HUMANS AND HYENAS

Humans and Hyenas examines the origins and development of the relationship between the two to present an accurate and realistic picture of the hyena and its interactions with people. The hyena is one of the most maligned, misrepresented and defamed mammals. It is still, despite decades of research-led knowledge, seen as a skulking, cowardly scavenger rather than a successful hunter with complex family and communal systems.

Hyenas are portrayed as sex-shifting deviants, grave robbers and attackers of children in everything from African folk tales through Greek and Roman accounts of animal life, to Disney's *The Lion King* depicting hyenas with a lack of respect and disgust, despite the reality of their behaviour and social structures. Combining the personal, in-depth mining of scientific papers about the three main species and historical accounts, Keith Somerville delves into our relationship with hyenas from the earliest records from millennia ago, through the accounts by colonisers, to contemporary coexistence, where hyenas and humans are forced into ever closer proximity due to shrinking habitats and loss of prey. Are hyenas fated to retain their bad image or can their amazing ability to adapt to humans more successfully than lions and other predators lead to a shift in perspective?

This book will be of great interest to students and scholars in the environmental sciences, conservation biology, and wildlife and conservation issues.

Keith Somerville is a Member of the Durrell Institute of Conservation and Ecology at the University of Kent, UK, where he is a professor at the Centre for Journalism. He is a Senior Research Fellow at the Institute of Commonwealth Studies, a Fellow of the Zoological Society of London, UK, and a Member of the IUCN CEESP/SSC Sustainable Use and Livelihoods Specialist Group.

ROUTLEDGE STUDIES IN CONSERVATION AND THE ENVIRONMENT

This series includes a wide range of inter-disciplinary approaches to conservation and the environment, integrating perspectives from both social and natural sciences. Topics include, but are not limited to, development, environmental policy and politics, ecosystem change, natural resources (including land, water, oceans and forests), security, wildlife, protected areas, tourism, human-wildlife conflict, agriculture, economics, law and climate change.

Natural Resources, Tourism and Community Livelihoods in Southern Africa
Challenges of Sustainable Development
Edited by Moren T. Stone, Monkgogi Lenao and Naomi Moswete

Leaving Space for Nature
The Critical Role of Area-Based Conservation
Nigel Dudley and Sue Stolton

Power in Conservation
Environmental Anthropology Beyond Political Ecology
Carol Carpenter

Rethinking Wilderness and the Wild
Conflict, Conservation and Co-existence
Edited by Robyn Bartel, Marty Branagan, Fiona Utley and Stephen Harris

Humans and Hyenas
Monster or Misunderstood
Keith Somerville

For more information about this series, please visit: www.routledge.com/Routledge-Studies-in-Conservation-and-the-Environment/book-series/RSICE

HUMANS AND HYENAS

Monster or Misunderstood

Keith Somerville

Routledge
Taylor & Francis Group
LONDON AND NEW YORK

First published 2021
by Routledge
2 Park Square, Milton Park, Abingdon, Oxon OX14 4RN

and by Routledge
52 Vanderbilt Avenue, New York, NY 10017

Routledge is an imprint of the Taylor & Francis Group, an informa business

British Library Cataloguing-in-Publication Data
A catalogue record for this book is available from the British Library

Library of Congress Cataloging-in-Publication Data
Names: Somerville, Keith, author.
Title: Humans and hyenas : monster or misunderstood / Keith Somerville.
Description: Agingdon, Oxon ; New York, NY : Routledge, 2021. |
Series: Routledge studies in conservation and the environment |
Includes bibliographical references and index.
Subjects: LCSH: Hyenas. | Hyenas--Effect of human beings on--History. |
Human beings--Effect of environment on--History. | Human-animal relationships--History.
Classification: LCC QL737.C24 S66 2021 (print) |
LCC QL737.C24 (ebook) | DDC 599.74/3--dc23
LC record available at https://lccn.loc.gov/2020044544
LC ebook record available at https://lccn.loc.gov/2020044545

ISBN: 978-0-367-43642-1 (hbk)
ISBN: 978-0-367-43641-4 (pbk)
ISBN: 978-1-003-00478-3 (ebk)

Typeset in Bembo
by Taylor & Francis Books

CONTENTS

FOREWORD

Due to their appetite for livestock and the occasional human, the great predators are the first wildlife to disappear when people get the technology to kill them efficiently, and the hardest to conserve today. Hyenas compound those sins by also eating human corpses, adding a supernatural dimension that makes them one of the most despised and misunderstood of all animals, a great irony given that the spotted hyena in particular is one of nature's most interesting, complex and, to those who know them, loveable creatures. They have been maligned throughout recorded history, and because they were probably one of early man's most dangerous predators and most direct competitors over kills, they have probably held that distinction since the dawn of mankind. With this volume, Keith Somerville has produced the definitive history of the tortured relationship between *Homo sapiens* and the Hyaenidae, his exhaustive research documenting our mutual history from the distant Pleistocene past to today's headlines and latest scientific studies.

Humans have been eliminating other species since we evolved into hunters nearly two million years ago, culminating in the Pleistocene extinctions of the last 20,000 years. Those saw the disappearance of scores of large mammal species, gradually at first in Eurasia and then very rapidly in the New World soon after humans arrived; we had eaten all of Australia's megafauna millennia earlier. Only in Africa and parts of Asia did the largest mammals survive, having had millions of years to gradually evolve behavioural defences against gradually evolving human predation. Today, however, firearms, poison, and the human population growth which resulted from colonialism have reduced Africa's wildlife to a small fraction of the numbers present when Europeans arrived. Kenya's great naturalist Leslie Brown estimated that 90% of its wildlife present at the beginning of the colonial period had disappeared by 1940, and that 90% of the remainder was gone by the 1970s. Reliable aerial census data show that it has declined nearly 90% since then. Thus, with more wildlife remaining than most African countries, Kenya has perhaps 0.1% of the large mammal fauna that were present in the

1890s. This book, like Prof. Somerville's previous ones on the ivory trade and the history of lions and humans, documents that cataclysm in excruciating detail.

Predators were far more abundant when our ancestors first domesticated wild grazers 8–10,000 years ago. The early herders must have developed effective antipredator measures as soon as they had livestock to protect, probably similar to those still used today by traditional African pastoralists. Had they not kept the first livestock in stout bomas (corrals) at night and tended them closely when out grazing by day, large carnivores would have ended the experiment in domestication as soon as it started. Ironically, hyenas are the easiest predator against which to defend livestock: as they cannot jump, a strong boma merely one meter high effectively excludes them. However, ready availability of cheap but lethal agricultural pesticides have made those efforts unnecessary and the great predators, along with the rest of African wildlife, are fast disappearing.

Lions, tigers and bears have a large and devoted constituency in the West, but hyenas attract only a small fan base; even today they are portrayed in popular entertainment as stupid, ugly villains. I was ejected from a major US conservation funding organisation for telling the representative from Disney, in a private email, that *The Lion King* had made finding money for hyena conservation even more difficult than it already was. Fortunately, however, lions are more popular than ever, and measures which protect livestock from lions also conserve hyenas.

This universal hatred is inexplicable to anyone who works with spotted hyenas, in the wild or in captivity; to know them is to love them. They are exceptionally social, have strong personalities and as infants they readily bond with humans. Wolves and most other wild animals raised by people become aggressive as they mature, but spotted hyena infants reared by a human regard him/her as their beloved, albeit dominant, mother for life. For this reason, and because their complex society is quite similar to ours, it may have been an accident of history that we domesticated wolves rather than hyenas.

Many scientists have devoted their careers to understanding and conserving the Hyaenidae. Due to space constraints and because they are so unique, I will mention here only the early history of spotted hyena research. The first study was by Aristotle, who debunked their mythical hermaphroditism 2,300 years ago. The next was by Leonard Harrison Matthews, who in 1935 addressed those ancient stories by dissecting over 100 spotted hyenas in then-Tanganyika, finding that females indeed have very peculiar external genitalia but are normal mammals internally. The female has no vulva; in its place is a scrotum that appears to contain testicles. Her clitoris is greatly enlarged and fully erectile, nearly indistinguishable from the male's penis and in the same position; through it she urinates, mates and gives birth. Aristotle, it seems, had examined striped hyenas, anatomically normal mammals.

Not until Hans Kruuk's classic study in Serengeti and Ngorongoro in the 1960s did we begin to understand spotted hyena behaviour and ecology. By following them at night, Hans discovered that, rather than being scavengers dependent upon lions' leftovers, they are highly efficient social predators, their abundance where protected making them a major ecological influence on prey populations. He found that that they live in permanent, highly territorial social groups of up to 100 animals which he termed clans. He documented that females are strongly dominant over males, and among the many social behaviours he described was the greeting ceremony: when two spotted hyenas meet after a separation, they stand head to tail, each carefully inspecting the other's groin and, regardless of their sex, one or both getting an erection, in this case a purely social signal unrelated to romance.

Ten years later and initially unbeknownst to each other, Gus Mills in the arid Kalahari and I in Kenya's lush Masai Mara studied their social ecology, finding similar patterns in spite of major differences in both hyena and prey density. Female youngsters remain in their mother's clan for life and 'inherit' her dominance rank, while males leave their natal clan upon reaching adulthood. The social system revolves around female dominance: high ranking females and their progeny drive subordinate individuals off a kill as it is consumed, and with their preferential access to food, individuals in those matrilines are more likely to survive, and thus produce more descendants than lower ranking families. A clan, therefore, comprises related females, their offspring and adult males which immigrate from other groups.

Their bizarre sexuality makes the spotted hyena an intriguing model in which to study the basic biology of sex differences in mammals, including humans. In 1984, Stephen Glickman, Julian Davidson and I created the Berkeley Hyena Project, raising 20 infants in large paddocks in the hills above the University of California campus. Steve recruited endocrinologists, embryologists, behavioural biologists, and neuroscientists to unravel a complex and still incompletely understood series of hormonal and physiological events leading to the female's unique anatomy and aggressive nature. When they were old enough to breed, we found that, unlike other carnivores, spotted hyena infants, usually twins, are born with their eyes open, sharp front teeth, good coordination and a drive to fight violently within minutes of birth. In the wild this often culminates in the death of one twin. As adaptations to winning that primal battle, the gestation period is unusually long and the neonates unusually large, making delivery through the narrow birth canal of the clitoris a dangerous moment for both first time mothers and their offspring. Thus far the evolution of these reproductive peculiarities eludes full explanation; there is nothing ordinary about spotted hyenas.

After dominating the earth for 60 million years, the last of the great mammals follow the dinosaurs into oblivion as we watch. This time, the blame lies not with an asteroid but with us, unable to control our own numbers, our destructive nature, our short-sightedness, our unwillingness to co-exist with the rest of Creation. Hyenas and other large carnivores are killed in retaliation for eating livestock and by bushmeat snaring, but the ultimate cause of wildlife decline is the exploding human population, converting wilderness to farms, and degrading rangeland. Dry bush savannah comprises more than half of sub-Saharan Africa, and a century ago supported abundant wildlife plus people and their cattle. Today, over 50% has been heavily degraded by decades of overgrazing, the soil washed away, the wildlife nearly gone. With the 1.2 billion people of Africa on a trajectory to reach 4–6 billion by the end of this century, the outlook for wildlife and wilderness is not encouraging.

The future of African animals lies with African people, but exponential population growth and development make it murky at best. Prof. Somerville's history of the hyena-human relationship documents the decline of three species in excruciating detail, a scholarly case study of the devastating impact the too-intelligent ape has had on the species with which it shares the planet; there are millions of similar stories as life on earth is crushed beneath eight billion humans. This volume joins his previous ones on the destruction of elephants and lions, clarion calls for humanity to wake up, grow up, and assume responsibility for repairing the destruction it has wrought.

Dr Laurence G. Frank, Living with Lions Project Director and research associate in the Museum of Vertebrate Zoology, University of California, Berkeley

ABBREVIATIONS

ANC	African National Congress (South Africa)
BP	Before present (present being identified as 1950)
BNP	Bwabwata National Park (Namibia)
BSAC	British South African Company (Southern Rhodesia)
CAMPFIRE	Communal Areas Management Programme for Indigenous Resources (Zimbabwe)
CAR	Central African Republic
CBNRM	Community based natural resource management
CDV	canine distemper virus
DICE	Durrell Institute of Conservation and Ecology
IIED	International Institute for Environment and Development
DRC	Democratic Republic of Congo
DWNP	Department of Wildlife and National Parks (Botswana)
FAO	Food and Agriculture Organization (UN)
GLTA	Greater Limpopo Transfrontier Area
GLTP	Greater Limpopo Transfrontier Park
GMA	game management areas (Zambia)
GR	Game Reserve
HALO Trust	Hazardous Area Life-support Organization
HNP	Hwange National Park
HWC	Human Wildlife Conflict
IUCN	International Union for the Conservation of Nature
KAZA	Kavango-Zambezi Transfrontier Area
KTP	Kgalagadi Transfrontier Park (formerly Kalahari Gemsbok National park)
KCP	Kwando Carnivore Project, Namibia
KWS	Kenya Wildlife Service

MET	Ministry of Environment and Tourism (Namibia)
MMNR	Masai Mara National Reserve
MPLA	People's Movement for the Liberation of Angola, Portuguese: Movimento Popular de Libertação de Angola.
NCA	Ngorongoro Conservation Area (NCA)
NP	National Park
Sanparks	South Afgrican National Parks
SSC	Special Survival Commission (IUCN)
SULi	Sustainable Livelihoods Specialist Group (IUCN)
UNITA	National Union for the Total Independence of Angola, Portuguese: União Nacional para a Independência Total de Angola
WCS	Wildlife Conservation Society
WMA	wildlife management area (Botswana and Tanzania)
WWF	Worldwide Fund for Nature
Zimparks	Zimbabwe National Parks
ZSL	Zoological Society of London

ACKNOWLEDGEMENTS

In researching and writing the book I have been supported, helped and advised by the leading scientific experts on hyenas and conservationists. I am hugely grateful to Dr Laurence Frank of the University of California and the Laikipia Predator Project for his encouragement, careful reading of draft chapters and for the foreword. Dr Amy Dickman of WildCRU (Oxford) and the Ruaha Carnivore Project; Arjun Dheer for his incisive comments on draft chapters and superb typo-spotting skills; Dr Sarah Edwards for her advice and for showing me the brown hyenas and a den at Africat, Okonjima, Namibia; Marcus Baynes-Rock of the University of Notre Dame, for reading and commenting on draft chapters and the data on the hyenas of Harar; Dr Mordecai Ogada of Conservation Solutions Afrika; Julian Rademeyer of the Global Initiative Against Trasnational Organized Crime; Dr Vivienne Williams of the University of Witswatersrand, for help with South African wildlife markets and muti; Emsie Verwey of the Skeleton Coast Brown Hyaena Project and Wilderness Safaris; Derek de la Harpe, Neil Midlane and Henry Parsons of Wilderness Safaris; Sara Fernandes Elizalde, for advice about hyenas in Angola; John O'Brien of Shamwari Reserve, Eastern Cape; Dr Ross Barnett; Brennan Peterson-Wood of the Conservation Research Africa in Malawi; Dr Ludwig Siege, formerly of the Ethiopian Wildlife Authority; Dr Julian Kesby; Dr Velizar Simeonovski; Dr Hans Bauer of WildCRU; the palaeontologist Dr Ross Barnett; Dr David Anderson of the University of Warwick; and, Craig Taylor-Hilton of IUCN for the use of the maps.

I would like to thank Emma Milnes and Sarah Broadhurst of the Zoological Society of London's Library for their skilled and patient assistance and Chief Curator at the ZSL for his advice on hyenas in zoos.

Rosie Anderson and her colleagues at Routledge for turning this manuscript into a book and for her immense patience, and Sophie Blocksidge for her copyediting.

Lastly, as ever, this wouldn't have been written without the love and support – not to mention grammatical advice and translations from German – from my wife Liz and son Tom.

INTRODUCTION

Late one night in October 1981 in Lilongwe, Malawi, I was driving home from a local bar. Outside the post office in the old town I saw what I thought was an ungainly dog crossing the road. As I got closer, the animal stopped and looked at me. It wasn't a dog but a large spotted hyena, the first I'd ever seen other than in TV wildlife documentaries. Having watched me, perhaps hoping I'd crash and provide it with a free meal, the hyena ambled off down the road. For the next eight months that I worked in Malawi, I didn't see another hyena but heard them occasionally at night, especially in and around Kanengo College, north of Lilongwe and surrounded by bush and small villages. Hyenas frequented the town at night, foraging among refuse, preying on dogs or any livestock not locked up, and even attacking drunks who had fallen into storm drains or the destitute forced to sleep on the streets.

On one occasion, a woman walking near the Lilongwe wildlife sanctuary was badly mauled by a hyena – the national newspaper, the *Daily Times*, reported that it was suspected of being rabid, but many local people told me they believed the hyena was bewitched. It could have been rabid, just as likely it was opportunistically attacking someone walking alone near a wooded area, where hyenas were known to rest during the day. Other places favoured by these urban or semi-urban hyenas were culverts under roads or tunnels that linked the storm drains that channelled away torrential rains during the wet season.

In subsequent trips to Africa as a journalist and then a researcher on human-wildlife conflict over a period of 40 years, I often saw spotted hyenas and, in South Africa and Namibia, brown hyenas. From the start, I found them fascinating and when I read Kruuk's ground-breaking work on spotted hyenas in the Serengeti and Ngorongoro,[1] I was won over and became a hyena supporter. The more I visited Africa, talked to people there about hyenas, saw hyenas depicted in the media, literature, and film, and even in reputable documentaries, as thieving, cowardly but vicious, ugly, or just plain disgusting animals, the more I found them worthy of in-depth research and a more accurate image. The traducing and demonisation of hyenas in films like the *Lion King* (both versions) and *Mowgli: Legend of the Jungle* were the final straw and I determined to write about the history of human-hyena relations and the origins of the myths and misrepresentations.

A flavour of the myths and the effects hyenas have on people can be found in the following quotes from leading carnivore specialists, conservationists and hunters:

> Perhaps the most important challenge facing those of us committed to the conservation of this group of animals is to overcome the very strong negative feelings many people have towards hyaenas. Until they are viewed in a more positive light it will be difficult to effectively implement management plans for hyaenas.[2]
>
> *Gus Mills and Heribert Hofer*

> Hyenas inspire horror in people. Often this human reaction is covered up by laughter (especially in Africa); in people's minds, hyenas are inexorably linked with garbage cans, corpses, faeces, bad smells, and hideous cackles.[3]
>
> *Hans Kruuk*

> If Africa has a voice, it is the hyena. Deep in the blackness of night it gropes through the bush in rising and falling echoes that come from nowhere, yet everywhere, insane choruses of whoops, chortles, chuckles, giggles, shrieks, and howls that have a way of reaching into the guts of a man as he sits by a lonely, dying fire and of raising the hackles of ancient, long-forgotten apprehension.[4]
>
> *Peter Hathaway Capstick*

> Hyenas are loathed, vilified, feared. Derided, persecuted, and, where people have the wherewithal, eradicated. Time and again I encounter negativity when I tell people about my research. They can't resist telling me how hyenas are disgusting or ugly…This is a little unusual for a large carnivore; other species of this order evoke awe, admiration, and adoration.[5]
>
> *Marcus Baynes-Rock*

> Like everyone else in Africa, I was brought up to believe that hyenas were nasty, cringing, crafty scavengers living partly on scraps from kills made by others, such as lions, and partly on refuse and human corpses.[6]
>
> *Elspeth Huxley*

She added that to call a man a *fisi*, Swahili for hyena, was a deadly insult in Kenya.[7]

Things have changed a little through the research by Kruuk, correcting the depiction of the spotted hyena as purely a scavenger with no hunting skills or courage, and subsequent studies by Frank, Mills, Hofer, Delia and Mark Owens, Holekamp, Dloniak, Abi-Said, Singh, Baynes-Rock, Yirga, Edwards, and others (which will be quoted and referenced in later chapters) shedding new, much-needed, and fascinating light on hyenas. But there is much more work to be done to improve the understanding and representation of hyenas and ensure they do not remain disdained and oft forgotten in conservation policy, let alone the media and popular culture. What follows is an honest and detailed account of the many millennia of human-hyena interactions. It will utilise scientific studies of hyena biology but is not a zoological/biological study, it concentrates on their coexistence and conflict with people. It will not shy away from the dark side of hyenas (such as devouring human corpses left in the bush or buried in graves), but will put hyenas into context and tell the whole story of their relationship with humans. For simplicity and in line with Kruuk, I have used hyena, not hyaena throughout, except in quotes by others.

The focus of this book will be the three hunting and scavenging species of hyenas – the spotted (*Crocuta crocuta*), the brown (*Parahyaena brunnea*), and the striped (*Hyena hyanea*). The fourth hyena species, the aardwolf (*Proteles cristata*) will be mentioned only in passing, as its ecology and behaviour differ significantly.

Chapter 1 deals with the basic details of size, appearance, diet, foraging and social behaviour, breeding, and the competition and conflict between hyenas, other carnivores and people. Chapter 2 looks at the origins and evolution of hyenas and humans, charting their interactions, conflict, competition and the fear engendered in humans by hyenas. Chapters 3 and 4 deal with the effects on human-hyena interaction of human population growth, and agricultural and technological advances up to the end of the 16th century in Africa, West, Central and South Asia. Chapter 5 covers the period from 1600 to the beginnings of colonialism at the end of the 19th century. Chapter 6 covers the colonial period in Africa and Asia, which saw the greatest diminution of habitats for wildlife and large-scale killing of wildlife, especially carnivores, to make way for livestock, for commercial gain and for sport. Chapter 7 deals with coexistence and conflict in the contemporary era and the modern images of the hyena. The final chapter, 8, will deal with the myths, demonisation and modern representation of the hyena, from early humans through to *The Lion King*.

Notes

1 Hans Kruuk (1972) *The Spotted Hyena A Study of Predation and Social Behaviour*, Chicago: University of Chicago Press.
2 Gus Mills and Heribert Hofer (1998) *Status Survey and Conservation Action Plan Hyaenas*, Gland, Switzerland: IUCN/SSC Hyaena Specialist Group, http://www.carnivoreconservation.org/files/actionplans/hyaenas.pdf accessed 4 September 2019 p. vi.
3 Kruuk, 1972, p. 6.
4 Peter Hathaway Capstick (1977) *Death in the Long Grass*, London: Cassell, p. 273.
5 Marcus Baynes-Rock (2015) *Among the Bone-Eaters. Encounters with Hyenas in Harar*, Pennsylvania: University of Pennsylvania Press, p. 2.
6 Elspeth Huxley and Hugo van Lawick (1984) *Last Days in Eden*, London: Harvill Press, p. 43.
7 Ibid.

1

DRAMATIS PERSONAE

The spotted, striped and brown hyenas

Class Mammalia
Order Carnivora
Suborder Feliformia
Family Hyaenidae
Subfamily Hyaeninae
Spotted *Genus* Crocuta *Species* Crocuta crocuta
Striped *Genus* Hyaena *Species* Hyanea hyaenea
Brown *Genus* Parahyanea *Species* Parahyanea brunnea

Brown, striped hyena and spotted hyenas belong to the subfamily *Hyaeninae*, a subfamily that includes all living and extinct bone-cracking hyenas.[1] Aardwolves are the only members of the Hyaenidae subfamily *Protelinae* and are insectivores. The hyenas are surprisingly varied in their ecologies and social behaviours[2] and are distributed across most of Africa (except the most arid areas of the Sahara, high mountains and dense Congo Basin and West African rainforest); the Middle East (from Turkey to southern Arabia); and West, Central and parts of South Asia. Population numbers are hard to assess and throughout, the estimates given are within broad ranges based on spoor, scat, camera trap and observational surveys. These all have a margin for error.[3]

The striped hyena is found across the hyena range, excepting the southern half of Africa. It is the only hyena found outside Africa. The IUCN Red List places it in Afghanistan; Algeria; Armenia; Azerbaijan; Burkina Faso; Cameroon; Chad; Djibouti; Egypt; Ethiopia; Georgia; India; Iran, Islamic Republic of; Iraq; Israel; Jordan; Kenya; Lebanon; Libya; Mali; Mauritania; Morocco; Nepal; Niger; Nigeria; Oman; Pakistan; Saudi Arabia; Senegal; Syria; Tajikistan; Tanzania; Tunisia; Turkey; Turkmenistan; Uganda; Uzbekistan; Western Sahara; Yemen; with a possible presence in Benin; Central African Republic; Eritrea; Guinea;

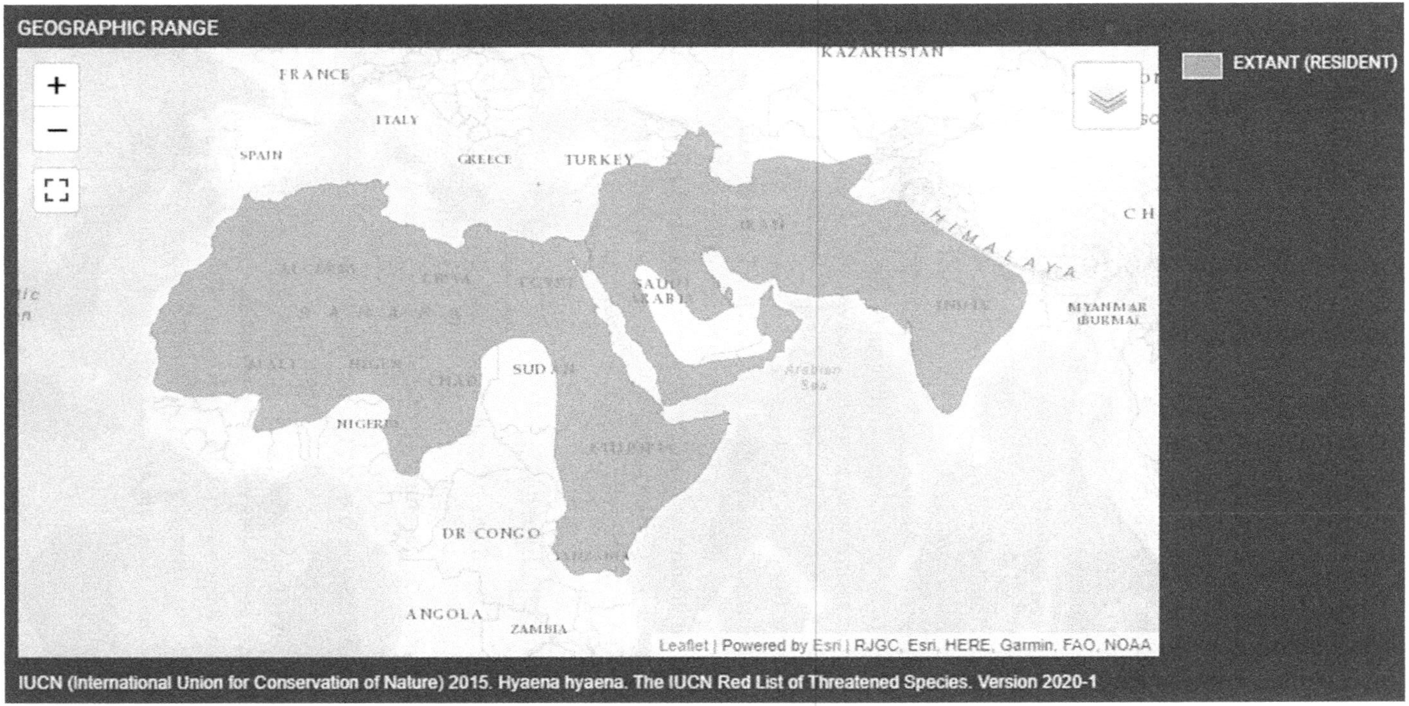

MAP 1.1 Striped hyena IUCN Red List

Kuwait; Qatar; Somalia; Sudan; United Arab Emirates. It is listed as "near threatened", with a population globally estimated at 5,000–9,999.[4] It is very possible that the difficulty in counting striped hyenas accurately means the population could be larger, as some conservationists believe.[5]

The spotted hyena is only found in Africa. The IUCN Red List records it in Angola; Benin; Botswana; Burkina Faso; Burundi; Cameroon; Central African Republic; Chad; Congo; Democratic Republic of Congo; Côte d'Ivoire; Djibouti; Equatorial Guinea; Eritrea; Eswatini; Ethiopia; Gambia; Ghana; Guinea; Guinea-Bissau; Kenya; Malawi; Mali; Mauritania; Mozambique; Namibia; Niger; Nigeria; Rwanda; Senegal; Sierra Leone; Somalia; South Africa; South Sudan; Sudan; Tanzania; Uganda; Zambia; Zimbabwe; and possibly Algeria and Togo. It is listed as of "least concern" but decreasing in numbers. A "tentative estimate of the total global population is between 27,000 and 47,000". The largest known populations occur in the Serengeti ecosystem in Tanzania and Kenya (7,200–7,700 in the Tanzanian sector and 500–1,000 in the Kenyan sector); in Kenya's Tsavo East and Tsavo West NPs (3903 +/- 514)[6]; and the Kruger NP in South Africa (1,300–3,900).[7] Yirga et al.'s suggestion that there could be 28,620 spotted hyenas in Ethiopia's Tigray province would make that the largest population and would require a recalculation of the overall spotted hyena population,[8] but it remains to be verified.

The brown hyena is found in Angola; Botswana; Namibia; South Africa; and Zimbabwe; with possible presence in Eswatini (Swaziland) and Mozambique. It is listed as "near threatened", with a population between 4,365 and 10,111. Botswana has the largest population at about 3,900 animals, followed by Namibia with between 566–2,440 and South Africa with 800–2.200; no reliable population size estimates are available for Zimbabwe, Angola, Mozambique or Eswatini.[9]

Striped hyena (*Hyaena hyaena*)

This hyena weighs between 26–34kg (females) and 26–41kg (males), and is 1.00–1.10m in length, 0.66–0.75m in height.[10] It is dog-like, with a sloping back, black vertical stripes on a background of grey or beige, pointed muzzle and ears and a mane along its back.[11] It occurs in open habitat or light thorn bush country, avoiding altitudes above 2,500m and dense forest.[12] It favours rocky areas, ravines and hills, preferably with caves in which to rest during the day. Mainly nocturnal and shy of humans, it does nevertheless reside in some areas (Palestine/Israel, for example) close to human settlements.[13] They generally occur at low densities.

Although the research has increased in recent years, with research on behavioural ecology by Wagner,[14] on status and conservation in Lebanon by Abi-Said[15] and in Rajasthan by Singh[16] and a few earlier papers by Kruuk[17] and Leakey et al.,[18] it is comparatively little studied.[19]

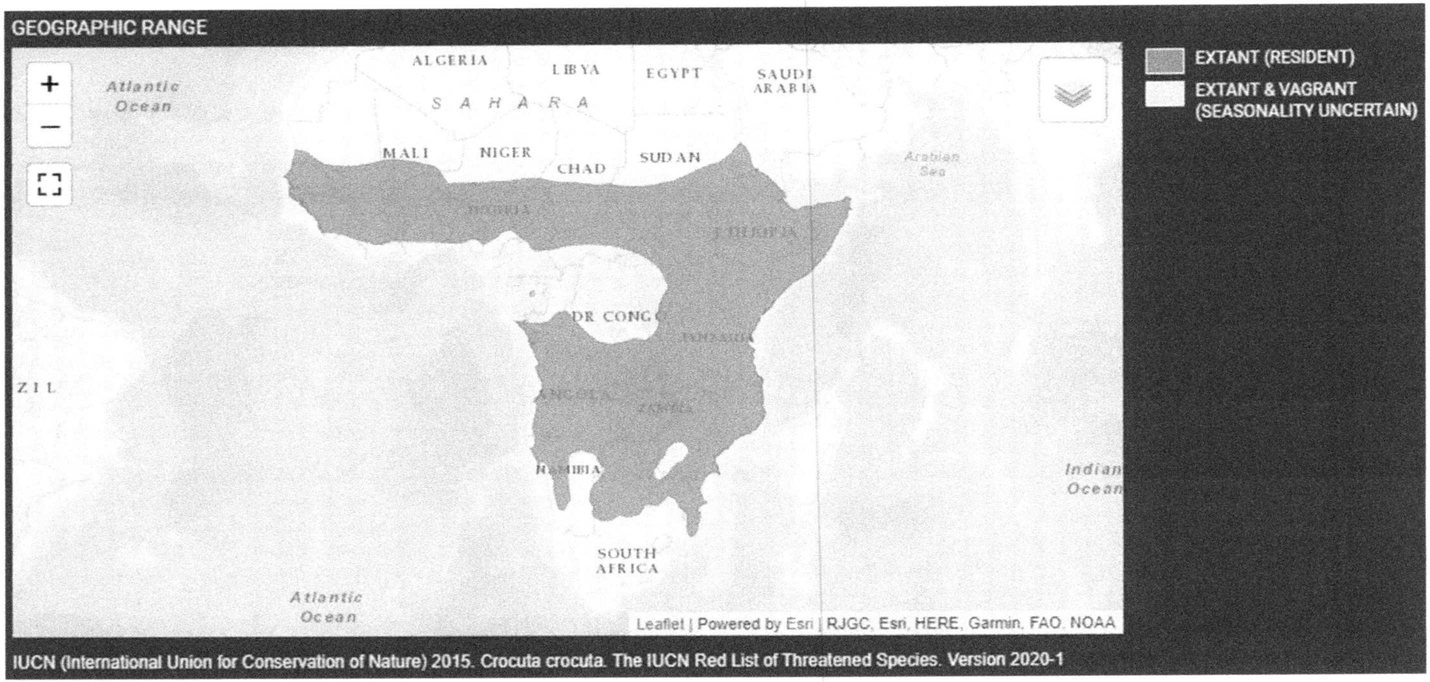

MAP 1.2 Spotted hyena IUCN Red List

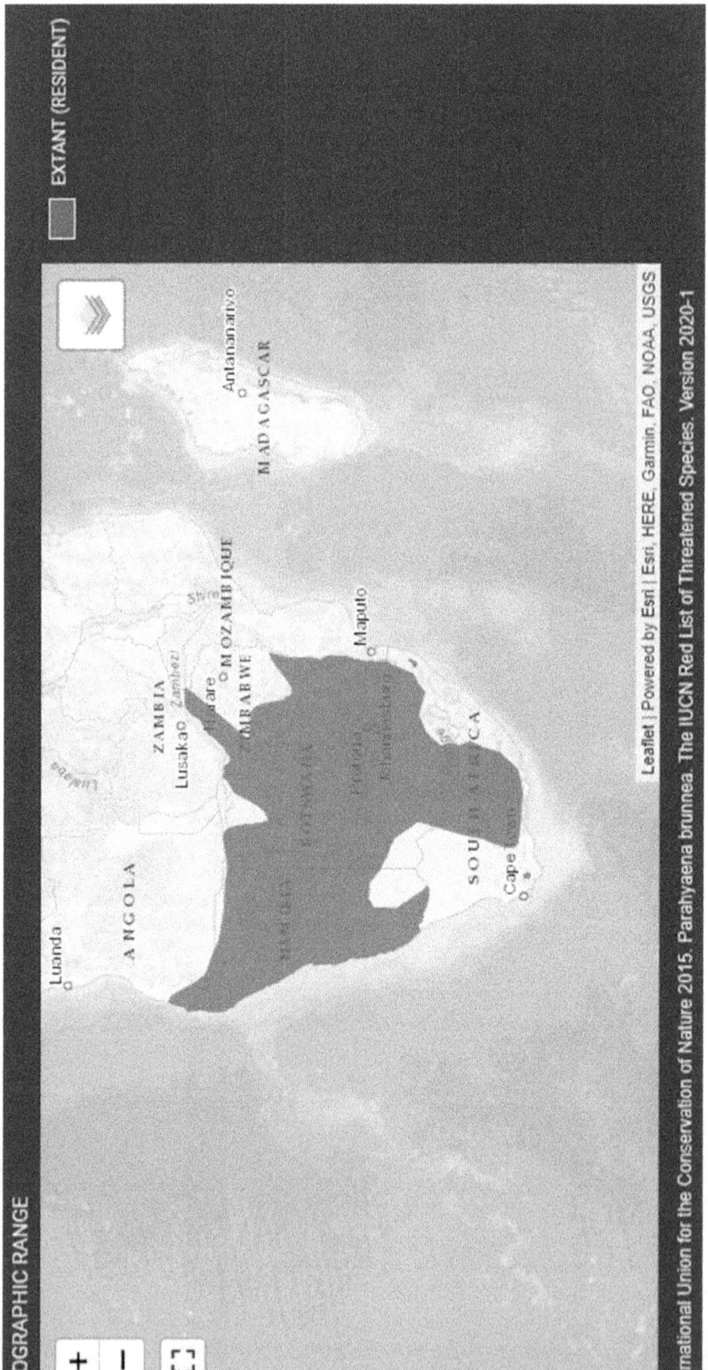

MAP 1.3 Brown hyena IUCN Red List

Diet

Striped hyenas have a diverse diet including carrion, medium and small vertebrates, invertebrates and fruits and vegetables (melons, dates and other fruits and vegetables matter), as well as utilising waste produced by humans.[20] They are capable of killing medium-sized ungulates and livestock.[21] Kruuk said they were "inveterate scavengers of human waste" and that in the Serengeti, they fed on "domestic refuse such as fruits...bread, boiled potatoes and any animal offal such as bones, pieces of leather".[22] He also referred to reports that they had killed and eaten people.[23] Studies in Kenya, Israel and Lebanon document that they take human remains from graveyards.[24] Faecal analysis of their diet in Tanzania showed the following: zebra in 4% of samples; wildebeest 22%; kongoni 4%; topi 2%; Grant's gazelle 24%; impala 2%; Thomson's gazelle 42%; dik-dik 12%; hare 10%; lizard or snake 32%; birds 44%; vegetable matter 36% – most would have been scavenged but the smaller mammals, reptiles and birds may have been hunted.[25] Bhandari et al. examined scats in two lowland districts of Nepal. The analysis showed the following percentages – wild boar 26.2; hare 16.23; rhesus monkey 15.58; muntjac deer 11.03; langur monkey 8.44; domestic goat 5.15; domestic sheep 2.59; and dog 2.59.[26]

Foraging and hunting

Striped hyenas are solitary foragers and don't regularly feed in large groups.[27] The exceptions being where there are particularly rich food sources, including man-made feeding stations established to observe feeding behaviour, at which large groups may gather.[28] Although this does not alter the size of groups in which they live, it means they are less solitary than thought.[29] They are almost totally nocturnal in their foraging. The sub-species found in North Africa, India and Syria are slightly larger than the others and periodically hunt and kill larger mammals. Leakey believed that striped hyenas in Kenya were not successful hunters of large mammals,[30] and were inefficient hunters of smaller mammals like hare, porcupine and small rodents, reptiles, birds and insects.[31]

In its Middle Eastern and West Asian range, hyenas scavenge the remains of camels, wild goats, domestic sheep and goats and donkeys, with some hunting of domestic animals (including dogs) and small-to-medium-sized mammals. Wild prey accounts for a small percentage of food (3.3%, according to Leslie).[32] The animal carcasses scavenged are either the kills of larger carnivores[33] or those which have died from accidents, disease or old age.

Social behaviour and breeding

Striped hyenas were thought to be solitary, but are now known to form small, loose groups to breed.[34] A survey in Lebanon found groups ranging from three to nine adults and juveniles.[35] Striped hyenas are polyandrous, with females mating with more than one male. Females establish "the minimum defendable territory with enough resources to provide food for herself and her offspring". Males live singly or in small groups with larger home ranges, to be able to mate with females in this range.[36] Females may mate with multiple males when they are in season.[37] Wagner et al. in their Kenyan study found that group-living males fathered 69% of the resident female's offspring, which "supports the interpretation of coalition formation as a strategy to defend mating opportunities".[38] It is not clear how well-defined territories are, but "the low degree of home range overlap between these groups does indicate that group members operate within common, exclusive ranges".[39] There have been no observations of them being aggressively defended. Any groupings that form among males relate to breeding. Similarly, females only associate to mate.[40] In Laikipia, Kenya, Wagner found that the population groups that formed in this way averaged three males to a female and they were stable entities.[41]

Striped hyenas do not have the range of vocalizations of spotted hyenas, but Kruuk identified "a number of different vocalizations, all of which bore a close resemblance to some of the spotted hyaena sounds" – the main sounds were whining by cubs before suckling, giggling by adults when frightened, yells when competing over food, "a soft long drawn-out lowing sound when frightened in a defensive position" and growling when fighting over food or play-fighting.[42] The raising of the mane was an important visual means of communication – notably in aggressive confrontations between hyenas, in encounters with spotted hyenas, when running from a human or chasing prey.[43] Kruuk noted that when members of the same family meet, they often sniffed each other's face and neck, followed by sniffing each other's anal region.[44] They communicate with other hyenas through scent marking, using paste from glands near the anus on twigs, grass stems and other chosen signal posts to mark ranges or communicate presence.[45]

Gestation takes 90–1 days and the litters vary in size from one to five, but usually two to three.[46] Cubs are blind until about seven days old and suckle from the mother, beginning to eat meat and other food brought to the den by the mother from 30 days old.[47] Males in the home range bring food to the cubs.[48] The cubs continue to suckle after taking solid food, usually for four or five months. Sexual maturity varies between the ages of at least 15 months up to three years. Adult hyenas leave their mother's range as they reach maturity.

Competition with other carnivores

Striped hyenas are attracted to kills made by other carnivores or to carcasses of animals that died from other causes. They avoid encounters with tigers, lions and spotted hyenas, and do not try to appropriate their kills, only feeding from remains after the larger carnivore has departed. Hyenas will drive cheetah from their kills and try to kill cheetah cubs. They do the same, not always successfully, with leopards. In the Middle East, the Caucasus, Central Asia and India, hyenas compete with wolves and usually dominate a single wolf, but not a group of four or more.[49] In Israel's Negev Desert, researchers found evidence of a striped hyena accompanying a wolf pack on a hunt[50] Hyenas will attack and kill domestic dogs, but if confronted by a pack of dogs or by humans, they may play dead to deter an attack.

Conflict with humans

Humans are the major source of hyena mortality, with persecution through shooting, trapping and poisoning in retaliation for the killing of livestock, and through traffic accidents, with 20–30 hyenas killed annually on roads in Palestine and Israel.[51] It is not unusual for striped hyenas to die after taking poisoned bait intended for wolves or leopards, or for hyenas to be trapped by hunters trapping wolves for their fur.[52] In the Caucasus, stories abounded in the 19th and 20th centuries of striped hyenas killing people, especially children,[53] leading to persecution. In Bihar, India, striped hyenas were blamed for the deaths of 19 young children in 1974.[54] In much of North Africa and the Middle East it is "loathed as a grave robber, and therefore severely persecuted through poison-baiting and trapping".[55] It is disliked by farmers across its ranges for its habit of raiding melon, date and other crops.

Spotted hyena (*Crocuta crocuta*)

Spotted hyenas are heavily built, dog-like animals with massive jaws and a spotted coat. They vary in size from 1.2–1.8m long, 77–81cm tall and weigh 40–86kg; females are 10% heavier than males.[56] It is the most abundant large carnivore in sub-Saharan Africa, with a wide range of habitat from semi-desert, savannah and open woodland to dense forest, montane forest and peri-urban areas.[57] Adult female spotted hyenas are socially dominant over all adult immigrant males.[58] A particular feature of this species is that the secondary sexual organs are very similar in males and females. The female clitoris is of the same size and shape as the penis and she possesses pseudo-testes.[59] The urogenital canal runs through the clitoris.

Diet

Spotted hyena are hunters and scavengers, eating little vegetable matter, consuming the offal, meat, hides and bone of animals they hunt or scavenge, along with waste from human settlements. They consume almost all of an animal killed or scavenged,[60] leaving only the horns and larger bones of large carcasses. They hunt medium-to-large animals, but will take smaller prey according to availability. A hyena can kill an adult wildebeest three times its own weight, but larger prey such as zebra and buffalo require several hyenas for a successful kill. Hunts are usually initiated by a single hyena with clan members joining, often attracted by the calls of the initiator.[61]

Kruuk in his ground-breaking study of the spotted hyena recorded that in the Serengeti the prey killed or scavenged varied in proportion according to woodland and plains – with wildebeest adults making up 17.3% on the plains and 39.1% in woodlands; wildebeest calves about 2.1% and 4.7%; zebra adults 17.5% and 18.1%; zebra foals 1.9% and 2.0%; gazelle adults 36.2% and 16.3%; and gazelle fawns 12.7% and 14.3%. In Ngorongoro Crater, adult wildebeest made up 49.1%; wildebeest calves 16.7%; adult zebra 12.4%; zebra foals 5.7%, gazelles (adults and fawns) 7.3%; and others 9.0%.[62] The percentage killed against scavenged varied with wildebeest at 62.5% (adult) and 68.8% (calf) in the Serengeti but 98.2% and 98.6% in Ngorongoro, and 33.0% of zebra foals in Serengeti killed but 100% in Ngorongoro.[63] In some areas of Ethiopia lacking wild prey, more livestock are killed or scavenged, and more human waste consumed. In the Kgalagadi Transfrontier Park, spanning South Africa and Botswana, gemsbok were killed most often. In Kruger NP, kudu, impala and warthogs were preferred over wildebeest and zebra.[64]

Foraging and hunting

While mainly nocturnal, "the activity pattern of the spotted hyena...appears to be the most flexible, with activity often extending into periods of daylight, or even occurring at midday".[65] Spotted hyenas, contrary to popular belief, often hunt alone or sometimes in small groups of two to five. They detect prey by sight, sound and smell (the latter key to detecting carrion). Having found ungulates, they will run at a moderate speed through a herd trying to identify weak, old or very young animals before chasing the prey at speeds up to 60km/h.[66] Kruuk said that in most cases one hyena started a hunt and others joined in. Hyenas are more successful in catching calves or foals when there are two or more hyenas to ward off the mother.[67] Kruuk has noted, in his latest work on carnivores, that at times hyenas hunted zebra in packs of 10 or more and assembled before the pursuit began.[68] Hyenas will pursue the chosen prey for several kilometres, the longest hunt recorded being 24km.[69] Once the animal has been brought down, often by the hyenas tearing at its belly or back end, it is killed and rapidly consumed. At large kills, with many hyenas present, there is a free-for-all, though with dominant females likely to feed first.[70] Frank says:

> on a kill the cubs of dominant females have precedence over all other hyenas. They get regular access to meat at a younger age and grow faster, larger, and with the 'self confidence' so important in dominance relations among females and cubs.[71]

In areas where hyenas live close to people, they will kill livestock, even getting into bomas to do so, and scavenge waste dumps. In urban areas they will forage at night around residential areas, abattoirs, markets and any place where they might find waste or even domestic/feral dogs and, on occasions, people sleeping in the open or incapacitated (by hunger, illness or alcohol).

Social behaviour and breeding

Spotted hyenas live in fission-fusion societies – "large communities that have their own territory, but within which the members are not permanently together.

Individuals may come and go in temporary small groups, they switch between groups or they may stay on their own for a time."[72] The social groups, known as clans, vary in size from a few animals to 90, and in a few cases over 100. Within a clan:

> all members know one another individually, rear their cubs together at a communal den, and defend a common territory, yet clan members spend much of their time alone or in small subgroups" and with very large clans, the whole membership is not often together.[73]

Clan size is governed by prey density and habitat availability, with suitable habitats "saturated with a mosaic of hyena territories".[74] Territories, den sites and food sources are fiercely defended. Clans fight and kill hyenas caught intruding into another's territory or over food sources along the boundaries.

Clans have well-established dominance hierarchies, with females and their female offspring comprising the dominant groups, with adult male members subordinate. Most males leave their natal clans as they by adulthood, trying to join other clans as immigrants. The offspring of a dominant female will lose rank on moving clan. The most recent immigrant male is at the bottom of the clan hierarchy. There will be dominance relationships between the females, with inheritance of dominance through behaviour learned from a mother with high status, rather than genetic inheritance.[75] In small clans, there may be a single dominant female and her offspring and perhaps one immigrant male. In medium to large clans, clan members may be apart for extended periods and engage in "reunion displays" when sub-groups come back together with other clan members.[76] The sub-groups of clans will be made up of related females and potential mates.[77] Clan members will come together to defend territories and to scent mark the boundaries of the territory, with pasting from glands near the anus providing strongly scented markers on grass stems or twigs. The paste of an individual is believed to be recognisable as being from a particular animal and helps with bonding/networking within clans.[78]

Vocalizations are an important source of communication. Group members appear to recognize the calls of clan members at a great distance. They use vocal signals in direct interactions – such as growls, giggles and shrieks. They emit a large array of distinct sounds:

> deep groans to call their cubs out of dens, high-pitched squitters to beg for food or milk, and cattlelike lowing sounds to bring group-mates to a common state of high arousal…[they] are often referred to as laughing hyenas because their 'giggle' vocalizations sound much like high-pitched, hysterical human laughter…long-distance vocalizations of spotted hyenas, called 'whoops,' are the sounds heard most commonly at night [and are]…rallying calls to gather scattered clan members together to defend the territory boundaries, food resources, the communal den, or clan-mates in danger…[or] recruit hunting partners.[79]

Breeding is non-seasonal and a female may mate with several males when she is in oestrus. Females give birth to one to four young after a 16-week gestation period.[80] Females usually give birth in a den they have moved to alone, and transfer their young to the communal den at two to five weeks old. It is not unusual for there to be 30 young of varying ages at a large clan's den.[81] Dominant females will breed more regularly, with shorter birth intervals and a higher rate of cub survival. Cubs will establish dominance relationships within litters and at the communal den learn their places in the clan hierarchy,[82] and sibicide (killing of one cub by another) is common, with a cub that kills its siblings, as Mills and Hofer and Frank have noted, gaining great benefits, notably more food and maternal investment in the dominant, surviving cub, which seems to have led to the occurrence of "high neonatal aggression leading to sibicide".[83] Mothers nurse their young until they are about 11–14 months old. In small clans and during periods when the hyenas' prey may migrate away from the clan territory, the young may be left alone for extended periods by the mother as she commutes as much as 40–50km to follow migrating herds. Brown et al. have been investigating the killing of cubs by dominant females, presumably to give advantage to their young over progeny of subordinate females. Within a Maasai Mara clan that had been studied for 30 years, of 99 observed deaths of cubs, 21 could be attributed to infanticide, always by female killers and usually at the communal den site.[84]

Competition with other carnivores

Hyenas will chase cheetah and leopards from kills. Wild dog packs will generally keep hyenas off their kills, unless outnumbered, but usually leave sufficient bones and other material for them to scavenge. Wild dogs in sufficient numbers can chase hyenas from their kills. A pack of hyenas will attempt and often succeed in chasing lions from a kill and will protect their kills. The numbers involved are important, with hyenas needing to outnumber lions substantially. They rarely drive lions from a kill if large males are present.

Hyenas have a conflictual and complex relationship with lions. For centuries, this was presented as hyenas scavenging from lion kills, living on their leavings. This was challenged in the 1960s with the work of Dr Hans Kruuk, who documented the hunting capabilities of hyenas and reported hyenas often had their kills scavenged by lions. Kruuk estimated that of the kills he saw fought over by hyenas and lions, 53% had been killed by hyenas and 33% by lions, the rest being scavenged by both. In Ngorongoro Crater he estimated that fought-over kills had an even higher hyena kill rate of 84%, with only 6% killed by lions.[85] Conflict between them is intense and packs of hyenas will mob individual or small groups of lionesses but are deterred by large males. Lions will attack hyenas and often kill or injure them.[86]

Leopards are usually muscled off their prey by hyenas but they may sometimes fight back and a large male leopard might hold its own against a single hyena. Spotted hyenas usually dominate brown hyenas (though occasionally a large brown hyena will stand up to a single spotted one) and striped hyenas.

Conflict with humans

Conflict derives chiefly from hyena predation on domestic stock and human retaliation or pre-emptive killings.[87] This frequently involves presuming a hyena eating from the carcass of an animal that died from disease or old age in fact killed it. But it is undeniable that spotted hyenas do kill livestock and can also be man-eaters, especially children, the sick, the elderly and people sleeping outside or incapacitated in some way. Humans are responsible for hyena mortality through shooting, spearing and poisoning, but also through unintentional snaring when hyenas are caught in snares designed to trap ungulates to provide meat for people. Some die as a result of being hit by vehicles on roads. Scavenging from human settlements also leads to conflict, though, as will be shown, it is often tolerated, especially in Ethiopia. They adapt to living close to humans if food is available, even to the extent of losing some fear and, as in the extreme case of Harar (Ethiopia), feeding from human hands and tolerating the presence of numerous people.

Brown Hyena (*Parahyanea brunnea*)

The brown hyena is a medium-sized, dog-like animal with long legs and a shaggy brown coat becoming to a white, cape-like pelage around the neck and shoulders. The lower limbs have white stripes on a brown background. There is little variation in size among adults and they weigh 28–47kg, stand 79cm at the shoulder and are 1.2–1.6m from head to tail. They are found in arid, semi-desert and dry scrub or woodland areas of southern Africa, favouring rocky terrain with good bush cover to establish dens.[88] In many areas, such as central and southern Namibia, they are the dominant carnivore, while in other areas they are subordinate to spotted hyenas, lions and wild dogs.[89] They largely nocturnal and very secretive in their habits,[90] and have only in recent years been the subject of a growing number of studies, starting with Mills.[91] Their main ranges are in the central and southern Kalahari; most of Namibia; parts of southern Angola, Zimbabwe; and drier areas of South Africa away from main area of human habitation. They have been reintroduced in some areas, particularly fenced reserves in the Eastern Cape.[92]

Diet

They have an extremely diverse diet ranging from animals they hunt (a maximum of 16% of their intake, including springhares, aardvarks, vulnerable young of medium and small ungulates, small mammals, reptiles, birds and eggs)[93] through carrion to a variety of vegetable foods, especially melons. They will feed on large ungulate carcasses, such as gemsbok, hartebeest and wildebeest, but these vary in availability, especially in areas where ungulate species migrate. This source is about 17% of their food intake.[94] They can go for months without water, getting moisture from the blood of animals they hunt and from tsamma melons and other succulent plants.[95]

In the southern Kalahari, Mills detailed the following food intake:

> remains made up 39% of all food consumed, wild fruits such as the tsamma and gemsbok cucumber made up 29%. The remainder is made up of insects, birds eggs, reptiles and small mammals or birds. Small mammals that were hunted made up only 4% of the diet – mainly springbok lambs, springhares, bat-eared foxes, jackals and korhaans…Wildebeest generally made up the major part of the large animal remains consumed (55%), followed by Gemsbok (19%), red hartebeest (16%) and eland (10%). Springbok made up 98% of medium-sized mammals consumed. Small mammal remains were a mix of steenbok, bat-eared fox, jackal and springhare.[96]

Foraging and hunting

Brown hyenas are solitary foragers but may gather in small groups where there is a big carcass or large quantity of food – including around feeding sites where vegetable and animal waste is put out near observation hides. Brown hyenas travel long distances at night foraging for carcasses, hunting small animals or searching for fruits and plants. Brown hyenas in the Kalahari averaged 31.1km per night foraging. The longest distance recorded by Mills for one hyena was 54.4km over 13h 55m.[97] They will spend 80% of the hours of darkness seeking food.[98] They are not adept hunters and Mills recorded that of 128 hunts observed in the southern Kalahari, only six were successful.[99] In that region, the remains of lion kills were most often available to hyenas. Leopard kills were a source, but where there were trees were nearby, they carried their kills into trees away from scavengers.[100] In central Namibia, in areas of dry woodland, brown hyenas would wait near trees where leopards had stashed prey and would drive leopards off prey being consumed on the ground.[101]

Brown hyenas are more successful hunters along the coast of Namibia, where they hunt seal pups and scavenge carcasses where Cape fur seals haul up in large numbers on the sands – one of the largest colonies is at Cape Cross, where 80–100,000 seals can be found along the shore. Brown hyenas will kill seal pups on land, where they are not very mobile, frequently killing large numbers at one time and only eating a portion of each pup. Many female hyenas have their dens in easy reach of the colonies. The ability to predate the seals and scavenge the carcasses of seals, whales and other marine animals along the coast is an important part of their diet and enables females to raise cubs more successfully.[102]

Social behaviour and breeding

Brown hyenas live in clans ranging in size from a solitary female with cubs to small groups containing several females, each with offspring of varying ages. Clans as large as 14 have been observed,[103] but this is rare. One den at Okonjima was home to 12 hyenas, which was the biggest clan in the area.[104] The size of groups is determined by food availability and breeding success.[105] Where there are adult females and males in a clan, it appears that they have an "alpha male and alpha female, who share equal status".[106] The Owens' study of brown hyenas of the Central Kalahari showed that the alpha female shared equal rank with the alpha male and dominated natal males until they emigrated. The alpha female was able to secure the longest feeding time of any clan member and "although subordinate clan females were lower in rank than the alpha male, the hierarchy was non-linear and reversals did occur. Dominance can vary with context and clan females sometimes dominated the alpha male at a carcass."[107]

Females tend to stay in their natal clan and become breeding adults. Most males disperse from their natal clans. About 33% of the adult population is nomadic, mostly males.[108] Within groups, foraging remains solitary but they may gather at a large food source.[109] The home ranges of clans are usually occupied only by clan members and when members from different clans meet the result is usually aggression with clan members coming together to deter or attack intruders. Fights are usually limited to neck-biting, and perhaps ear ripping, accompanied by yells or growling.[110] Neck-biting, shaking of a subordinate animal by a dominant one and muzzle wrestling are employed in hierarchy disputes within clans. Hierarchy contests occurred when a new hyena joined a group (almost always a male) or when the status of one hyena was not well-established, according to the Owens.[111]

The territories are generally stable and are delineated by scent marking along the boundaries, either by defecating or pasting, with brown hyenas leaving a smeared black paste and a thick, white lipid paste. Nomadic males may enter territories to mate but do not become part of clans.[112] Territory size varies greatly according to the nature of the habitat, food availability, density of hyena and other predator populations and proximity to humans. Territories in the southern Kalahari may vary from 235–480sq km, on the Namib Desert coast around 220sq km, but as low as 49sq km in an agricultural region of the Transvaal.[113] There may be some overlap in territories in some areas, with conflict only occurring with incursions deep into one clan's territory, particularly near dens.

They are non-seasonal breeders with a gestation period of 97 days and a litter size ranging from one to five. They give birth in natal dens. These dens will be visited by other clan members to socialize and bring food.[114] The cubs suckle for 12–15 months. After about four months cubs are moved from the natal to the communal den. These dens may be moved regularly to get away from collections of parasites, avoid predators and because of things like cave-ins. All adult females in a clan may mate and give birth, having mated with males that have joined the clan or nomadic ones. Females within a clan may suckle each other's cubs and clan members will bring food to the den for the cubs. Females give priority to suckling their own cubs first.[115] The Owens recorded two males with half-siblings in the communal den helping feed them. These males migrated at 40 months of age as males usually did. In one case they observed, the cubs being fed were orphaned and a natal male was one of three clan members helping rear them to maturity.[116]

Once weaned, by the age of 15 months or a little before, cubs are effectively sub-adults, having been foraging for themselves from about 10 months old. Infant mortality is low, because of the cooperative feeding. Cubs born at a clan's communal den have a much better survival rate than those born in solitary dens of nomadic females, because of clan feeding of the young.[117] In the southern Kalahari, Mills estimated infant survival at 86%.[118] Sub-adult males disperse from their natal clans. They then have a higher mortality rate, especially when they moved out of protected areas into grazing land, where they may be killed as suspected stock raiders.[119] Conflict with larger carnivores, such as lions and spotted hyenas, is another cause of mortality.

The pasting mentioned and defecation at specific sites (such as the crossing of tracks and paths used by the hyenas, as I saw in Okonjima) are the main forms of communication within clans and to demarcate territory. They have no long distance calls, unlike spotted hyenas, and a limited vocal repertoire with a variety of whines, a yell (often when fighting each other) and growls.[120]

Competition with other carnivores

The brown hyena benefits from the presence of lions, which provide carcasses from which to scavenge. Lions may kill brown hyenas, but there is not the level of competition between them that there is between lions and spotted hyenas, chiefly because brown hyenas do not attempt to drive lions from kills and give way to lions on all occasions. Brown hyenas may compete with spotted hyenas for carcasses, but the spotted species is larger and more aggressive. Mills observed a single spotted hyena displacing several browns at a carcass on a number of occasions.[121] Spotted hyenas have been known to kill brown hyenas. There have been recorded examples of a brown hyenas successfully stealing from a spotted hyena,[122] including film at Okonjima of a brown taking a carcass from a spotted hyena.[123] Brown hyenas take kills from leopards and may force them to take refuge in trees.[124] On occasions, brown hyenas will not challenge an adult leopard but wait until the leopard has finished feeding and scavenge the remains.[125] Brown hyenas are easily driven from carcasses and kept away from kills by packs of wild dogs, but are capable of dominating cheetahs.

Conflict with humans

Brown hyenas are persecuted by livestock farmers across most of their range. They are blamed, often unjustly, for killing livestock. Usually, when they are seen on livestock carcasses, the animal has died of disease, natural causes or been killed by another carnivore. Brown hyenas may take young animals or sick ones but rarely healthy ones. Despite this, death at the hands of people is a major cause of mortality for adult brown hyenas, with animals shot, poisoned, trapped or hunted with dogs.[126] Brown hyenas may occasionally be shot by trophy hunters, who wish to tick an animal off a list.[127] Loss of habitat, with the expansion of areas for grazing, the death in bushmeat snares and deaths on roads are the other main anthropogenic threats.

Notes

1 Jennifer E. Smith and Kay E. Holekamp (2018) Spotted Hyenas, in J. C. Choe (ed), *Encyclopedia of Animal Behavior*, 2nd edition, Amsterdam: Academic Press/Elsevier, 190–208, p. 190.
2 Heather E. Watts and Kay E. Holekamp (2007) Hyena societies, *Current Biology*, 17, 16, no page numbers, https://www.cell.com/current-biology/pdf/S0960-9822(07)01497-2.pdf accessed 15 January 2020.
3 Personal communication with Dr Laurence Frank.
4 M. AbiSaid and S. M. D. Dloniak (2015) *Hyaena hyaena. The IUCN Red List of Threatened Species*, https://www.iucnredlist.org/species/10274/45195080 accessed 10 June 2020.
5 Person communication with Dr Laurence Frank.
6 P. Henschel et al. (2020) Census and distribution of large carnivores in the Tsavo national parks, a critical east African wildlife corridor, *African Journal of Ecology*, 2020, 00, 1–16. DOI: 10.1111/aje.12730 accessed 28 July 2020, p. 1.
7 T. Bohm and O.R. Höner (2015) *Crocuta crocuta. The IUCN Red List of Threatened Species*, https://www.iucnredlist.org/species/5674/45194782 accessed 10 June 2020.
8 Gidey Yirga et al. (2016) Spotted hyena (Crocuta crocuta) concentrate around urban waste dumps across Tigray, northern Ethiopia, *Wildlife Research*, 42, 7, 563–569, pp. 566–7.
9 I. Wiesel (2015) *Parahyaena brunnea. The IUCN Red List of Threatened Species*, https://www.iucnredlist.org/species/10276/82344448 accessed 10 June 2020.
10 San Diego Zoo, https://animals.sandiegozoo.org/animals/striped-hyena accessed 10 June 2020.
11 Gus Mills and Heribert Hofer (1998) *Status Survey and Conservation Action Plan Hyaenas*, Gland, Switzerland: lUCN/SSC Hyaena Specialist Group, http://www.carnivoreconservation.org/files/actionplans/hyaenas.pdf accessed 4 September 2019, p. 21.
12 Ibid.
13 Ibid., p. 22.
14 Aaron Parker Wagner (2006) Behavioural Ecology of the Striped Hyena (Hyaena hyaena), a dissertation submitted in partial fulfilment of the requirements for the degree of Doctor of Philosophy in Biological Sciences Montana State University Bozeman, Montana April 2006.
15 Mounir R. Abi-Said (2006) Reviled as a grave-robber: the ecology and conservation of striped hyaenas in the human-dominated landscapes of Lebanon, PhD Thesis, Durrell Institute of Conservation and Ecology, University of Kent.
16 Priya Singh (2008) Population density and feeding ecology of the striped hyena (Hyaena hyaena) in relation to land use patterns in an arid region of Rajasthan, thesis Submitted to The Manipal University In partial fulfilment for the degree of Master of Science in Wildlife Biology and Conservation.
17 Hans Kruuk (1976) Feeding and social behaviour of the striped hyaena (Hyaena vulgaris Desmarest) *East African Wildlife Journal*, 14, 91–111.
18 L. N. Leakey et al. (1999) Diet of Striped Hyaena in Northern Kenya, *African Journal of Ecology*, 37, 3, 314–26.
19 David E. Leslie (2016) A Striped Hyena Scavenging Event: Implications for Oldowan Hominin Behavior, *Field Notes: A Journal of Collegiate Anthropology*, 8, 1, 122–38; see also, Wagner, 2006, p. 1.
20 Randeep Singh et al. (2014) Population density of striped hyenas in relation to habitat in a semi-arid landscape, western India, *Acta Theriologica*, 59, 521–527, p. 521.
21 Mills and Hofer, 1998, pp. 22–3.

22 Kruuk, 1976, p. 93–4.
23 Ibid.
24 James T. Pokines and Julian C. Kerbis Peterhans (2007) Spotted hyena (Crocuta crocuta) den use and taphonomy in the Masai Mara National Reserve, Kenya, *Journal of Archaeological Science*, 34,1914–1931, p. 1916; Liora Kolska Horwitz and Patricia Smith (1988) The Effects of Striped Hyaena Activity on Human Remains, *Journal of Archaeological Science*, 1988,15, 47, 471–481, p. 472.
25 Kruuk, 1976, p. 95.
26 Shivish Bhandari et al. (2020) The diet of the striped hyena in Nepal's lowland regions, *Ecology and Evolution*, 3 April, 1–10, https://doi.org/10.1002/ece3.6223 accessed 14 June 2020, p. 6.
27 Wagner, 2006, p. 118–21.
28 See, for example, D. W. Macdonald (1978) Observations on the behaviour and ecology of the striped hyena, *Hyaena hyaena*, in Israel, *Israel Journal of Zoology*, 27, 4,189–198.
29 Hila Shamoon and Idan Shapira (2019) Limiting factors of Striped Hyaena, Hyaena hyaena, distribution and densities across climatic and geographical gradients (Mammalia: Carnivora), *Zoology in the Middle East*, 65, 189–200, p. 191.
30 L. N. Leakey et al., 1999, p. 315.
31 Horwitz and Smith, 1988, p. 472.
32 Leslie, 2016, 124–5.
33 Mills and Hofer, 1998, pp. 23.
34 Aaron P. Wagner, Laurence G. Frank and Scott Creel (2008) Spatial grouping in behaviourally solitary striped hyaenas, Hyaena hyaena, *Animal Behaviour*, 75, 1131–1142, p. 1138.
35 Mounir Abi-Said and Zuhair S. Amr (2012) Camera trapping in assessing diversity of mammals in Jabal Moussa Biosphere Reserve, Lebanon, *Vertebrate Zoology*, 62, 1, 145–152, p. 149.
36 Ibid.
37 Smith and Holekamp, 2018, p. 197.
38 Aaron P. Wagner et al. (2007) Patterns of relatedness and parentage in an asocial, polyandrous striped hyena population, *Molecular Ecology*, 16, 4356–4369, pp. 4365–6.
39 Wagner, 2006, p. 118.
40 Leslie, 2016. p. 127.
41 Wagner, 2006, p. 118.
42 Kruuk, 1976, p. 104.
43 Ibid.
44 Ibid., p. 105.
45 Ibid., p. 106.
46 Mills and Hofer 1998, p. 24.
47 Ibid.
48 Ibid.; see also Kruuk, 1976. p. 106.
49 Mills and Higer, 1998, pp. 24–5; V. G. Heptner and A. A. Sludskii (1992) *Mammals of the Soviet Union: Carnivora (hyaenas and cats), Volume 2*, Smithsonian Institution Libraries and National Science Foundation. https://archive.org/details/mammalsofsov221992gept/page/10 accessed 24 July 2019.
50 Vladimir Dinets and Beniamin Eligulashvili (2016) Striped Hyaenas (Hyaena hyaena) in Grey Wolf (Canis lupus) packs: cooperation, commensalism or singular aberration? *Zoology in the Middle East,* 62, 1, 85–87, pp. 85–6.
51 Mills and Hofer, 1998, p. 25.
52 Heptner and Sludski, 1992, pp.23–4.
53 Ibid., p. 46.
54 Mills and Hoger, 1998, p.25.
55 Abi-Said, 2006, p. 33.
56 San Diego Zoo, https://animals.sandiegozoo.org/animals/spotted-hyena accessed 16 June 2020.

57 Laurence G. Frank, Kay E. Holekamp and Laura Smale (1995) Dominance, Demography, and Reproductive Success of Female Spotted Hyenas, in A.R.E. Sinclair and Peter Arcese (eds), *Serengeti II Dynamics, Management and Conservation of an Ecosystem*, Chicago: University of Chicago Press, 364–384, p. 365.
58 Smith and Holekamp, 2018, pp. 192–3.
59 Mills and Hofer, 1998, pp. 29–31.
60 Mills and Hoger, 1998, p. 32.
61 Watts and Holekamp, 2007, no page numbers.
62 Hans Kruuk (1972) *The Spotted Hyena A Study of Predation and Social Behaviour*, Chicago: University of Chicago Press, pp. 80–1.
63 Ibid., p. 88.
64 Mills and Hofer, 1998, p. 33.
65 Joseph M. Kolowski (2007) Daily Patterns of Activity in the Spotted Hyena, *Journal of Mammalogy*, 88, 4, 1017–1028, p. 1017.
66 Ibid.
67 Kruuk, 1972, pp. 151 and 172–4.
68 Hans Kruuk (2019) *The Call of the Carnivores. Travels of a Field Biologist*, Exeter: Pelagic Publishers, p. 52.
69 Mills and Hofer, 1998, p. 33.
70 Kruuk, 1972, pp. 124–6.
71 Personal communication from Dr Laurence Frank.
72 Hans Kruuk (2002) *Hunter and Hunted. Relationships between carnivores and people*, Cambridge: Cambridge University Press, p. 21.
73 Smith and Holekamp, 2018, pp. 197–8.
74 Ibid.
75 Ibid.; Lily Johnson-Ulrich and Kay E. Holekamp (2020) Group size and social rank predict inhibitory control in spotted hyaenas, *Animal Behaviour*, 160, 157–168, p. 158.
76 Jennifer E. Smith et al. (2008) Social and ecological determinants of fission-fusion dynamics in the spotted hyaena, *Animal Behaviour*, 76, 619–636, pp. 619–20.
77 Ibid., p. 20.
78 Sarah Edwards (2015) *Human-wildlife conflict issues on commercial farms bordering the Sperrgebiet and Namib-Naukluft National Parks borders, southern Namibia*, PhD thesis, Royal Holloway, University of London, 8 September 2015, p. 58.
79 Smith and Holekamp, 2018, p. 200.
80 Ibid.
81 Mills and Hofer, 1998, p. 36; Kowolski, 2007, p. 1018.
82 Smith and Holekamp, 2018, p. 200.
83 Mills and Hofer, 1998, p. 37; personal communication with Dr Laurence Frank.
84 Ally K. Brown, Kay E. Holekamp, Eli D. Strauss (2020) Infanticide by females is a leading source of juvenile mortality in a large social carnivore, https://www.biorxiv.org/content/10.1101/2020.05.02.074237v1.full accessed 27 August 2020.
85 Kruuk, 1972, pp. 129–30.
86 Ibid., p. 132–3.
87 Mills and Hofer, 1998, p. 34.
88 Ibid., p. 26.
89 Bryan Shorrocks (2007) *The Biology of African Savannahs*, Oxford: Oxford University Press, p. 104.
90 Mills and Hofer, 1998, p. 3.
91 Michael G. L. Mills (1981) *The Socio-Ecology and Social Behaviour of the Brown Hyaena* Hyaena brunnea, *Thunberg, 1820, in the Southern Kalahari*, DSc Thesis, University of Pretoria; M. G. L. Mills (1982) Hyaena brunnea, *Mammalian Species*, 194, 23 November 1982, pp. 1–5.
92 Ibid., p. 1.
93 Mark and Delia Owens (1985) *Cry of the Kalahari*, Glasgow: Fontana/Collons, p. 74.
94 Shorrocks, 2007, p. 104.

95 Owens, 1985, p. 74.
96 Mills, 1981, pp. 43 and 46.
97 M. G. L Mills (1990) *Kalahari Hyenas Comparative Ecology of Two Species*, Caldwell, New Jersey: The Blackburn Press, p. 73.
98 Mills and Hofer, 1998, p. 27.
99 Ibid.
100 Mills, 1981, pp 54–55.
101 Personal communication from Dr Sarah Edwards.
102 Ingrid Wiesel, Sabrina Karthun-Strijbos and Inga Jänecke (2019) The use of GPS telemetry data to study parturition, den location and occupancy in the brown hyaena, *African Journal of Wildlife Research*, 49, 1, 1–11, pp. 2–3.
103 Watts and Holekamp, 2007.
104 Personal communication from Dr Sarah Edwards.
105 Mills, 1998, p. 75.
106 Sarah Edwards (2019) Understanding the shaggy-haired scavenger, *Africa Geographic*, 2 August 2019, https://africageographic.com/stories/the-brown-hyena-2/#agtravel-3 accessed 23 June 2020.
107 Delia Owens and Mark Owens (1996) Social dominance and reproductive patterns in brown hyaenas, Hyaena brunnea, of the central Kalahari, Animal Behaviour, 51, 535–551, p. 546.
108 Edwards, 2019.
109 Mills and Hofer, 1998, p. 28.
110 Personal communication from Dr Sarah Edwards; Mills and Hofer, 1998, p. 28.
111 Owens, 1985, pp. 78–9.
112 Mills, 1981, p. 4.
113 Mills and Hofer, 1998, p. 28.
114 Wiesel et al., 2019 p. 2.
115 Mills and Hofer, 1998, p. 28.
116 Owens, 1996, p. 543.
117 Ibid., p, 548.
118 Mills, 1981, p. 75.
119 Ibid.
120 Mills, 1981, p. 163.
121 Mills, 1981, pp. 54–55.
122 Ibid., p. 219.
123 Personal communication from Dr Sarah Edwards.
124 Owens, 1985, pp. 134–5.
125 Personal communication from Dr Sarah Edwards.
126 Mills and Hofer, 1998, p. 29.
127 Westbury et al., 2017 no page numbers.

2

HUMANS AND HYENAS FROM THE PLEISTOCENE TO THE HOLOCENE

Time periods

Eocene – 56 to 33.9 million years ago, henceforth mya
Oligocene – 33.9 million to 23.03mya
Miocene – 23.03mya to 5.33mya
Pliocene – from 5.33 to 2.6mya
Quaternary period – the last 2.6 million years
Pleistocene epoch – part of the Quaternary, stretches from 2.6mya to 11,700 mya
Early Pleistocene – from 2.6mya to c781,000 years ago
Middle Pleistocene – from c781,000 to 126,000 years ago
Late Pleistocene – from 126,000 to 11,700 years ago
Holocene – from 11,700 to present.

The evolution of hyenas from the first carnivorous mammals

To examine the evolution of modern hyenas and their relationship with humans, a good starting point is the appearance of early mammalian prototype-carnivores. These evolved from the first mammals and took over the niche in the food chain previously occupied by carnivorous dinosaurs, which disappeared with the extinction of dinosaurs 65mya. Many early mammalian carnivores were chiefly scavengers, but some hunted insects and small vertebrates. Some early meat-eating mammals were marsupials and precursors of the carnivorous and omnivorous marsupials that evolved in South America and Australia; marsupials of the northern hemisphere died out, apart from North American opossums.[1]

The mega-marsupial hunters of South America were huge, the size of a grizzly bear, weighing up to 200kg. One, *Proborhyena gigantea*, was found in South America during the Eocene (56 to 33.9mya) and up to the Oligocene (33.9mya to

23mya).[2] Despite its name it was not an ancestor of the modern hyena, even though, like hyenas, it had teeth capable of crushing large bones and combined hunting with stealing prey from other predators and scavenging carcasses. There were other wolf-like *borhyaenids* in the region, which died out in the later megafauna extinctions between 130,000 and 8,000BCE.

The creodont was another early meat eater, which developed from the squirrel-like *Cimolestes*, which had appeared around the time of the dinosaur extinction 65mya. Creodonts produced a variety of species – dog-like, bear-like, cat-like and hyena-like.[3] The latter were part of the creodont branch named hyaenodontids, which included animals the size of a weasel up to that of a striped hyena. Macdonald wrote, "the most fearsome of the hyaenodontids was a wolf-sized animal by the name of *Hyaneodon horridus*," found in North America around 25mya. Some creodonts had teeth designed for crushing bones. The largest hyaenodontid was the *Megistotherium*, which inhabited the Sahara (before it became a desert) 20–25mya. It may have weighed 800kg[4] and been "the largest mammalian land predator ever".[5]

Creodonts competed with and were succeeded as the dominant meat-eaters by the ancestors of modern carnivores, which also evolved from *Cimolestes*.[6] The order Carnivora evolved over millions of years ranging from small animals catching insects and small vertebrates to massive hyenas and feliform carnivores, such as sabre-toothed and scimitar-toothed cats (some were ancestors of the modern *Felidae*, but many were not true cats but members of other feliform families)[7] preying on huge herbivores. Many of these emerging members of the carnivore guild developed carnassial teeth suited to cutting flesh from prey, which enabled them to more efficiently process carcasses. Some also had the bone-crushing teeth that are still to be found in modern hyenas. The evolving carnivores replaced the creodonts and became the top land predators across the northern hemisphere by 30–20mya, mainly because of their dietary adaptability, as many were able to eat fruit and vegetables, as well as meat, and so survive declines in prey species in a way that the more specialized flesh-eating creodonts could not.[8]

The evolution of these carnivores coincided with climatic changes and the expansion of savannah and lightly wooded areas, starting 60–55mya, notably in Africa. Grasses and grazing animals appear to have evolved in tandem.[9] By 30–20mya the ancestors of most of the large mammals of modern African savannahs had appeared, helped by volcanic and geological activity that formed the Rift Valley and generated soils producing abundant grasses. The hoofed mammals found in Africa included pigs, giraffes, buffalo, wildebeest, hartebeest, reedbuck, gazelle and smaller antelopes,[10] providing prey species and so enabling the evolution and dispersal of carnivores, with the eventual appearance of the hyenas *(Hyaenidae)* as well as dogs, wolves, foxes and jackals (*Canidae*) and cats (*Felidae*). Many species which developed in Africa dispersed into the Middle East, Europe and Asia and some as far as the Americas. Others, like the wolf, developed outside Africa.

The modern family of carnivores closest to the primitive carnivores in appearance and diet is the *Viverridae* (genets, civets, linsangs and binturong etc)[11], from which the hyenas evolved. The viverrids produced cat-like and dog-like branches.

The cat-like branch produced the *Hyaenidae*. The true hyenas developed from species of *Viverridae* in heavily forested areas of Eurasia 22–25mya. They were arboreal and civet-like.[12] The earliest examples found are *Progenetta*, a small viverid-like hyaenid, found in European Miocene fossil deposits,[13] and *Plioviverrops*, civet-like and partly arboreal, eating small animals, carrion, insects and fruit. They evolved into terrestrial mammals; the aardwolf is the surviving descendent of this branch. The other modern hyena species evolved from the bone-crushing branch of feliform viverrids. Over 60 taxonomically confirmed species of hyaenids have been identified in Africa, Eurasia and North America, with the first appearing in the Middle Miocene,[14] but it is believed that in total they exceeded that as "in the fossil record… hyaenids are both diverse and abundant, and nearly 100 species have been named", existing for 25 million years from the Miocene.[15]

Another group of hyena-like animals, which probably evolved from the dog-like branch of viverrids, and are described by Macdonald as "a mysterious group",[16] were the percrocutoids, of which fossil remains of *Percrocuta australis* and *Percrocuta tobieni* have been found at sites in South Africa. Werdelin believes that while they have similar dentition to hyenas, they "are likely to be phylogenetically [in terms of evolutionary development and diversification of a species] distant from hyaenids".[17] They may have competed and coexisted with early hyenas around 15mya, but the percrocutoids disappeared by 5mya, partly because the true hyenas developed bone-cracking teeth and could out-compete their rivals. Hyenas thrived in Africa and Eurasia, in the latter until the arrival of competing dog species.

The dog-like hyaenids, ranging from jackal to wolf-size, are the most widely represented in the fossil record.[18] The carnivore fauna excavated sites in Turkey have shown that there were numerous hyenas there: of the carnivore fossils found on the central Anatolian plateau of Turkey, 68% were *Hyaenidae* and 12% *Felidae*.[19] *Hyaenictitherium wongii*, found at late Miocene sites in China, appears to have been common across Eurasia in dry, open areas without thick forest from China to Greece. Another hyena species, *Hyaenictitherium intuberculatum*, was also present. These hyenas coexisted with and are likely to have scavenged from the kills of sabre-toothed cats, which were found from China to Western Europe.[20] *Lycyaena chaeretis* and *Lycyaena dubia*, hunting rather than scavenging hyenas, have been found to have been present in many of these areas with finds in Spain, Greece and Turkey, in savannah/open woodland and where there were large numbers of equid and bovid herbivores.[21]

Around 10mya a broad shift happened in climate, vegetation and habitats for animals, with the decline of dense forests and expansion of grassland, bush savannah and open woodland. This encouraged the evolution of fleeter-footed antelope, gazelle and other grazers/browsers which thrived in open countryside, and fast predators or cooperative hunters like lions and spotted hyenas (which survived while most other hyena species disappeared). It also affected the development of hominids and ultimately the emergence of the hominin branch which evolved into the *Homo* genus and eventually modern humans, *Homo sapiens*.

The arrival of hominids and hominins as competing carnivores

The evolution of *Homo sapiens* from hominid and hominin predecessors paralleled carnivore evolution in Africa. Hominid ecology, like that of hyaenids from the Pliocene to the Holocene, was affected by changes in climate, vegetation and geology. In the early period when forest was the dominant vegetation across Africa, hominid distribution and survival relied heavily on "potable water, and animal-based food resources and plant-based food resources, with the plant diet providing more food than animal diet".[22] The early hominids were chimp-sized and needed safe sleeping sites, confining them to forests. They were preyed on by sabre and scimitar-toothed cats. Although there is no definitive evidence that they were preyed on by hyaenids, it is likely they scavenged hominid remains and were powerful enough to have killed hominids.

Geological and climate changes, which produced drier upland areas and the spread of open woodland and grassland, encouraged the evolution of hominids adapted to such terrain in East and Southern Africa. Such habitats also helped the evolution of hyenas. Spotted hyaenas were present in Africa just after 4.0mya, dispersing into Asia and first appearing in Europe in the Early Pleistocene. Brown and striped hyaenas appear to have originated in Africa around 3.0mya.[23] The fossil evidence leads archaeologists and anthropologists to believe that in East Africa the hominid line derived from arboreal ancestors who inhabited forests that disappeared during climate changes. These key changes in environment, which influenced human and carnivore evolution (especially lions and hyenas), took place first during the Miocene (15.5–12.5mya) and then the Pliocene-Pleistocene period between 3.0 and 2.5mya. They involved lower, seasonally restricted rainfall; tectonic changes creating highland areas; and related erosion, leading to increased diversity of soil and vegetation types. Volcanic activity enriched soils, providing more nutritious grasses over wide areas. The improved availability and quality of plant food supported a wide range of mammalian herbivores.

The two periods of cooling and drying were followed by what Owen-Smith calls "turnover pulses" in large mammalian herbivores, taking place between 10 and 5mya and at around 2.5 mya.[24] The first saw the development of the modern bovid genera and the first australopithecines, which evolved from the hominids. The arrival of the australopithecine hominins living in savannah and open woodland at the time of the second turnover pulse encouraged changes in their meat-eating and meat acquisition, just as hyenas were adapting to the new habitat, and, "As far back in human evolutionary history as evidence permits us to see, the presence of hyenas in ancestral human landscapes is always implied, if not positively attested to by their bones and teeth marks".[25] The human-hyena relationship appears to have evolved in Africa between the two pulses, around 4.4mya. Bone finds in the Middle Awash in Ethiopia includes the remains of six individual *Ardipithecines* (hominins from the Late Miocene and Early Pliocene) and evidence of the presence of hyenas. Baynes-Rock says that almost all the remains of the *Ardipithecines* from that place and time "have been heavily 'ravaged' by hyenas, probably *Crocuta dietrichi*, a precursor to spotted hyenas".[26] There is also evidence of hominins and hyenas occupying the same landscape at Laetoli in Tanzania. Three hominin prints were left in volcanic ash and mud there, which also preserved prints from hyenas, likely a *Crocuta* species as large as the modern spotted hyena.[27] Five species of hyena were present in the region at this time.[28]

According to Vinuesa et al.:

> [The] most likely evolutionary scenario for the genus *Crocuta* includes an African origin, being already recorded in this continent by *Crocuta dietrechi* at 3.85–3.63 Ma. Subsequently, this genus would have first dispersed into Asia [*Crocuta ultima*] by the earliest Pleistocene (ca. 2.2 Ma) and into Europe [*Crocuta spelaea*] around the early-middle Pleistocene (ca. 0.8 Ma).[29]

They add:

> ...there are no remarkable differences between the two extinct studied species, *C. ultima* and *C. spelaea*, except for the somewhat larger brain size of the former. Their overall brain morphology is more similar to that of *C. crocuta* than to that of the other extant bone-cracking hyenas...However, the two studied extinct species similarly differ from extant C. Crocuta...

with the modern spotted hyena having a more developed brain, suggesting it was the more intelligent and adaptable of the *Crocuta* species.[30] This is may explain its ability to survive in a more diverse range of habitats and its flexibility in hunting and foraging behaviour.[31]

By the second turnover pulse, archaeological evidence indicates meat-eating by hominins including:

> unmistakable evidence for at least a partial focus on tool-assisted consumption of medium to larger-sized mammals at 2.5–2.6 Ma (millions of years ago)... This fundamental shift does not simply represent a change in diet, but also a change in...habitat preferences, activity patterns, population size and structure, social behavior, predator avoidance, technology, and cognitive capabilities.[32]

It may also have "forced increased and novel interactions between hominins and carnivores, including competition for these carcasses...and enhanced predation risk from sympatric carnivores".[33] It is highly probable that after 2.6mya, hominins entered into the carnivore guild, through their increased meat consumption. Research into faunal remains at FLK Zinj (Olduvai) and Kanjera (Western Kenya) suggests that "Oldowan hominins at both locations acquired meat through hunting and scavenging".[34] The evidence cited by Oliver et al indicates that by 2.0mya, and probably even before that date, hominins at Kanjera specialized in hunting vulnerable juvenile ungulates rather than relying purely on "chance encounters and scavenging of carnivore-kills". In wooded areas around Olduvai, ambush hunting by hominins would have been a likely strategy.[35] This does not rule out scavenging (passive and aggressive) as complementary means of meat acquisition. Excavations and modern studies of hyena scavenging strongly suggest that spotted hyenas and striped hyenas "would have competed with Oldowan hominins for early access to carcasses for meat and late access for bone marrow"[36]. The study by Leslie in Kenya at Oldowan excavation sites suggests that "early hominins would have been able to scavenge mammal remains from the Pliocene phylogenetic counterparts of striped hyenas more easily than from those more closely related to spotted hyenas".[37]

The second turnover period involved the thinning out of bovid species, with those unable to adapt becoming extinct, and the divergence from other hominids of "robust australopithecines" adaptable to the open environments and the evolution of the more advanced hominins.[38] The growth in open habitat prompted changes in social organization through the development of social/pack hunting by hominins and other carnivores,[39] as used on occasions by modern spotted hyenas; but little is known about the precise social behaviour of extinct hyenas.[40] The opening up of savannah areas and decline in dense woodland had an effect on the fauna of eastern and southern Africa, with herbivore and carnivore megafauna disappearing (heavy sabre and scimitar-toothed cats and the mega-hyaenas), causing a decline in the number of hyena species.[41] Open canopy woodland, low tree and shrub bush, and grassland came to cover about 40% of Africa's land surface. This supported a diversity of herbivores and is an explanatory factor for why Africa avoided the more substantial megafaunal extinctions which affected North America and Eurasia. Owen-Smith estimates that during the megafaunal extinctions at the end of the Pleistocene between 15,000 and 10,000 years ago, the Americas lost 75% of genera, Europe and Australia 45% and Africa only 13% [42]. One should add, as Frank says, that African animals coevolved with hominins as the latter developed hunting and evolved evasive and defensive strategies that enabled them to exist as hominins became social hunters armed with simple weapons. In Eurasia and the Americas, experienced hunters with dogs and fire arrived in areas populated by large herbivores unused to human hunters, hastening extinctions started by climate changes.[43]

The diversity of African grazers and browsers which survived the period of extinctions enabled large predators like spotted hyenas to flourish, and helped hominids and then hominins to multiply and evolve. The hominins that emerged from the forests as they contracted and savannah expanded included early predecessors of humans, like *Australopithecus ramidus* (aka *Ardipithecus ramidus),* whose remains have been found in Ethiopia, and *Australopithecus afarensis* in Ethiopia and Tanzania. During the Plio-Pleistocene pulse, australopithecines split into two divergent lineages, both bipedal and utilising stone tools. One lineage brought forth *Paranthropus robustus* and *Paranthropus boisei*, which disappeared around 1mya. The other evolved into *Homo sapiens*. Owen-Smith believes that the success of the latter may have been related to the ability to adapt to a wider variety of foodstuffs, with increasing consumption of animal protein, obtained by scavenging, and perhaps the killing of slow, small or weak animals.[44] One adaptation to savannah living, another parallel with hyenas, was greater social organisation and the expansion in the size of groups. Without comparatively safe arboreal sleeping quarters, hominins needed to band together to survive. They began to use basic tools and weapons for food acquisition, processing, competitive scavenging and defence. As social organisation and tool use became more advanced, the capability developed to competitively scavenge with hyenas and to compete with hyenas, lions and other predators in hunting larger prey.

The skull of the first known hominin, *Sahelanthropus tchadensis*, was discovered at Lake Chad and dates from 7–6mya.[45] Around 2.6mya hominins began to fashion tools from stone, rather than just using unworked stones and branches.[46] They were named *Homo habilis* and identified as part of a stone-tool making culture from 2.5–1.5 million years ago by Louis Leakey, following tool and fossil discoveries at Olduvai Gorge. The making of tools and weapons, such as spearheads, improved hunting and defensive capabilities and made the hunting of large herbivores and competitive scavenging possible with lower risk.[47] Around 1.8mya, the species *Homo ergaster* evolved, with the ability to make and use more sophisticated tools. *Homo ergaster* developed into the more modern human species, evidence for this coming from Ethiopia's Awash valley, dated at 160,000 years ago.[48] There is new evidence from the Atlantic coastal mountains of North Africa of the evolution of *Homo* species. Recent finds from Jebel Irhoud in Morocco suggest that the evolution of *Homo sapiens* occurred around 315,000 years ago (fitting in with fossil evidence in East Africa of *H sapiens* originating 400,000–200,000 years ago).[49]

As Palmqvist et al. have noted, the evolving nature of the carnivore guild, especially of social hunters and scavengers like spotted hyenas and lions, may have forced *Homo* species towards behavioural and technological improvement in what is called the Acheulian tool-making period (1.6mya to 200,000years ago).[50] The relationship between the members of the carnivore guild, in which they influence each other's evolution, especially in behavioural terms, is termed co-evolution. As Stiner argues, "Virtually every major period in human evolutionary history seems to provide examples of co-evolutionary processes involving animals…we must conclude that humans are exceptionally prone to forming co-evolutionary bonds with other species."[51] This certainly seems to be the case with humans and social carnivores such as hyenas. They appear to have developed a co-evolutionary relationship through "predator-prey relations, predator-predator competition… Species may compete…by depleting shared resources…by competing directly over them…or both".[52] Stiner dates this competition and probable evolutionary influence as starting at least 2mya in Africa and spreading as hominins migrated into Eurasia and beyond, as did hyenas.[53]

For meat-eaters, the evolution of prey species had a clear effect on physical and social development. In East and Southern Africa, as Yamaguchi et al. explain, "From the late Miocene [23.03 to 5.333mya] through the Pliocene [5.33 to 2.58 mya], African mammalian faunas experienced a great change: 76% of the land mammals were new, and of these c. 53% were found in Africa alone".[54] The East African carnivore guild "underwent extensive changes in species richness and taxonomic composition during the course of the Plio-Pleistocene. Peak species richness occurred in the early late Pliocene (about 3.6–3.0 Ma), with subsequent gradual loss of richness to the present".[55] During the transition from Miocene into Pliocene 5.33mya, the number of *hyaenid* species fell in Africa and Eurasia from 59 to 32.

The decline in numbers of carnivore species occurred as hominins were becoming more efficient and aggressive as scavengers, and more adept as hunters. Werdelin and Lewis believe the evidence of increasing human size, brain capacity and technological skill mean that during the period of carnivore guild shrinkage, "hominins of the Plio-Pleistocene must have evolved effective anti-predator strategies in general and effective strategies against kleptoparasitism to compete successfully with the diversity of large-bodied carnivorans present in eastern Africa, particularly during the late Pliocene".[56] This could have been a contributory factor in the disappearance of early hyena species, and why the spotted hyena, with its ability to hunt most of its food and scavenge when the opportunity arose, survived and thrived. The survival success of the spotted hyena may also have been linked to the feeding pattern on its own kills and carcasses foraged or scavenged from other predators, in which meat and marrow-bearing bones were consumed or removed, leaving little for hominin scavengers. Pobiner's research at Ol Pejeta in Kenya found that, "In stark contrast to the lion-consumed samples, none of the bones from the carcasses consumed by hyenas had any bulk flesh left for a scavenger. Less than 30% of the bones retained even flesh scraps".[57]

As human brain capacity increased and social behaviour became more sophisticated, humans evolved a range of strategies for defence against and competition with predators.[58] Russell et al describe five aspects of hominid/ hominin interactions with wildlife:

> 1. Hominids as regular scavengers of remains of kills by other carnivores. 2. Carnivores as scavengers of remains stored by hominids in their shelters or campsites. 3. Carnivores as prey of hominids. 4. Hominids as prey of carnivores. 5. Coexistence, commensalism and domestication.[59]

The need for defence against predation or theft of kills by hyenas, growing hominin consumption of animal protein and increased brain capacity/intelligence led to "significant behavioural transformations. New subsistence strategies based on obtaining animal protein emerged, which generated changes in the ecological relationships between hominins and the other predators".[60] The move from being prey to scavenging carnivore kills, defending carcasses from scavenging by carnivores and killing of carnivores all attest to an evolving relationship, though not all aspects of vulnerability evident in early interactions disappeared as the ability to compete with carnivores increased.

The disappearance of giant hyenas, sabre and scimitar-toothed cats on both sides of the Mediterranean in the Middle Pleistocene "are contemporaneous with the first long-term hominin settlements. At that time, hominins hunted in groups and relied on new effective weapons; these two improvements allowed them to slaughter larger gregarious preys and to handle encounters with dangerous competitors".[61] The development of group hunting, as in other carnivores like hyenas, lions, wolves and wild dogs, permitted hominins to have a wider selection of prey than just small mammals, birds and sick or injured animals. The hunting strategy may have developed out of the group living of hominins and have reinforced the need for groups in order to maximise hunting and foraging capabilities.[62] Group hunting and foraging would also have had the advantage (as shown when hyenas defend kills from lions or drive lions from their kills) of being able to defend kills or steal carcasses from other predators.

Hominins in southern Africa developed tool-making in the 1.7–1.1mya period and in North Africa around 1.5–1.1mya. Around 1.0mya, man is believed to have started utilising fire for warmth, protection from predators at night, cooking and the production of fire-hardened wooden weapons.[63] Herbivore remains in fossil finds from after 1.8mya rose from about 15–25% of assemblages to 45%.[64] The increase in meat consumption at a time when the expansion of open grasslands made ambush harder suggests the development of more efficient thrown or fired weapons. *Homo erectus*, living about 1.8mya–300,000 years ago, has been shown to have used stone weapons for hunting. Fossil evidence also shows increasing hominin body size as *Homo erectus/ergaster* evolved, with improved stamina for hunting, increasing the ranges over which prey could be hunted in grasslands and greater strength for communal defence against predators and in competing for food.[65] At some stage snares, pitfalls and game pits would have been added to the array of hunting technologies, as shown in bone assemblages and artefacts from early San settlements in southern Africa.[66] There would have been competition with hyenas over carcasses of animals preyed on by humans and over scavenged kills (particularly lion kills).[67] Schaller and Lowther perceptively noted:

> [such] social carnivores as the wolf (*Canis lupus*), African wild dog (*Lycaon pictus*), spotted hyena (*Crocuta crocuta*), and lion (*Panthera leo*) possibly resemble the early hominids more closely in their social systems than any living non-human primate…some of the selective forces which influenced the social carnivores also had an effect on societies of hominids.[68]

In areas with high hyena populations there would likely have been increased competition between them and hominins for carcasses in savannah and open woodland areas, such as in northern Tanzania (close to Olduvai and Laetoli, where so many hominin and predator/prey bones have been found).[69] This area is a particularly good one for study, as its modern carnivore-human interactions and hunting/scavenging patterns can be used to suggest Pleistocene and early Holocene hominin hunting/scavenging and competition with hyenas and other predators, as "the broad structure of ecosystems in East Africa has not changed radically in the past several million years".[70] The savannah and open woodlands that support substantial numbers of small, medium and large ungulates and diverse predators, as well as humans, developed about 3mya. The hyaenid presence has changed little in that period, with the spotted, brown (the brown only now in southern Africa) and striped being present, but earlier hyaenids, notably *Pachycrocuta*, having disappeared around 1.5mya.[71] Blumenschine notes that hyenas would have been major competitors for carcasses in this region at the time that hominin development was enabling them to be more efficient scavengers, but adds that hyena kills or lion kills appropriated by hyenas would have provided little opportunity for passive scavenging because of the efficiency of hyenas in consuming carcasses swiftly and dealing with large bones in a way that lions, leopards or wild dogs could not.[72] Competition between humans and hyenas for carcasses would have been most intense when it came to large carcasses and recently abandoned felid kills. It is very clear that hominins and hyenas had "overlapping niches throughout much of their evolutionary history. Both evidently exploited fauna by hunting and scavenging, processed carcasses and bones to obtain valuable nutritional resources".[73]

Hominin efficiency in gaining access to these kills would have increased as communities expanded, developed cooperative scavenging strategies and used stone and projectile weapons. The growing advantage to early *homo* species in the scavenging power balance with hyenas is demonstrated by Schaller and Lowther's experiments in the Serengeti, in driving hyenas from kills.[74] Binford et al.'s work in Kruger National Park had the interesting result of suggesting that bone assemblages found at Olduvai, which had shown hyena tooth marks on bones assembled by hominins and were interpreted as meaning that hyenas gnawed the bones after they had been processed by humans, may have been scavenged by humans after the bones had been cracked, defleshed or gnawed by hyenas. This could mean that early humans were on occasions scavenging bones largely devoid of flesh from hyenas to obtain marrow.[75] Bunn and Ezzo argue that Koobi Fora, Olduvai and other bone assemblages, with evidence of defleshing by hominins, indicate the likelihood that this was not just passive scavenging of carcasses abandoned by other predators but also "active, confrontational scavenging to acquire large animals" and active scavenging and hunting of smaller animals.[76] This would clearly have involved hominins in conflict with predators such as hyenas. Bunn and Ezzo believe that *Homo rudolfensis* and *Homo erectus* were large enough for several of them to have driven hyenas or lions from carcasses.[77]

As the skills of the *Homo* species advanced and the population expanded they dispersed, spreading to Eurasia and southern Asia. Their superior size, tool-making capabilities and intelligence enabled them to out-compete less advanced hominins and may have contributed to the extinction of *australopithecines* and *paranthropus* by 1.5mya.[78] The evolutionary process saw changes in physiology, notably bone/skull structure, culminating in the appearance about 300,000 years ago, or possibly earlier, as evidenced by the Moroccan fossils, of *Homo sapiens*. This period saw advances in technology with the production of more sophisticated bladed tools and weapons.[79] Hafted weapons with sharp points appear in the African archaeological record 100,000–200,000 years ago[80]. Evidence uncovered by Mary Leakey suggests "groups of early hominids were never very large, but comprised a sufficient number of active males to form hunting bands and to protect the females and young in case of attack".[81]

One of the leading authorities on modern carnivores, George Schaller, takes the view that hominin ability in competing for carcasses contributed to aggressive patterns of behaviour towards predators and they might have used weapons to intimidate or attack predators.[82] Other studies also support the ability of both social hominids and then hominins to obtain meat through "the wresting of animals killed by other predators"[83]. King argues strongly that, "At the least, it would have been advantageous if early hominids could have defended their own kills against such competitors as hyenas, wild dogs, big cats" and driven other predators from their kills. He continues, "The carnivore data indicate that early hominids could have accomplished all of these ends, which today are accomplished by dholes, about half the size of early hominids, and by hyenas".[84]

Prior to the Plio-Pleistocene mammal decline, at least 30 hyena species had inhabited Africa, Europe and Asia and many were believed, on the basis of fossil evidence, to be wolf-like, fast hunters. One, *Ictitherium viverrinum*, was jackal-sized and its diet is likely to have included meat, insects and fruit.[85] Many of these species became extinct due to climate change and competition from dogs, which migrated to Eurasia around 7–5mya across the Bering Straits from America.[86] The dogs thrived and 22 species are to be found in Africa and Eurasia, compared with four hyena species, of which only the striped hyena is still found in Eurasia. The striped hyena lost out to the dogs in western, central and southern Europe during the Pleistocene and was restricted to its current range in Turkey, the southern Caucasus, Central, West and South Asia, North, East and West Africa.[87] The modern *hyaena* and *parahyaena* lineages are believed to have developed from hyenas known to have been present in eastern and southern Africa from about 6.5mya.[88]

Remains of early hyena species (such as *Hyaena algeriensis*) dating to the Miocene have been found at Wadi el Hammam in Algeria, while Early Pleistocene remains of spotted hyenas and striped hyenas have been found at Makapansgat, Swartkrans and Kromdraai in South Africa, and of brown hyenas at Olduvai in Tanzania.[89] Spotted hyena fossils from the same period have been found at Omo in Ethiopia and East Turkana in Kenya. Fossils from a number of extinct hyenas (*Hyaena bellax, Hyaena abronia, Hyanea preforfex*, and *Leecyaena forfex*) have also been found at sites in South Africa.[90] In the Early Pleistocene, a total of nine hyena species lived in Africa, which coincided with an abundance of sabre-toothed cats and mega-herbivore prey providing substantial numbers of carcasses from which to scavenge.[91] Only one hyena species, *Chasmaporthetes*, made the trip from Eurasia to North America around 3mya during the Pliocene. It was a fast, almost cheetah-like carnivore, but it failed to compete in the long-term with wolves, coyotes and other dog species and was extinct by 1.5mya.[92]

In Africa and parts of Eurasia, where hyenas proved able to compete more successfully with dogs than in America, they survived and were for a long period the dominant bone-crushing carnivore species. They were predominantly scavengers rather than specialised hunters (the spotted hyena an exception to this pattern), feeding largely on the carcasses of mega-herbivores. *Pliocrocuta* was the most widespread large bone-crushing hyaenid during the Pliocene and early Pleistocene, the earliest fossils found in China dating to 5–5.6mya.[93] *Pachycrocuta* was the largest hyena of all time. Originating in Africa, it outlasted *Pliocrocuta perrieri* and the cheetah-like pack hunter *Chasmaporthetes lunensis*. Remains of the huge *Pachycrocuta brevirostris*, were found in 1845 in the French region of Le Puy. It was a massive beast and was called by its discoverers the short-faced hyena.[94] This species evolved over 3mya simultaneously in Africa and Asia, migrating to Europe by about 2mya, as climate, vegetation and prey changes favoured dispersal. This coincided with human and lion migration from Africa into Europe and Asia. There is fossil evidence that these hyenas lived in the same habitat as humans. *Pachycrocuta* and human remains have been found next to a fossil elephant carcass in Spain, "and more than half the *Homo erectus* bones from China's famous Dragon Bone Hill showed gnaw marks that demonstrated how effective these carnivores were at dismantling human bodies".[95]

Pachycrocuta brevirostris was only as tall as a spotted hyena, but bulkier and low-slung, with hugely powerful jaws.[96] It was a scavenger, eating meat and large bones from carcasses of sick or old animals, but primarily those killed and only partially eaten by predators like sabre-toothed cats. Its size and evident power suggests the possibility of killing large prey, but it is doubtful that it was a swift runner like spotted hyenas. Its teeth have been described as sledgehammers that "could splinter the marrow bones of an elephant".[97] *Pachycrocuta* was present in the faunas of Africa and Eurasia, as confirmed by fossil records from Hungary, Czech Republic, Germany, Greece, Italy, France and the Mendip Hills in Britain.[98] It was clearly a powerful member of the carnivore guild, "as the largest hyaena ever recorded, it was certainly one of the top predators of that guild",[99] being larger than other predators except the lion-sized sabre-toothed cat *Homotherium latidens*. [100] It was the only hyena present in Europe at this time. It was joined and seemingly out-competed by the spotted hyena (*Crocuta crocuta*) at the end of the Early Pleistocene (around 781,000 years ago).[101] Evidence from the Early Pleistocene in southern Asia indicates that among the dominant carnivores were *Pachycrocuta brevirostris* and *Crocuta crocuta*. [102] *Pachycrocuta* appears to have lived in Asia from about 3.0–0.2mya.[103] *Pachycrocuta* was present in eastern and southern Africa (from 3.0–1.5mya). It is thought that *Pachycrocuta* was gregarious and may have formed extended family groups. *Crocuta crocuta spelaea* and *Crocuta crocuta* fossils also suggest they lived in social groups from early in their existence.[104] *Pachycrocuta* reached as far as East and Southeast Asia, from northern Korea to Thailand and Cambodia, but was eventually replaced by an Asian variant of the spotted hyena, *Crocuta crocuta ultima*, by 200,000 years ago in Southeast Asia.[105]

There is some evidence that *Pachycrocuta* preyed on or at least scavenged the bodies of hominins. This comes from the study of the large number of fossil remains of *Homo erectus* at Zhoukoudian in China, known as Peking Man. Controversy surrounded evidence from these early *homo* remains, with some experts believing that evidence of the breaking open of skulls and large bones of Peking Man suggested they practised cannibalism. But others believed that these broken-open bones could have been scavenged and crushed by the powerful jaws of hyenas.[106] The Chinese paleoanthropologist Pei Wenzhong, who codirected the early Zhoukoudian excavations, believed that the skulls had been chewed by hyenas.[107] That none of the *H. erectus* skeletons found was complete, and that they made up a small percentage of the bone assemblage at the Zhoukoudian cave site, reinforced the idea that these could have been the remains of carcasses scavenged or even hunted by *Pachycrocuta*. Examination of the *H. erectus* remains by Boaz and Ciochon supported the hyena thesis with extremities (hands and feet) missing and canine and bone-crushing tooth marks on many bones.[108] They believed that:

> Bite marks on the brow ridge above the eyes indicate that this protrusion had been grasped and bitten by an animal in the course of chewing off the face. Most animals' facial bones are quite thin, and modern hyenas frequently attack or bite the face first...[109]

The hyena method of biting through the face or tearing the face off or biting off extremities is also consistent with reports from the first European travellers and hunters in Africa and modern observations of injuries inflicted by hyenas (see Chapters 5, 6 and 7).

There is also evidence from the Middle Pleistocene from Spain of likely hyena (species unidentified) consumption of humans. The finds at Cova Negra (Valencia) have been dated from the second half of the Middle Pleistocene to Late Pleistocene and include 24 cranial, dental and postcranial remains of a juvenile Neanderthal, with two measurable carnivore tooth pits on the cranial surface compatible with bears and hyenas.[110] The authors of that study note that "Carnivore damage is much more common on Neanderthal bone surfaces than was previously thought", indicating hunting or scavenging of Neanderthals by large carnivores, including hyenas.[111] Hart and Sussman believe that the evidence that has been found at such sites supports the view that hominids were preyed on by ancient predators, including ancestors of or early forms of the modern spotted hyena.[112] These finds suggest that hyena-on-human predation was more than just sporadic, though we perhaps shouldn't go quite as far as Hart and Sussman in suggesting that hyenas were "consuming our ancestral kin like popcorn".[113]

Bone assemblages in excavated sites used by both hominins and hyenas often contain ungulate or other prey bones with evidence of both human and hyaenid alteration and "often confound analysis as to which species was the accumulator and which was the scavenger. This mutual theft or scavenging of food may have originated millions of years ago".[114] There remains considerable debate over early *Homo* species' roles as scavengers and hunters at the time of the first human dispersal out of Africa and the status of the genus *Homo* in Europe as carnivores during the late Early Pleistocene.[115] In the Iberian Peninsula, fossil finds in the Orce area of Spain suggest that:

> ...large scavenging carnivorans, such as the giant, short-faced hyena *Pachycrocuta*, were responsible for an intense bone modification activity during the late Early Pleistocene. Evidence of anthropic [human-related activity or influence] action on bones is also recorded at these sites.[116]

In the Terrassa area of north-eastern Spain studied, fossilised faeces from *Pachycrocuta* were found along with the bones of numerous herbivore species, with a huge proportion of hippos, fallow deer-like cervids the second most abundant species, and rhinos and horses also well represented. Other, less frequent finds include primates, bovids, pigs and giant deer.[117] These would all have been probable prey for human hunters and prey or sources of carcasses for hyenas. Madurell-Malapeira et al. suggest that the hyena tooth marks on the bones of herbivores suggested reliance on scavenging rather than hunting.[118] Some level of human-hyena competition would have been unavoidable and the study cited concludes:

> the presence of stable populations of saber-toothed felids in the late Early Pleistocene of Europe...probably opened an ecological niche for scavengers. This niche was probably occupied by both *Pachycrocuta* and early *Homo*, as Megantereon, given its great killing capabilities...was probably unable to consume all of the meat that it obtained by hunting large ungulates...these early human populations likely competed for these carcasses directly with very powerful carnivorans...such as *Pachycrocuta*. [119]

Given the size and strength of *Pachycrocuta* it would have been highly confrontational and the ability of hominins to engage successfully in intraguild competition would have been crucial to ensure access to carcasses.[120]

Competition would have been high in East and Southern Africa and even higher between hominins and other predator/scavengers in the late Villafranchian and Epivillafranchian (3.5–1.0mya) ecosystems of Southern Europe because of the lower productivity of vegetable foods and the need for protein and fats to cope with lower winter temperatures.[121] As Rodríguez and Mateos believe, competition for carcasses occurred but was not such that it led to the decline of either hyenas or humans and they coexisted for a prolonged period due to the fact that:

> their niches did not overlap completely and that neither of the two competitors overcame the other. Existence of both hyenas and hominins at low population densities, as suggested by our results, might be also a factor explaining the coexistence of both scavengers.[122]

The low hominin population levels at a time of hominin dispersal and growth in populations elsewhere may be a result, as Madurell-Malapeira et al. suggest, of the presence of large predators, such as sabre-toothed felids and *P. brevirostris*, which were "a limiting factor for food acquisition for these hominins".[123] When the sabre-toothed *Megantereon* and other large carnivores disappeared during the extinction events, hominins (part of the carnivore guild but omnivorous)[124] were better placed to survive and multiply/disperse than *Pachycrocuta*. But *Crocuta* survived in Europe for a period, and still survives in Africa because of its flexible hunting and scavenging behaviour and its ability to adapt to different diets and environments.

Globally, the survival of mega-hyenas was threatened in the Middle Pleistocene when the Earth's climate system changed substantially, with warm (interglacial) and cold (glacial) cycles causing substantial shifts in seasonal rainfall and temperature. This led to major changes in vegetation and fauna, with many mega-herbivores disappearing. This, in turn, was a driver of the extinction of many large predators – such as sabre and scimitar-toothed cats. The loss of the carcasses of large herbivores killed by those huge predators could, along with competition from an increase in spotted hyena numbers and the arrival in Europe of the lion (*Panthera leo)*, have been factors in the disappearance of *Pachyocrocuta.* [125] The "complex change in the inter-actions between flesh-eating and bone-consuming species of the [carnivore] guild"[126] could have been crucial to the inability of *Pachyocrocuta* to adapt in the way that the spotted hyena could because of its flexibility as a hunter and scavenger.

The major climatic and vegetation changes that occurred with the glacial and inter-glacial periods had a huge effect on ungulate prey numbers and through that on both hunter and scavenger members of the evolving carnivore guild. As Palombo et al. argue:

> it is reasonable to suppose a rough correlation between climatic changes and carnivore dispersal and turnover. In addition, changes in herbivore guild should remove keystone prey…[altering] the pre-existing prey–predator equilibrium, and led to new internal dynamics, causing evolutionary change in carnivores…[127]

and changes in the balance of power between humans and competing hunters and scavengers.

Despite the spotted hyena's hunting prowess and its ability to survive in quite diverse habitats, as demonstrated by its presence across wide areas of Africa, Asia and Europe, the Asian spotted hyena became extinct around the end of the Pleistocene, possibly surviving, according to some records, as late as 32,000–19,000 years ago in northern India, northern Thailand, and in central China until c16,700 years ago.[128] Early humans and hyenas coexisted in the same areas and archaeological finds suggest that in China they used the same caves, though at different times, for shelter from around 120,000 years ago, when it is believed *Homo sapiens* migrated into the Central China and surrounding areas.[129] There is no clear evidence of a relationship between the arrival of modern humans and the disappearance of the hyena, although Bacon et al. hypothesise that the disappearance "corresponds to two important events which could have been fatal to the hyena populations: the rapid shift to warmer and wetter climatic conditions (estimated over one or two centuries), and the occupation of growing populations of hunter-gatherers".[130] The changes around 11,500 years ago, leading to the Asian summer monsoon, growth in forest cover and the consequent reduction of open savannahs in the region, could have rendered the hyenas vulnerable to competition from tigers, leopards, dholes, wolves and the incoming human hunter-gatherers.

The fast-running or cursorial dog-like hyena *Chasmaporthetes* filled a very different niche to *Pachycrocuta* and mainly scavenging hyenas. Its ability to hunt fast prey may have been similar to the hunting technique of spotted hyenas. Appearing around 4.9mya and becoming extinct by around 780,000 years ago, at the start of the Middle Pleistocene, *Chasmaporthetes* was a genus of hyena distributed across Eurasia and Africa. It was the only hyena genus to cross the Bering land bridge to North America. Tseng at al. believe that the Old Crow Basin discoveries indicate that *Chasmaporthetes* "may have ranged throughout much of North America from its initial arrival during the Pliocene, to the Early Pleistocene," though not in large numbers anywhere, as suggested by the paucity of fossil finds.[131] This was not to last, though. Changes in the fauna and competition from *canids* spelled the end for *Chasmaporthetes*, which only survived until 1.5mya. The first large wolf-like species to appear in North America was *Canis armbrusteri*, a likely precursor to the to the dire wolves (now extinct) and the surviving gray wolves. *Chasmaporthetes* became extinct in the Americas within 600,000 years of the arrival of *C. armbrusteri* shortly before 1.4 to 0.85mya.[132]

By the Middle to Late Pleistocene, the scene was being set for the disappearance of most of the hyena species, with only the four modern species surviving into the Holocene. The modern striped hyena (*Hyaena hyaena*) is thought to have developed from the African *Hyaena makaponi* which lived from 3.6mya to 2mya, overlapping with the striped descendant which is first recorded about 2.5mya. The striped hyena then dispersed rapidly according to Rohland et al.,[133] into West Asia, parts of Central Asia and India from Africa around 0.13mya, with further dispersal in the Neolithic period less than 12,000 years ago. The modern brown hyena (*Hyaena brunnea*, formerly *Parahyaena brunnea*) developed from *Parahyanea howelli*, first recorded in Kenya 4.1mya. The likely evolutionary path of the spotted hyena is from other members of the *Crocuta* genus, *Crocuta dietrechi* or *Crocuta eturono*, found in Africa 4–3mya.[134]

The *Crocuta* genus in Africa dispersed into Europe and became the dominant member of the genus. They, or their immediate predecessors, were part of carnivore dispersals from Africa around starting about 3.5mya, and reaching parts of East Asia and also Pakistan[135] but with major dispersal events at 2.4mya, 1.3–1.5mya (when they moved into Europe and there was separation, according to Rohland et al., of northern and southern spotted hyena populations in Africa) and then 0.36mya.[136] Barnett has described the dispersal of *Crocuta* hyenas from Africa, noting periods of expansion and contraction of populations and clear connections between populations verified by DNA analysis. He says that "Africa has acted as a kind of hyaena pump, occasionally disgorging spotted hyenas into the rest of the Old World", concurring with the projection of other researchers that the first dispersal started about 3.5mya and with further dispersals, the last two around 1.5mya and 300,000 years ago.[137]

The European spotted or cave hyena (*Crocuta crocuta spelaea*) developed as a Eurasian species, found from Spain to Siberia, and is considered a sub-species of the spotted hyena which diverged from it 0.36mya after the third dispersal of *Crocuta* to

Eurasia.[138] The cave sub-species were larger than modern spotted hyenas – having been estimated to weigh 102kg,[139] compared with 44–90kg. There is controversy still over whether the Eurasian *Crocuta crocuta* was a separate sub-species or whether it was a geographic variation on the African spotted hyena, but they are clearly related and Europe's hyenas developed from animals that dispersed from Africa. Varela et al. argue that rather than being a sub-species, let alone a separate species (designated *Crocuta spelaea*), "the taxonomic status of the Pleistocene European spotted hyenas was revised using ancient DNA. That analysis confirmed the existence of genetic flow between the African and Eurasian populations during the Pleistocene". They treat the cave hyena as being *Crocuta crocuta* and a European version of the spotted hyena.[140] To avoid confusion between the European and African populations, I will refer to the European Pleistocene spotted hyena as a cave hyena, as is common in much of the literature.

Major archaeological finds in the Bohemian Karst region of Czech Republic indicate that during the Middle and Late Pleistocene hyenas were common in the area. Cave deposits from the Middle Pleistocene show the presence of *Pachycrocuta* and then the cave hyena and evidence of bones scavenged or from hunted prey.[141] Many of the caves in the region have bones from the Late Pleistocene of cave hyenas, *Crocuta crocuta spelaea*, suggesting the caves were dens. The caves may also at some stage have been used as hibernation sites for cave bears (*Ursus spelaeus*). The remains of cave bears showing signs of disarticulation and consumption by cave hyenas have been found at the sites, indicating scavenging of dead bears.[142] The bone assemblages in the caves indicate that the hyenas preyed on or scavenged Przewalski's horse, woolly rhinoceros, steppe bison, reindeer, red deer, Irish elk, European ass and the alpine species chamois and alpine ibex. Competing carnivores were the cave bear, cave lion (*Panthera leo spelaea*), wolf and wolverine, and early humans.

Homo antecessor (or *Homo erectus antecessor* – 1.2mya–500,000 years ago) reached Europe around 780,000 years ago.[143] *Homo heidelbergensis* and *Homo neanderthalensis* emerged in Eurasia between 600,000 and 350,000 years ago in the Middle Pleistocene. Modern humans (*Homo sapiens*) arrived in Europe during the Later Pleistocene between 45,000 and 40,000 years ago.[144] In the Middle to Late Pleistocene (300,000 to 30,000 years ago), many of the caves in the Bohemian Karst were used by early humans as well as hyenas, though there is no evidence to suggest of cohabitation or eviction of one species by the other.[145] The bone assemblages of Przewalski's horse compare closely with spotted hyena hunting patterns in East Africa, with much of the prey animal consumed where it was killed but large leg bones frequently carried away, sometimes back to den sites.[146] Remains of Przewalksi's horse make up the largest numbers of remains (16–51%), woolly rhinos the second largest in the hyena bone assemblages (25–30%), reindeer (7–15%) and bison (1–6%).[147] Bone assemblages from open-site, as opposed to cave, excavations at Westeregeln in central Germany show that cave hyenas in the Middle to Late Pleistocene (300,000–30,000 years ago) consumed large quantities of both woolly rhinoceros and woolly mammoth and Przewalski's horse, with indications that rhinoceros (especially juveniles) were not just scavenged but hunted by the hyenas.[148] The hyenas were regular hunters of the wild horses, and are thought to have hunted them in packs, as modern spotted hyenas do with zebras.[149]

There is evidence from fossil sites in Spain of human killing of and consumption of hyenas during the Late Pleistocene. The ulna bone of a cave hyena found at the Sala de los Huesos site (Cáceres, Spain) exhibits cutmarks and other carnivore damage in the form of pits and furrowing. Hidalgo says the bone shows "Evidence of human processing and consumption of carnivores during the Pleistocene", adding that this is quite unusual. The specimen Hidalgo examined was dated approximately around 120,000 and "represents a clear case of hominid butchering of a hyena carcass".[150] Examination of the bone assemblage in which it was found, in what was clearly a hyena den, showed the possibility that the hyena died and was scavenged by humans or that they killed and partially dismembered it. There is also evidence, from Arcy-sur-Cure in France, of hyena consumption of human flesh and bones – whether hunted or scavenged is not clear. The remains from there show typical hyena gnawing patterns on the maxilla bone of a Neanderthal human.[151]

Humans and hyenas (cave hyenas, spotted hyenas and striped hyenas) coexisted and competed for prey or carcasses during the Late Pleistocene in Eurasia and Africa and may at times have preyed on each other when opportunities arose or certainly scavenged from each other's kills or carcasses. The shrinking of the range and then extinction of the Eurasian cave hyena occurred at the end of the Pleistocene and early years of the Holocene (approximately 50,000–10,000 years ago), with the spotted, brown and striped hyena surviving in Africa and the striped hyena population outside Africa being restricted to south-eastern Europe; Turkey; the Caucasus; and parts of the Middle East, Central, West and South Asia. This was part of the global megafauna extinction, which removed many of the larger species of mammals from the scene, especially in Europe, parts of West Asia and North America.[152] In the western Palearctic region (covering Europe, Turkey, North Africa, the Middle East and the most westerly, temperate parts of West Asia), over a third of mammalian megafauna became extinct.[153] These included mega-herbivores which had been preyed upon or scavenged by cave hyenas and scavenged by striped hyenas – notably the Irish Elk, the woolly rhinoceros and the mammoth. Przewalski's horse disappeared from western and central Europe and was increasingly restricted to Central Asia. *Homo neanderthalensis* (Neanderthal man) also disappeared at this time, leaving *Homo sapiens* as the only human species.

The cave hyenas are most likely to have disappeared from much of Western Europe by 29,000 years ago. Stiner believes their final disappearance from Europe might have been as late as 13,000–11,000 years ago,[154] though this remains to be confirmed by carbon-dated remains. Ross Barnett is sceptical that they survived as late as that,[155] though does note that the radiocarbon dating from the most recent bone discovered "is not the same as the latest animal. Time and again, the animals of the Pleistocene have surprised scientists by turning up in time periods later than originally thought possible".[156] Stuart and Lister are more cautious and believe the earlier dates for extinction are the most reliable. They suggest that the

> latest calibrated dates for spotted [cave] hyaena from north-west and southern Europe at about 31 ka [ka = thousand years ago]...the period between 33.0 and 26.5 ka saw growth of the ice sheets...Spotted hyaena populations in Eurasia may have been impacted by seasonally very cold temperatures, and/or indirectly by a reduction in prey abundance...

around 30,500–28,500 years ago.[157] In Central Asia, West Asia and South-Western Asia, the spotted hyena survived longer but underwent a significant decrease in numbers during the last glaciation and disappeared around 13–11ka BP.[158] The shrinking of its range with total extinction in Eurasia and then in North Africa meant that the spotted hyena was limited in its range to sub-Saharan Africa, with some surviving in southern Egypt and northern Sudan.

Climate change, and the consequent changes in vegetation, reduced the more open habitats favoured by hyenas and their prey in Eurasia, but particularly western and northern Europe. This seemed to have reduced the habitat suitable for the cave hyena. This is surprising in some ways, given the known adaptability of modern spotted hyenas to diverse habitats, including forests, in Africa. The changes did, though, reduce megafaunal prey and carcass availability, and split cave hyenas off geographically from the main spotted hyena species, which survived in Africa, where loss of prey species was less pronounced, as savannah or open woodland habitat remained. Many existing hyena prey species – notably red deer, roe deer and wild boar – survived in Europe and could theoretically have maintained a reduced hyena population, but other factors intervened and prevented this.

Stiner explains the disappearance as being a result of two main factors which are closely linked:

> Europe may have experienced catastrophic loss of the kinds of habitats most suited to spotted hyenas, and a corresponding increase in mixed woodlands. Under these circumstances, spotted hyenas would have been inferior competitors to wolves, the latter being as much at home in forests as in open lands, and in highlands as in lowlands.[159]

Hyena specialist Arjun Dheer is sceptical of the idea of an assumed inferiority as a major influence, given the ability of modern spotted hyenas to live and thrive in environments with competitors such as lions and wild dogs.[160] But despite their flexibility in food acquisition, which has seen them survive across most of sub-Saharan Africa, even in the face of massive human expansion and habitat loss, the advance of dense forest at the same time as, and no doubt influencing, the advance of the wolf as an apex predator is likely to have had some role in sounding the death knell for European hyenas, and the role of the *H sapiens* cannot be discounted as another factor. Striped hyenas were also progressively pushed out of Europe, surviving the longest in south-eastern Europe, but now only found on the fringes of south-eastern Europe in the Caucasus and Turkey.

Homo sapiens had started to disperse from Africa 50,000–45,000 years ago and their arrival in Europe came at the time of climate change. Their arrival and the expansion of their range and numbers were followed by the disappearance of Neanderthals, many herbivore prey species and cave hyenas. The combination of climate change, wolf expansion and human activity "could have negatively and severely impacted large mammal populations", including cave hyenas, particularly because of likely loss of prey and increased and novel carnivore guild competition from wolves and humans.[161] They had, prior to the start of their decline, been present in Europe for at least a million years and were found from the Iberian peninsula to the Ural mountains of eastern Russia.[162] They are only missing in the fossil record from the Arctic zone of Eurasia and montane regions.[163] Bone assemblages at caves in Yorkshire and Devon indicate that cave hyenas lived and hunted in Britain. The fossils found show that large numbers of hyenas used the dens and they brought back body parts from animals as large as the hippopotamus.[164] The British cave hyenas disappeared around 31,000 years ago.[165] The evidence of *homo* co-habitation of regions of Europe with hyenas is provided by the Chauvet cave paintings, dating from around 32,000–30,000 years ago, which include a Palaeolithic painting of a hyena, appearing on one of the cave walls in Grotte de Chauvet.[166] There are scores of cave lions, but just one hyena depicted there.[167] Prior to this one image being found, there was only one representation of a hyena from this period and region – a small ivory sculpture from La Madlaine, France.

This lack of images has also been emphasised by Nikolai Spassov and Todor Stoytchev in their examination of images of hyenas:

> After looking through thousands of large mammal depictions in the rock art of Europe's Upper Palaeolithic published during the last century, our interest was raised by the inexplicably low number of Cave Hyaena zoomorphic pictograms. It is in discrepancy with the numerous fossil finds and the large area of Late Pleistocene *Crocuta crocuta spelaea*. During the Late Pleistocene this species inhabited all of Europe except the northern part of the continent.[168]

The few depictions there are can all be found in southern France at Chauvet, Lascaux and a few other sites. Why so few? Spassov and Stoytchev reject the idea that as a scavenger the hyena was somehow taboo and suggest that as there are a few images:

> It seems more probable that its lower rank in the 'animal worship' hierarchy in the imaginations and beliefs of the Ice Age hunters and artists was caused by entirely mundane, everyday reasons: the species has the unpleasant exterior of a scavenger, it's not a typical and looked-for prey as artiodactyls, it's not a serious enemy or rival like cave lions and bears, and it lacks the impressiveness of the mammoth or woolly rhino.[169]

The idea of hyenas as merely scavengers has a strongly European perspective, according to Marcus Baynes-Rock.[170]

In Africa, spotted hyenas had coexisted and competed with humans for longer, and the glacial period effects were not so severe there as in Eurasia, nor was there such a great loss of herbivore diversity. As Baynes-Rock puts it:

> While lions were still comparatively abundant in the upper Pleistocene and persisted in Europe into classical times hyenas in Eurasia were on the way out. After millions of years of persistence in the face of human presence and harsh conditions, hyenas could not adapt successfully to the modern human presence when combined with encroaching forests and competition from wolves…

which were better able to hunt in a variety of habitats, but particularly the advancing forests. The *Crocuta* species disappeared from Europe and Asia just prior to ten thousand years ago;[171] they survived a little longer in the Middle East before being confined to Africa.

Until the final restriction of *Crocuta* to Africa, they, and striped hyenas (which still survive there), were common in much of the Middle East, notably Palestine around the Galilee area. There is evidence that spotted hyenas competed with humans during the Middle Paleolithic period (c300,000 to 30,000 years ago). Fossil finds from the Manot caves in Galilee show the presence across that period of spotted hyenas and humans, both hunting the antelope and deer species found in abundance in the region before it became more arid.[172] Both hyenas and humans carried carcasses back to den sites. These periods, Middle Paleolithic in Eurasia and North Africa and the Middle Stone Age in sub-Saharan Africa, "are widely considered to offer the time frame in which behaviorally modern humans evolved".[173] Masseti and Covarelli estimated that during the late Pleistocene and the early Holocene periods, the striped and spotted hyenas were found in the Middle East and parts of Asia. Finds from various prehistoric sites in Palestine, where its remains have been found in levels ranging from 500,000–450,000 years ago to about 12,500 and 10,200 years ago, suggest a long period of residence there by both species.[174] The spotted hyena completely disappeared from Asia in the Upper Paleolithic period lasting from 43,000 to 26,000 years.[175]

The level of competition and conflict between hyenas and humans in the Palestine and the Middle East is hard to assess. Orbach and Yeshurun explain that, despite some niche partitioning of prey selection (gazelles for humans and fallow deer for hyenas) being essential if competition was not to lead to the elimination of one of the carnivores in a region, there was overlap in prey selection between them, but not a damaging level of competition.[176] Bones from caves in the Misliya area inland from Haifa in Israel from the early part of the Middle Paleolithic suggest by then an ability of human inhabitants to regularly hunt medium and large game, with substantial numbers of meat-bearing bones in the assemblages.[177] These assemblages showed no sign of the bones being from other carnivore kills or of having been modified by carnivores such as hyenas.[178]

By 10,000 years ago, human hunting abilities had improved, with evidence of projectile weapons being used in the Middle East as *Homo sapiens* migration from

Africa continued. Stone and antler points for weapons were found in the Manot caves.[179] This may have worked to prevent too much overlap in prey selection with hyenas, as the ability to use projectile weapons enabled hominins to hunt solitary prey in woodland, while hyenas appear, from the fossil record, to have preferred hunting ungulate herds in more open areas. This interpretation is supported by the evidence that Pleistocene spotted hyenas hunted medium-sized ungulates, which the Manot cave record supports, while humans were more flexible in their hunting/foraging and there was further partitioning of the diet they each relied on by humans "diversifying and broadening their diet to include vegetal food and small game".[180] When the spotted hyena declined and disappeared in this region, the increasing aridity of the region and shrinking of the prey base may have been a cause (though spotted hyenas have adapted to arid regions of southern Africa), with another factor being the introduction of domesticated animals into the region by people, who would then have seen spotted hyenas as a greater threat than the scavenging striped hyenas, though the latter are still persecuted today in some areas by pastoralists who blame them for killing livestock.

The survival of the spotted, brown and striped hyena in Africa, when all but the striped had disappeared in Eurasia, is undoubtedly a product both of the high percentage of savannah (including arid savannah and semi-desert) and open woodland habitat, and the large herbivore prey base that remained. This enabled spotted hyenas to prey on them and scavenge carcasses from other predators (including humans) and for the brown and striped hyenas to find sufficient carcasses to survive. Spotted hyenas were the most widespread of the survivors, living across the savannah, semi-desert, open woodland habitats and even tropical and montane forest, using caves, warthog and aardvark burrows as dens for their clans, which could be from a few animals up to 130 in number, according to prey availability and competition from other hunters (especially lions and humans).[181]

Brown hyenas, although generally solitary foragers, also use communal dens (usually involving related females and their offspring). Early evidence of this was found at the excavations at Swartkrans, Sterkfontein and Kromdraai in South Africa (where remains were also found of a possible ancestor of the brown hyena – *Hyaena bellax*, dating to the Early Pleistocene[182] – and of earlier, extinct hunting hyenas of the genera *Hyaenictis* and *Eryboas*, dating from the Pliocene).[183] The cave bone assemblages described by the South African researcher, Brain, included both bones clearly assembled by hyenas and cracked or chewed by them, and bones that show evidence of early human hunting or scavenging, and of processing of meat from the bones and their marrow for food.[184] Hominin remains found at the caves include *Australopithecus africanus*, dated to between 2.6 and 2.0mya, and bones thought to be *Homo habilis* or a similar early *Homo* species, perhaps *Homo gautengensis* dating from 1.8–1.5mya.[185] Brain also found from the examination of hominin bones that some had been modified by action of teeth, and so that hominins were falling prey to carnivores, including hyenas, who Brain described as "potential culprits".[186] Excavations at Gladysvale, 13 km northeast of the Sterkfontein caves, yielded a rich Plio-Pleistocene fauna, including specimens attributed to *Australopithecus africanus*, and also fossil hairs of probable human origin in a brown hyaena coprolite (fossilised faeces) from Gladysvale cave in South Africa in calcified cave sediment dated between 257,000 to 195,000 years ago.[187]

The late Pleistocene and early years of the Holocene saw extinctions of mammals in parts of Africa, due both to climate and vegetation change and the growth of human numbers and consequent hunting with more advanced weapons, as well as the first appearance of domestic stock, which competed for grazing and water.[188] The development of pastoralism would also have initiated a new phase in human-carnivore relations, with pastoralists defending stock vigorously from predation. In North Africa – and this may explain the restriction of the spotted hyena to sub-Saharan Africa – zebra, warthog, hartebeest, roan antelope, reedbuck, eland and other important prey species disappeared, most of them having gone by the dry period that followed the last Saharan moist period from about 14,000 to 4,000 years ago. These species had disappeared after the previous moist period (from 40,000–30,000 years ago), but repopulated parts of the Sahara and North Africa when the final moist period started.[189] Their eventual disappearance would have deprived spotted hyenas of their prey, but not necessarily affected the striped hyena, which was mainly a scavenger and for whom smaller, surviving mammals would have provided sufficient carcasses. Spotted hyenas are more water dependent than either striped or brown hyenas, and growing aridity in North Africa may have influenced the shrinking of the spotted hyena range.[190] The remaining populations or desert-adapted oryx, addax and gazelles (plus Barbary sheep in mountainous areas) would not have supported large populations of social carnivores, though lions survived there longer than spotted hyenas.[191]

Notes

1 David Macdonald (1992) *The Velvet Claw. A natural history of the carnivores*, London: BBC Books, pp. 13–4.
2 Roman Uchytel (no date) *Prehistoric fauna*, https://prehistoric-fauna.com/Proborhyaena-gigantea accessed 27 March 2019.
3 C.A.W. Guggisberg (1961) *Simba The Life of the Lion*, London: Bailey Bros and Swinfen, p. 15.
4 Macdonald, 1992, p. 23.
5 Guggisberg, 1961, p. 23.
6 Macdonald, 1992, pp. 22–3.
7 Paul Z. Barrett (2016) Taxonomic and systematic revisions to the North American Nimravidae (Mammalia, Carnivora), *PeerJ*, 4, e1658, https://www.ncbi.nlm.nih.gov/pmc/articles/PMC4756750/ accessed 27 March 2019.
8 Ibid., p. 26.
9 Robin S. Reid (2012) *Savannas of Our Birth. People, Wildlife and Change in East Africa*, Berkeley, Calif.: University of California Press, 2012, p. 83.
10 Ibid., p. 83.
11 Macdonald, 1992, pp. 26–7.
12 Ibid., p. 119.
13 M.G. L. Mills (1990) *KJalahari Hyenas. Comparative Behavioural Ecology of Two Species*, Caldwell: Blackburn Press, p. 9.
14 Victor Vinuesa (2018) *Bone-Cracking Hyaenas (Carnivora, Hyaenidae) from the European Neogene and Quaternary: Taxonomy, Paleobiology and Evolution*, PhD Thesis, Universitat

Autonome de Barcelona, October 2018, https://ddd.uab.cat/pub/tesis/2018/hdl_10803_665133/vivi1de1.pdf accessed 9 April 2019, p. 61.
15 Lars Werdelin and Nikos Solounias (1991) The Hyaenidae: taxonomy, systematies and evolution, *Fossils and Strata*, 30, 31st May 1991, p. 1.
16 Macdonald, 1992, pp. 126–7.
17 Werdelin and Solounias, 1991, pp. 4–5.
18 Vinuesa, 2018, p. 73.
19 Sakir Ōkurti et al. (2015) Carnivores from the Late Miocene locality of Hayranlı (Hayranlı, Sivas, Turkey), *Turkish Journal of Zoology*, 39: 842–867, http://journals.tubitak.gov.tr/zoology/issues/zoo-15-39-5/zoo-39-5-14-1407-38.pdf accessed 29 March 2019, p. 862.
20 Alan Turner and Mauricio Anton (1997) *The Big Cats and their fossil relatives*, New York: Columbia University press, pp. 43–8.
21 Dolores Pesquero (2011) An exceptionally rich hyaena coprolites concentration in the Late Miocene mammal fossil site of La Roma 2 (Teruel, Spain): Taphonomical and palaeoenvironmental inferences, *Palaeogeography, Palaeoclimatology, Palaeoecology*, 311, 1–2, 15 October 2011, pp. 30–7.
22 Eileen M. O'Brien and Charles R. Peters (1999) Landforms, Climate, Ecographic Mosaics, and the Potential for Hominid Diversity in Pliocene Africa, in Timothy Bromage and Friedemann Schrenk, *African Biogeography, Climate Change, and Human Evolution*, New York: Oxford University Press, 115–137, pp. 134–5.
23 A. Turner (1990) The evolution of the guild of larger terrestrial carnivores in the Plio-Pleistocene of Africa, *Geobios*, 23, 349–368, p. 349.
24 Norman Owen-Smith (1999) Ecological links between African Savanna Environments, Climate Change, and Early Hominid Evolution, in Bromage and Schrenk, 138–149, p. 138.
25 Marcus Baynes-Rock (2012) Hyenas like Us: Social Relations with an Urban Carnivore in Harar, Ethiopia, PhD Thesis, Department of Anthropology Macquarie University, Sydney, p. 21.
26 Ibid.
27 M. D. Leakey and R. L. Hay (1979) 'Pliocene footprints in the Laetolil Beds at Laetoli, northern Tanzania', *Nature*, 278, pp. 317–23, p. 320.
28 J. C. Barry (1987) Large carnivores (Canidae, Hyaenidae, Felidae) from Laetoli, M.D. Leakey and J.M. Harris (eds), in *Laetoli: A Pliocine Site in Northern Tanzania*, Oxford: Clarendon Press, pp. 235–58, p. 240.
29 Victor Vinuesa et al. (2016) Inferences of social behavior in bone-cracking hyaenids (Carnivora, Hyaenidae) based on digital paleoneurological techniques: Implications for human–carnivoran interactions in the Pleistocene, *Quaternary International*, 413 Part B, 7–14.
30 Ibid.
31 Ibid.
32 Briana Pobiner (2015) New actualistic data on the ecology and energetics of hominin scavenging opportunities, *Journal of Human Evolution*, 80, 1–16.
33 Ibid.
34 James S. Oliver et al. (2019) Bovid mortality patterns from Kanjera South, Homa Peninsula, Kenya and FLK-Zinj, Olduvai Gorge, Tanzania Evidence for habitat mediated variability in Oldowan hominin hunting and scavenging behavior, *Journal of Human Evolution*, 131, June 2019, 61–75.
35 Ibid.
36 David E. Leslie (2016) A Striped Hyena Scavenging Event: Implications for Oldowan Hominin Behavior, *Field Notes: A Journal of Collegiate Anthropology*, 8, 1, 122–38, p. 122.
37 Ibid.
38 Owen-Smith, 1999, p. 138.
39 Nobuyuki Yamaguchi et al. (2004), Evolution of the mane and group-living in the lion (Panthera leo): a review, *Journal of Zoology*, 263, 329–342, p. 331.
40 Vinuesa, 2018, pp. 70–1.
41 Mills, 1990, p. 9.

42 Owen-Smith, 1999, p. 148.
43 Author's personal communication with Dr Laurence Frank concerning the megafaunal extinctions.
44 Owen-Smith, 1999, p. 148.
45 John Iliffe (2007) *Africans. The history of a continent,* Cambridge: Cambridge University Press, 2nd edition, p. 6; M. Brunet et al. (2005), A new hominid from the Upper Miocene of Chad, Central Africa, *Nature*, 418, 2002, pp. 145–51; C. P. E. Zollikofer et al. (2005) Virtual cranial reconstruction of. Sahelanthropus tchadensis, *Nature*, 434, pp. 755–9.
46 Christopher Ehret (2016) *The Civilizations of Africa. A History to 1800*, Charlottesville, Virginia: University of Virginia Press, pp. 16–7.
47 L. S. B. Leakey (1961) *The Progress and Evolution of Man in Africa*, Oxford: Oxford University Press, 9 and 40.
48 Iliffe, 2007, pp. 7–8.
49 Jean-Jacques Hublin et al. (2017), New fossils from Jebel Irhoud, Morocco and the pan-African origin of Homo sapiens, *Nature*, 546, 289–292, 8 June 2017, p. 289.
50 Paul Palmqvist et al. (2007), A re-evaluation of the diversity of megantereon (mammallia, carnivora, machairodontinae) and the problem of species identification in extinct carnivores, *Journal of Vertebrate Paleontology*, 27, 1, 2007, pp. 160–175.
51 Mary C. Stiner (2012) Competition Theory and the Case for Pleistocene Hominin-Carnivore Co-evolution, *Journal of Taphonomy*, 10, 3–4, 129–45, p. 129.
52 Ibid.
53 Ibid., pp. 130–1.
54 Yamaguchi et al., 2004, p. 331.
55 L. Werdelin and M.E Lewis (2013) Temporal Change in Functional Richness and Evenness in the Eastern African Plio-Pleistocene Carnivoran Guild. *PLoS ONE*, 8, 3, https://doi.org/10.1371/journal.pone.0057944 accessed 12 April 2019.
56 Ibid.
57 Ibid.
58 Stiner, 2012, p. 133.
59 Jordi Russell et al. (2012) New Insights on Hominid-Carnivore Interactions during the Pleistocene, *Journal of Taphonomy*, 10, 3–4, p. 127.
60 Hans Kruuk (2002) *Hunter and Hunted. Relationships between carnivores and people*, Cambridge: Cambridge University Press, pp. 103–4.
61 Ibid.
62 John L. Gittleman (1989) Carnivore Group Living: Comparative Trends, in John L. Gittleman (ed) *Carnivore Behaviour, Ecology, and Evolution*, Ithaca, N.Y.: Cornell University Press, 183–208, p. 186.
63 Francesco Berna et al. (2012) Microstratigraphic evidence of in situ fire in the Acheulean strata of Wonderwerk Cave, Northern Cape province, South Africa, *Proceedings of the National Academy of Sciences*, 109, 20, E1215–E1220, DOI:10.1073/pnas.1117620109 accessed 28 March 2017.
64 Barham and Mitchell, 2008, p. 146.
65 Ibid., p. 147.
66 G. Mokhtar (ed) (1990) *General History of Africa. II Ancient Civilizations of Africa*, London: James Currey/UNESCO, 1990, p. 353.
67 Robert J. Blumenschine (1986) *Early Hominid Scavenging Opportunities. Implications of Carcass Availability in the Serengeti and Ngorongoro Ecosystems*, Oxford: BAR International Series 283, p. 69.
68 George B. Schaller and Gordon R. Lowther (1969) The Relevance of Carnivore Behavior to the Study of Early Hominids, Southwestern *Journal of Anthropology*, 25, 4 (Winter, 1969), 307–341, p. 308–9.
69 Blumenschine, 1986, p. 97.
70 Ibid., p. 109.
71 Ibid., p. 119.

72 Ibid., p. 128.
73 S. W. Lansing et al. (2009) Taphonomic and zooarchaeological implications of spotted hyena (Crocuta crocuta) bone accumulations in Kenya: a modern behavioral ecological approach, *Paleobiology*, 35, 289–309, p. 289.
74 Schaller and Lowther, 1969.
75 Lewis R. Binford, M. G. L. Mills and Nancy M. Stone (1988) Hyena Scavenging Behavior and Its Implications for the Interpretation of Fauna1 Assemblages from FLK 22 (the Zinj Floor) at Olduvai Gorge, *Journal of Anthropological Archaeology*, 7, 99–135, p. 126.
76 H. T. Bunn and J. A. Ezzo (1993) Hunting and scavenging by Plio-Pleistocene hominids: nutritional constraints, archaeological patterns and behavioral implications. Journal of Archaeological Science, 20, 365–389, p. 388.
77 Ibid., p. 389.
78 Ehret, 2016, p. 17.
79 Ibid., pp. 20–1.
80 Reid, 2012, p. 93.
81 Mary D. Leakey (1976) A Summary and Discussion of the Archaeological Evidence from Bed I and Bed II, Olduvai Gorge, Tanzania, in Glynn L. Isaac and Elizabeth R. McCown, *Human Origins Louis Leakey and the East African Evidence*, Menlo Park, Calif.: Staples Press, 431–459, p. 434.
82 George B. Schaller and Gordon R. Lowther (1969) The Relevance of Carnivore Behavior to the Study of Early Hominids, Southwestern, *Journal of Anthropology*, 25, 4 (Winter, 1969), 307–341, p. 324; see also R. Dart (1956), The Cultural Status of the South African Man-Apes, *Smithsonian Institution Report for 1955*, 317–338, p. 329.
83 Alan Turner (1988) Relative Scavenging Opportunities for East and South African Plio-Plestocene Hominids, *Journal of Archaeological Science*, 15, 327–341, p. 327.
84 Glenn E. King (1976) Socioterritorial Units and Interspecific Competition: Modern Carnivores and Early Hominids, *Journal of Anthropological Research*, 32, 3, 276–284, p. 282.
85 Macondald, 1992, p. 121.
86 Ibid.
87 N. K. Vereshchagin and G. F. Baryshnikov (1984) Quaternary Mammalian Extinctions in Northern Eurasia in Paul S. Martin and Richard G. Klein, *Quaternary Extinctions. A Prehistoric Revolution*, Tucson, Arizona: University of Arizona Press, pp. 483–516, p. 496.
88 Vinuesa, 2018, p. 86.
89 Savage, 1978, p. 259.
90 Ibid., pp. 259–60.
91 Mills, 1990, p. 9.
92 Macdonald, 1992, p. 128.
93 Vinuesa, 2018, p. 85.
94 Brian Switek (2017) Paleo Profile: The Short-Faced Hyena. *Pachycrocuta brevirostris* was the largest hyena of all time, *Scientific American*, 20 January 2017, https://blogs.scientificamerican.com/laelaps/paleo-profile-the-short-faced-hyena/ accessed 27 March 2019.
95 Ibid.
96 Ibid.
97 Macdonald, 1992, p. 128.
98 Vinuesa, 2018, p. 78.
99 A. Turner and M. Anton (1996) The giant hyaena Pachycrocuta brevirostris (Mammalia, Carnivora, Hyaenidae), *Geobios*, 29, 4, 455–468, p. 455.
100 Ibid.
101 A. Turner (1992) Large carnivores and earliest European hominids: changing determinants of resource availability during the lower and middle Pleistocene, *Journal of Human Evolution*, 22, 109–126.
102 R. W. Dennell, R. Coard, A. Turner (2008) Predators and scavengers in Early Pleistocene southern Asia, *Quaternary International*, 192, 78–88, p. 78.
103 Ibid., p. 79.
104 Vinuesa, 2018, p. 74.

105 Anne-Marie Bacon et al. (2018) Testing the savannah corridor hypothesis during MIS2: The Boh Dambang hyena site in southern Cambodia, *Quaternary International*, 464, Part B, 15 January 2018, 417–439, https://www-sciencedirect-com.chain.kent.ac.uk/science/article/pii/S1040618217300368 accessed 7 May 2019; see also, G. Shen et al. (2009) Age of Zhoukoudian *Homo erectus* determined with ^{26}Al/^{10}Be burial dating, *Nature,* 458, 198–200.

106 Noel T. Boaz and Russell L. Ciochon (2001) The Scavenging of 'Peking Man', *Natural History*, March 2001, 110, 2, p. 46.

107 Ibid.

108 Ibid.

109 Ibid.

110 E. Camarós (2017) Hunted or Scavenged Neanderthals? Taphonomic Approach to Hominin Fossils with Carnivore Damage, *International Journal of Osteoarchaeology*, 27, 606–620, pp. 608–9.

111 Ibid., p. 616.

112 Donna Hart and Robert W. Sussman (2009) *Man the Hunted. Primates, Predators, and Human Evolution*, Philadelphia, PA: Westview Press, p. 5.

113 Ibid., p. 261.

114 James T. Pokines and Julian C. Kerbis Peterhans (2007) Spotted hyena (Crocuta crocuta) den use and taphonomy in the Masai Mara National Reserve, Kenya, *Journal of Archaeological Science*, 34, 1914–1931, p. 1915.

115 Joan Madurell-Malapeira (2017) Were large carnivorans and great climatic shifts limiting factors for hominin dispersals? Evidence of the activity of Pachycrocuta brevirostris during the Mid-Pleistocene Revolution in the Vallparadís Section (Valles-Pened es Basin, Iberian Peninsula), *Quaternary International*, 431, 42–52, p. 42.

116 Ibid., p. 43.

117 Ibid., p. 45.

118 Ibid., p. 49.

119 Ibid., pp. 50–1.

120 Jesús Rodríguez and Ana Mateos (2008) Carrying capacity, carnivoran richness and hominin survival in Europe, *Journal of Human Evolution*, 118, 72–88, p. 73.

121 Ibid., p. 84.

122 Ibid.

123 Madurell-Malapeira, 2017, p. 51.

124 Rodríguez and Mateos, 2008, p. 84.

125 Palombo, 2008.

126 Turner, 1992, p. 466.

127 Maria Rita Palombo, Raffaele Sardella and MicaelaNovelli (2008) Carnivora dispersal in Western Mediterranean during the last 2.6 Ma, *Quaternary International*, 179, 1, 176–189.

128 Ibid. See also V. Zeitoun et al. (2010) The Cave of the Monk (Ban Fa Suai, Chiang Dao wildlife sanctuary, northern Thailand), *Quaternary International*, 220, 160–173.

129 G. J. Shen et al. (2013) Mass spectrometric U-series dating of Huanlong cave in Hubei Province, central China: evidence for early presence of modern humans in eastern Asia, *Journal of Human Evolution*, 65, 162–167.

130 Bacon et al., 2018.

131 Z. Jack Tseng, Grant Zazula and Lars Werdelin (2019) First Fossils of Hyenas (Chasmaporthetes, Hyaenidae, Carnivora) from North of the Arctic Circle, *Open Quaternary* (online version no volume or issue number), https://www.openquaternary.com/article/10.5334/oq.64/ accessed 20 July 2019.

132 Ibid.

133 Nadin Rohland et al. (2005) The Population History of Extant and Extinct Hyenas, *Molecular Biology and Evolution*, 22, 12, December 2005, pp. 2435–2443, https://doi.org/10.1093/molbev/msi244 accessed 11 April 2019.

134 Vinuesa, 2018, p. 86.

135 Rohland et al., 2005.
136 Ibid.
137 Ross Barnett (2019) *The Missing Lynx. The Past and Future of Britain's Lost Mammals*, London: Bloomsbury, pp. 45–6.
138 Ibid.
139 Carlo Meloro (2007) Plio-Pleistocene large carnivores from the Italian peninsula: functional morphology and macroecology, PhD Thesis, Università degli Studi di Napoli "Federico II", http://www.fedoa.unina.it/1935/1/Meloro_Scienze_Terra.pdf accessed 15 April 2019, p. 28.
140 Sara Varela et al. (2010) Were the Late Pleistocene climatic changes responsible for the disappearance of the European spotted hyena populations? Hindcasting a species geographic distribution across time, *Quaternary Science Reviews*, 29, 17–18, 2027–2035, p. 2027.
141 Cajus G. Diedrich and Karel Žák (2006) Prey deposits and den sites of the Upper Pleistocene hyena *Crocuta crocuta spelaea* (Goldfuss, 1823) in horizontal and vertical caves of the Bohemian Karst (Czech Republic), *Bulletin of Geoscience*, 81, 4, pp. 237–276, p. 237.
142 Ibid.
143 Alice Roberts (2018) *Evolution. The Human Story*, London: Dorling Kindersley, p. 130.
144 Ibid., p. 189.
145 Diedrich and Žák, 2006, p. 246.
146 Ibid., p. 253.
147 Ibid., p. 265.
148 C. G. Diedrich (2012) Late Pleistocene *Crocuta crocuta spelaea* (Goldfuss, 1823) clans as Prezewalski horse hunters and woolly rhinoceros scavengers at the open air commuting den and contemporary Neanderthal camp site Westeregeln (central Germany), *Journal of Archaeological Science*, 39, 6, 1749–1765.
149 Ibid.
150 Antonio Rodríguez Hidalgo (2010) The Scavenger or the Scavenged? *Journal of Taphonomy*, 2010, 1, 75–6, p. 75.
151 Barnett, 2019, p. 49.
152 Varela et al., 2010, p. 2027.
153 A. D. Barnosky et al. (2004) Assessing the causes of late Pleistocene extinctions on the continents, *Science*, 306, 70–75, p. 71.
154 Mary C. Stiner (2004) Comparative ecology and taphonomy of spotted hyenas, humans, and wolves in Pleistocene Italy, *Revue de Paléobiologie, Genève*, 23, 2, 771–785, p. 771.
155 Personal communication with Ross Barnett, 20 July 2019.
156 Barnett, 2019, p. 57.
157 A. J. Stuart and A. M. Lister (2014) *Quaternary Science Reviews*, 96, 108–116, p. 114.
158 Hervé Monchot and Marjan Mashkour (2010). Hyenas around the cities. The case of Kaftarkhoun (Kashan-Iran), *Journal of Taphonomy*, 8, 1, 17–32. p. 18.
159 Stiner, 2004, p. 781.
160 Personal communications, 10 November 2019.
161 Varela et al., 2010, p. 2027.
162 Ibid.
163 Vereshchagin and Baryshnikov, 1984, p. 496.
164 C. K. Brain (1981) *The Hunters or the Hunted. An Introduction to Africa Cave Taphonomy*, Chicago: University of Chicago Press, p. 56.
165 Barnet, 2019, p. 39.
166 J. M. Chauvet, E. B. Deschamps and C. Hillaire (1996) *Dawn of Art: The Oldest Known Paintings in the World*, New York: H. N. Abrams, p. 110.
167 Baynes-Rock, 2012, p. 33.
168 N. Spassov and T. Stoytchev (2004) The cave hyaena in the Upper Palaeolithic rock art of Europe, *Historia naturalis bulgarica*, 159, 16, 159–166, p. 159.
169 Ibid., p. 164.
170 Personal communication with the author.
171 Baynes-Rock, 2012, p. 33.

172 M. Orbach and R. Yeshurun (2019) The hunters or the hunters: Human and hyena prey choice divergence in the Late Pleistocene Levant, *Journal of Human Evolution*, published January 2018, https://doi.org/10.1016/j.jhevol.2019.01.005 accessed 15 May 2019, p. 13.
173 Israel Reuven Yeshurun, Guy Bar-Oz and Mina Weinstein-Evron (2007) Modern hunting behavior in the early Middle Paleolithic: Faunal remains from Misliya Cave, Mount Carmel, *Journal of Human Evolution*, 53, 656–77.
174 Marco Masseti - Alessandra Covarelli Presence and distribution of Hyaena hyaena L., 1758, in Jordan, with particular reference to its occurrence in the southern desert, https://www.academia.edu/7979320/Presence_and_distribution_of_Hyaena_hyaena_L._1758_in_Jordan_with_particular_reference_to_its_occurrence_in_the_southern_desert?email_work_card=view-paper accessed 21 March 2020, p. 1.
175 Ibid.
176 Orbach and Yeshurun, 2019, p. 14.
177 Yeshurun et al., 2007, p. 678.
178 Ibid.
179 Orbach and Yeshurun, 2019, p. 14.
180 Ibid., p. 15
181 Brain, 1981, p. 57.
182 R. J. G. Savage (1978) Carnivora, in Vincent J. Maglio and H. B. S. Cooke (eds), *Evolution of African Mammals*, Cambridge, Mass: Harvard University Press, 249–267, p. 260.
183 Brain, 1981, p. 266.
184 Ibid., p. 140
185 See D. Curnoe (2010) A review of early *Homo* in southern Africa focusing on cranial, mandibular and dental remains, with the description of a new species (*Homo gautengensis* sp. nov.), *HOMO – Journal of Comparative Human Biology*, 61, 3, 151–177.
186 Brain, 1981, p. 83.
187 Lucinda Blackwell (2009) Probable human hair found in a fossil hyaena coprolite from Gladysvale cave, South Africa, *Journal of Archaeological Science*, 36, 6, 1269–1276.
188 Richard G. Klein (1984) Mammalian Extinctions and Stone Age People in Africa, in Martin and Klein op. cit., 553–573, p. 563.
189 Ibid., p. 555.
190 Person communication with Arjun Dheer.
191 Somerville, 2019, pp. 100–2.

3

HUMANS AND HYENAS IN AFRICA TO 1600CE

Pastoralism, cultivation and the environmental effects of human development

The growing influence of human activity over the natural world, population growth, improvements in food acquisition and, particularly, hunting methods, substantially affected the nature of human-hyena coexistence. Some forms of conflict (competition for carcasses from which to scavenge) declined and new ones developed. The key area of conflict began to focus on predation of livestock or other domestic animals by hyenas and their protection by humans. The growth in human cultivation of crops and expansion of settlements also influenced hyena interactions with people, as habitats previously only moulded by natural processes were now increasingly engineered by humans – forest clearance, the planting of crops in place of natural vegetation and the grazing of cattle sheep and goats on land that previously provided food and cover for wild ungulates, the major food sources for spotted, brown and striped hyenas.

As soon as humans had crops and domesticated animals to protect, they started to control wildlife where it was "perceived to harm the livelihoods, lives or lifestyles of people".[1] While the protection of human life from predators was nothing new, crop and livestock protection was, which affected not only hyenas but all predators and other wildlife inhabiting the land converted to human settlements, crop fields or grazing for livestock. The growth of pastoralism and farming, combined with technological advances in hunting capabilities, had growing effects on the balance of power between humans, their potential prey and the other members of the carnivore guild with whom they competed in an expanded range of ways. When wildlife was perceived by people as destroying or damaging crops, or killing livestock and domesticated animals like hunting dogs, "a common response has been to kill them".[2] This was, and still often is, seen as the cheapest and most practical strategy, avoiding the need for fences, more labour-intensive herding techniques and other measures to deter predators.[3]

From around 20,000BCE, what Reid has labelled the "soft boundary" between humans, wildlife and the environment, with people living in small communities gathering wild plant foods and hunting/scavenging for meat, began evolving. The soft boundary involved protection of people from predators, people hunting wild prey or scavenging, foraging for wild cereals, fruits and tubers and gathering of wood for fires. The wildlife side of the boundary involved predation by carnivores; competition within the carnivore guild (of which humans and hyenas were members) for live prey and carcasses; and browsing of plants by ungulates in effective competition with human gathering of wild plants, seeds, fruit and nuts. Humans and wildlife shared territory and used the same resources in similar ways. Over the next 5–10,000 years, it became an increasingly hard boundary through the domestication of animals and spread of the cultivation of plants for food, the latter developing from the regular harvesting of wild cereals, vegetables, tubers and fruits.[4]

Using Reid's model, human economic evolution first created a mixed boundary, as people began to cultivate plants and to domesticate ungulates for their milk and meat, while still hunting and foraging. Instead of having temporary shelters, more structured seasonal and then increasing numbers of permanent settlements were established. The hard border advanced with the clearing of woodland or bush, fencing of cultivated land, attempts to exclude wildlife from fields and the enclosure of livestock at night to protect them from predators. As cultivation and animal husbandry methods improved, the boundary between farmed/settled areas and wildlife habitat became ever more exclusionary.[5] While wildlife was progressively excluded from cultivated areas, and predators seen as mortal enemies excluded for the safety of people and livestock, it is evident from the long-term historical record, that some wildlife survived this process, and spotted and striped hyenas became adept at penetrating the boundary. Their ability to both hunt and scavenge, especially from the detritus of human settlements (including the bodies of humans deposited in the bush or in shallow graves), and the omnivorous diet of striped hyenas, including vegetable as well as animal matter, enabled them to benefit from waste produced by humans. The hard border for much wildlife was more permeable for hyenas.

Around 19,000–20,000 years ago, people began intensive exploitation of tubers and fish in Egypt's Nile Valley. By 15,000BCE, communities along the northern Nile were systematically harvesting wild cereals rather than just foraging for plant food.[6] To the west of the Nile, peoples of the Nilo-Saharan and Niger-Kordofan language groups, in north-west Sudan and Chad, pursued livelihoods based on foraging, hunting and fishing, and were producing more sophisticated spears, bows and arrows.[7] In the eastern Mediterranean and the area of the Tigris and Euphrates rivers (present-day Iraq), hunter-gatherers established more permanent settlements as they harvested wild cereals, nuts, pulses and fruit, supplemented with hunting and fishing.[8] The evolution from harvesting wild plant foods into early forms of arable farming based on cereals and pulses took place independently in different parts of the world. There is evidence that this form of food production had developed in the Middle East and West Asia by 9600BCE and spread to other areas, including South Asia and Africa, though with some evidence suggesting there could have been independent development in some regions of plant cultivation.

At this stage of human history, the spotted hyena was found across most of Africa, but had gone from Eurasia and was disappearing from North Africa. The brown hyena was limited to southern Africa's arid regions. The striped hyena was found in North Africa, the drier parts of West Africa with open landscapes (not in areas of thick forest) and the Sahel across into Sudan and the Horn of Africa, and in East Africa as far south as central Tanzania. Beyond Africa it was present in the Arabian peninsula, the Levant, Palestine, and from Turkey across Central and West Asia and into South Asia, but was only present on the fringes of eastern and southern Europe. As far as can be judged from the limited archaeological and written evidence, the striped hyena was widely distributed across its range in dispersed populations. As a shy, nocturnal forager, it can survive close to human settlements without being easily spotted. The striped hyena is still the most widely distributed carnivore in West Asia[9] and is the most widespread carnivore in India.[10]

Across the ranges of the hyenas, the growth of settled cultivation of plants and the domestication of animals for food created new areas of conflict between people, their stock, wild ungulates and predators. For combined hunter/scavengers like hyenas it also created new opportunities for hunting livestock and foraging from the waste produced by humans. The domestication of sheep and goats and later cattle combined dangers and huge opportunities for hyenas. Sheep and goats were the first animals domesticated, around 10,500 years ago in West Asia, in what was called the Fertile Crescent along the Tigris and Euphrates rivers. Wild ibex and mouflon were the likely ancestors of domesticated goats and sheep.[11] Cattle and pigs were domesticated slightly later. The new farming communities based on cultivation and livestock husbandry expanded and dispersed across the eastern Mediterranean and Mesopotamia, with pastoralism important in many areas.[12] These regions had widely distributed populations of striped hyenas (which survived the disappearance of the spotted hyena and the spread of human settlements)[13], for whom the chance to scavenge from human settlements[14] and livestock carcasses would have supplemented their natural diet of wild ungulate remains and vegetable matter. This brought them into conflict with people, who are likely to have thought that hyenas seen consuming a dead sheep, goat, cow or donkey had killed it in the first place.

The new food production methods spread out westwards into Anatolia and then southern Europe around 6000–700BCE, and into central, western and northern Europe around 5600BCE, where the cultivation of plant foods was supplemented by domestication of animals, including wild boar and wild cattle. Arable farming and pastoralism spread eastwards or in some areas developed independently in Asia, with evidence of cultivation and pastoralism around the Caspian Sea by 6000BCE.[15] Parts of the Indian sub-continent were suitable for the early development of agriculture-based societies while others, like the dense humid forests of the north-east, the Eastern and Western Ghats of central-southern India and other mountainous or thickly-forested regions, retained hunter-gatherer modes and basic shifting agriculture.[16] Settled agriculture developed 10,000–4,000 years ago, with agricultural settlement and population expansion really taking off around 4000BP (Before Present).[17] The earliest evidence from the region of domestication of animals comes from Mehgarh in Baluchistan in Pakistan in 7000BCE (with the growing of wheat, barley, pulses and the keeping of sheep, goats and cattle)[18], though some put it around 8,000 years ago.[19] Estimates of the start of rice cultivation vary between 7000BP[20] and 300BP;[21] Roberts puts it during the third millennium BCE in the Ganges Valley.[22] Farmers and pastoralists spread across India over a period of several thousand years as livestock husbandry, based particularly on zebu cattle and the cultivation of rice, gram and beans enabled populations to expand.[23]

The clearing of land and need to protect livestock will have involved increasingly strenuous attempts to exclude wildlife from cultivated fields and deter or kill predators. But well into the Common Era, wildlife remained diverse, widely distributed, with wild prey providing food for diversity of predators, including hyenas. As with the hyena species in Africa, striped hyenas across Asia appear to have adapted to life around human settlements, feeding opportunistically on waste produced by humans and on the carcasses of livestock that died from disease or other natural causes. More such carcasses would have been available in regions where religious beliefs prohibited the killing of cattle and consumption of meat. It is notable that the ideal striped hyena habitats, in areas with open woodland, tall grassland, hills with valleys that provide den sites, were often in areas where pastoralism and cultivation developed. Neither hyenas nor farming communities lived in dense, moist forests.[24]

Africa – the hyena heartland

Around 10000BCE, the climate across Africa grew hotter and wetter, flooding parts of the Nile Valley and pushing people on to higher plains. Iliffe says by 12,000–7,500 years ago, many communities had moved away from river valleys to higher grasslands and open woodland.[25] These were well-watered, due to the increased rainfall, and plant cultivation largely took the place of harvesting wild grains and vegetables/fruits. These changes may have been introduced into hunter-gatherer inhabited regions by migrants from areas of expanding population, where cultivation had already developed. In some regions this could have been independent, emerging from systematic foraging of wild cereals, pulses and tubers. The raising of sheep and goats, and then cattle, probably spread to the Nile Valley and beyond from the Middle East, where there is the earliest evidence of domestication,[26] though some local domestication may have occurred independently. Domesticated animals and their herders appear to have moved into the Nile Delta region between 6000 and 5000BCE as the climate became more arid in the Levant, encouraging migration to better watered areas.[27]

The first farming settlements clearly identified as villages in the Nile region date back to 5200–4500BCE at Fayum and Merimde Beni-Salama in the Delta.[28] Excavations of waste middens at these sites show the continued presence and hunting of wild ungulates such as hartebeest and gazelles, from which one can infer the likely presence of large carnivores, such as hyenas. Striped hyenas, by then the only hyena in North Africa, adapted to living close to human settlements, being mainly nocturnal in their foraging and able to survive by hunting small or weak animals (including young or sick livestock), eating carcasses of dead ungulates (wild or domesticated), foraging for fruit and vegetables and scavenging from human waste, including buried or abandoned human bodies. Spotted hyenas may have survived in southern Egypt and interior districts of North Africa, but in decreasing numbers before their disappearance in the face of human expansion. There is evidence that they may have occurred in the past in the Ahaggar and Tassili d'Ajjer regions of Algeria, but Mills and Hofer do not put any date on their presence there.[29]

Domestication of animals and crop cultivation spread across the coastal plains and foothills of the mountain ranges of North Africa. Here the spotted hyena had been exterminated or had died out as the prey base was depleted, but the striped hyena survived in widely distributed populations; interestingly, lions survived in these regions until the end of the 19 century.[30] Striped hyena numbers and ranges are impossible to asses at this time – observations and the ability to assess numbers made difficult by the animal's nocturnal habits, their wariness of people and their scavenging, which made them less worthy of note than lions, for example. They were and still are found in all the countries of North Africa, from Egypt across to Morocco; in the countries to the south on the Atlantic and Red Sea coasts; in the Sahel region, bordering the Sahara; and in much of West and East Africa.[31]

Clutton-Brock says that securely dated finds of domesticated cattle in North Africa indicate their presence from the seventh to sixth millenniums BCE, noting that it is not certain whether pastoralism reached there from West Asia via Egypt or whether it may have developed independently in some areas.[32] Some estimates put the expansion of pastoralism in North Africa and the Saharan littoral around 6500 years ago, when the drying of the Sahara began, though with some areas still suitable for nomadic pastoralism.[33] Those who had settled in more fertile areas of the Sahara were forced to move south and south-east into sub-Saharan Africa or back towards the Nile.[34] Striped hyenas do not live in fully desert areas but, like their brown counterparts in southern Africa, can survive in arid regions with sufficient food and water (often in the form of succulent plants and their fruits).

Livestock and arable farming in the North Africa and the Nile Valley reduced the need to hunt but created the need to defend livestock from predators. Pastoralists, who were dependent on their livestock for food and could not accept serious losses, would have resorted to killing or driving off hyenas and other predators. The cultivation of land reduced grazing for predators' wild prey, thereby increasing predation on livestock. As Bertram suggests, "Herbivores which would feed on a farmer's crops are either destroyed or kept out by fences" once more settled cultivation developed,[35] and carnivores would be killed in retaliation for livestock loses or pre-emptively to prevent future attacks. Hyenas may have been particularly well-placed to prey on stock around settlements, as they adapted to scavenging waste produced by humans.

Some pastoral communities in Africa, notably East Africa, developed considerable skill in hunting predators to protect their cattle and smallstock. In some this evolved into culturally important manhood rituals. The hunting of lions, such as by the Maasai, Barabaig and other East African communities, became part of the rites of passage to adulthood. The killing of leopards was also seen as a manly act that helped protect the community. This did not extend to hyenas, which were seen as vermin to be killed with no great prestige being gained. Dr Amy Dickman, who founded the Ruaha Carnivore Project in Tanzania and studied Maasai and Barabaig conflict with carnivores, told the author that huge prestige was attached to killing lions. Hunters who did so would be rewarded by the community and held in high esteem. There was no respect to be gained for killing hyenas, seen as ignoble and worthless pests.[36] The hyena and lion expert and researcher Dr Laurence Frank told me:

> Pastoralists really loathe hyenas and I doubt that they would ever hesitate to kill one given the opportunity, but, being nocturnal and wary, that does not arise often. I am not aware that killing a hyena is worth any kudos or celebration to pastoralists. In fact, it probably happens rarely, as they are hard to hunt; I would guess that in modern times the vast majority are poisoned, with no prestige attached.[37]

Although in many parts of Africa, hyenas were thought by many to be linked to the supernatural, as will be demonstrated later in the next chapter, among most East African pastoralist communities of Cushitic or Nilotic origin hyenas were despised but not seen as possessors of occult power or linked with witchcraft. Where such beliefs did exist in East Africa, they were more common among settled communities engaged in cultivation. Some communities which ascribed occult connections to hyenas would not kill them as they were seen as perilous to hunt or harm because of the association with sorcerers. But in some areas of sub-Saharan Africa (Harar in Ethiopia being one), hyenas were surprisingly tolerated and seen as nature's waste cleansers and recyclers. In a few places, hyenas received religious protection, such as those scavenging from human settlements in Ethiopia, which denned in sacred forests surrounding Ethiopian Orthodox churches.[38] Some communities, like the Tugen and Nandi of Kenya, had hyenas as totem animals and would not harm them. They believed hyenas to be links between this world and dead ancestors.[39]

Cultivation and pastoral cultures spread from the Nile Valley and fringes of the Sahara to Sudan. Livestock-keeping communities settled there about 6,000 years ago, spreading over the next 1,000–2,000 years to Eritrea, Ethiopia, Somalia and then Kenya. They kept goats, sheep and cattle.[40] In the west-central Sahara, people moved south and west into West Africa, particularly around the Niger river, where there is the first evidence of the cultivation of West African rice from plants gathered from the wild.[41] Archaeological evidence from 9500 to 5000BCE indicates people were raising animals and growing plants for food in the south-eastern Sahara, the savannahs of West Africa and the Ethiopian highlands.[42] Archaeological finds around Lake Chad even show the survival of some nomadic livestock-keeping in highland areas of the central Sahara with sufficient rainfall or other water sources, such as the Nubian aquifer system beneath the surface of the Sahara.[43] Ehret writes that by the eighth millennium BCE, Saharo-Sudanese people of the eastern Sahara were cultivating grains previously gathered from the wild and building thorn fences to protect homesteads and cattle from predators.[44] Settled and nomadic communities still hunted to supplement food production, and will have produced carcasses from butchering prey animals or domestic stock, as well as a result of disease or age. Hyenas are likely to have adapted swiftly to scavenging from humans, while also opportunistically hunting poorly protected stock. Ehret says that by 7000 BCE:

> …the typical Saharo-Sahelian extended family of the eastern Sahara region resided in a large homestead surrounded by a thick thornbush fence. The enclosed area acted as a cattle pen, protecting animals at night from human cattle raiders as well as predators principally hyenas, leopards and lions.[45]

They also developed metalworking skills. Iliffe suggests this started in the Nile Valley in the fifth millennium BCE, possibly introduced by migrants from North Africa or West Asia.[46] Iron-working spread from the Nile Valley into eastern Africa and the Horn; it is believed to have been introduced to the Great Lakes region by 900BCE and northern Nigeria by 500BCE.[47]

Despite growth in human populations as food availability increased, human densities remained low and there was abundant land into which communities could disperse. Until the 20th century CE, "Africa was an underpopulated continent".[48] Low population density enabled nomadism, transhumance and shifting cultivation because of a lack of competition for land and the absence of delineated and defended territories. Low population density, small settlements and transhumance limited the depletion of prey species and habitat through land clearance for grazing and cultivation. This reduced the anthropogenic effects on predators like hyenas. But where natural prey may have been thinned out by human hunting or agricultural expansion, they preyed on livestock and scavenged from the waste of human settlements or nomadic camps. Where the exclusion or depletion of prey species did occur, there was a decline in predator ranges and greater human-predator conflict, though this did not affect hyenas quite as it did lions and other predators, for example, because of their adaptable foraging behaviours.

Human and hyenas in East Africa and the Horn of Africa from the Holocene to 1600

The early Mashariki Bantu migrants who moved into East Africa 2–3,000 years ago in at least two waves, whose language forms the basis of many of the region's modern language groups, had words in their vocabulary showing knowledge of the wildlife, including for spotted hyenas (*mbui*), bushbuck (*songa*) and other forest-dwelling mammals.[49] This indicated not only human contact with hyenas but also suggested the flexibility of habitat use by hyenas, and that they were to be found in the forests of West, Central and North-Central Africa from which many of the migrants originated or through which they moved. In central Sudanic areas, the peoples were of Nilo-Saharan origin, and had their own languages that only mixed with Bantu after 1000BCE. The Central Sudanic peoples had words for spotted hyena (*na*) as well as lion (*ebi* or *ngboro*) and wild dog (*lu*), suggesting open savannah or lightly wooded terrain, which suited diverse carnivores.[50] The Bantu migrations into the Great Lakes, Rift and East Africa led to a greater pace of forest and bush clearance both for cultivation and to provide grazing for livestock.

In East Africa, pastoralism generally preceded widespread cultivation of crops, chiefly because human migration from the North and West was into grassland or woodland/savannah, where lower rainfall limited crop yields. Many pastoralists followed the rains and the grazing they produced on a seasonal basis, establishing temporary cattle camps as they moved their herds or flocks. Mobility limited conflict with wildlife in any one area and did not involve the permanent exclusion of herbivores or predators from farmed land, as settled farming did. Pastoralists came into conflict with wildlife at times of poor rains, when wild ungulates competed for food and water with livestock. The ungulates moved to follow the rains for the same reasons as pastoralists, while carnivores preyed on the migrant wild ungulates and the domestic stock of nomads that passed through their territories. Nomadic or commuting carnivores following the herds would also take livestock. It is likely as that hyenas fell into both categories, with packs having long-used den sites and territories but commuting long distances, while those dispersing from their natal clans or not part of clans were more likely to follow the herds.

For spotted and striped hyenas, waste produced by expanding human settlements or temporary camps was a supplementary food source, as was the custom of some communities to dispose of their dead by leaving them in the bush or in shallow graves. Many African communities, like the Hehe of southern Tanzania, are believed to have buried only members of the ruling families, "the bodies of commoners being left in the bush for hyenas and other scavengers".[51] The Maasai[52] and their Maa-speaking ancestors followed similar body disposal patterns, with only chiefs being buried and the bodies of ordinary people left in the bush. Among some peoples, such as the Tugen and Nandi of Kenya and some communities in Sudan, it was necessary for people's spirits to ascend into the afterlife that their bodies be consumed by hyenas.[53] The Hadza hunter-gatherers of northern Tanzania would take the bodies of the young who died and leave them in the bush, while older people would be placed inside their huts, which were burned.[54] This feeding on dead bodies and link with the afterlife in some belief systems may have given rise to the reputation of hyenas as grave-robbers and creatures linked to supernatural forces.

By 2000BCE, pastoralist communities keeping sheep, goats and cattle had spread along the Rift Valley and into the savannahs and woodlands between the Rift and the Indian Ocean coast.[55] The area was rich in wildlife, including buffalo, a wide range of antelope, zebra, wild pigs and predators like hyenas, lions, leopards, cheetah and wild dogs. Predators coexisted and competed with hunter-gatherer communities. Archaeological excavations in south Nyanza in Kenya indicate that although pastoralism was practised there by the second millennium BCE, the remains of wild animals from Neolithic sites show hunting persisted and "the pattern of wild animal taxa present is suggestive of a combination of widespread hunting of the more solitary species (e.g. oribi and, reedbuck) and selective hunting or scavenging of larger herd animals like zebra, hartebeest and topi".[56] Hunting and carrying parts of carcasses to temporary camps or more permanent settlements would undoubtedly have attracted hyenas to those areas.

There is a paucity of evidence from archaeological sites or oral history of the nature of the relationship of humans migrating into these regions with the wildlife, their hunting practices or strategies for protecting livestock. As Mutundu explains, "The mobility of both early pastoralists and later hunter-gatherers and their interactions… means that the archaeological records of these groups are not always easy to find or distinguish".[57] A problem associated with the study of pastoral economies in the region is the "difficulty of identifying and distinguishing archaeological sequences of hunter-gatherers with access to domestic stock or in the early phases of the adoption of herding, from those of pastoralists practising a subsistence strategy that included the utilisation of wild animals"[58]. Robertshaw also emphasised this: "Faunal bone assemblages from areas that adopted pastoralism suggest the decline but not disappearance in hunting for food".[59] Lamprey and Reid believe that what evidence there is suggests that in south-western Kenya, the region now occupied by the Maasai Mara National Reserve, there was a long history of pastoral-wildlife interactions.[60] As early as 2,000 years ago, in the Neolithic period, the Mara was occupied by pastoralists, rearing cattle and sheep and goats, who do not appear to have been active hunters.[61] Cushitic communities moved into the Horn of Africa and Kenya around 5,000–4,000 years ago, progressing south into northern Tanzania after 3000 BCE, living alongside the resident Khoisan hunter-gatherers.[62] Some communities in East and Southern Africa, the ancestors of the people who became known as Hadza and San/Bushmen, retained hunter-gatherer modes of subsistence and did not adopt cultivation or pastoralism.[63] Some of these communities were eventually assimilated by incoming farming or pastoral peoples, but many were pushed aside into areas unsuitable for crops or livestock, living by hunting and foraging in areas with sufficient wild plants to support them.[64] The human populations of these areas would have been accustomed to the presence of hyenas, which may conceivably have been viewed with greater tolerance than lions, leopards and cheetahs which were always seen as out and out predators, because of their waste disposal role.[65]

The climate of Africa became progressively drier and hotter from the fourth millennium BCE. Savannah and open woodlands expanded at the expense of denser forest, favouring pastoralism and providing more suitable conditions for large ungulates and social carnivores like hyenas and lions, and, of course, humans.[66] Crop farming did not disappear totally as some areas had sufficient rainfall for mixed farming. The cultivation of a greater variety of plant foods developed as Bantu-speaking peoples migrated from west-central Africa into the western Rift Valley, and eventually East and Southern Africa.[67] Gifford et al. provide evidence of wild ungulates around Nakuru-Naivasha in the last two millennia BCE from bone assemblages showing "a diverse range of herbivores living in close proximity to the neolithic pastoralists... [with] numbers of zebra, wildebeest...kongoni...impala, Grant's and Thomson's gazelle populations".[68] Evidence from excavated sites in this region suggests that cattle were slaughtered at the site for human consumption, while the wild ungulate remains found lacked skull bones and teeth, suggesting only parts of the carcasses had been carried back to the settlements.[69] The excavation at the settlement, which may have been a residential base camp, shows there was a midden for bones and other carcass remains from butchery.[70]

Farming, hunting and defensive capabilities improved when stone age technologies were replaced by ironworking.[71] Between 3000 and 1000 BCE, Bantu migrants from Central Africa arrived in the Great Lakes and Rift Valley region.[72] They assimilated ironworking skills from Nilo-Saharan peoples who had moved south and transferred to them more advanced methods of crop cultivation.[73] This combination of migration and technological advance enabled growth in human settlements, increasing the pressure on wildlife as livestock herds grew and more land was brought under cultivation.[74] Over the next three millennia, into the modern period, the expansion of cultivated areas, livestock herds and human settlements became the major threat to habitats, wildlife and predators which posed a threat to humans and their stock.

In the last millennium BCE and the early years of the Common Era (CE), the drying of the East African climate opened up more savannah in elevated areas of the Rift Valley and the neighbouring plateaus. This increased grassland habitat for wild herbivores and enabled the further expansion of grazing for livestock.[75] Areas such as Laikipia and Uasin Gishu, which became areas of extensive pastoral activity while retaining large and diverse wildlife populations, were opened up by climate change. They became famed for their wildlife and hunting opportunities when British traders, hunters and then settlers arrived from the mid-to-late 19th century. The savannah region stretched from these higher altitudes and plateau areas to the Loita and Mara plains and the Serengeti and the Crater Highlands in Tanzania. It supported the massive numbers and diversity of game, alongside the growing human population. By the time of the arrival of Europeans as traders, hunters and explorers in the interior of East Africa, much of the savannah was peopled by Maa-speaking pastoralists (from whom developed the Maasai and Samburu), though they had been preceded by other pastoral peoples.[76]

From around 300 to about 1000CE, ancestors of the Kalenjin-speaking peoples of the Rift Valley of Kenya, referred to as Sirikwa, moved into the region. They were pastoralists and had a hierarchy and societal structure based on male age-grades (as did the Maa-speakers who pushed them out centuries later). As males reached the age for transition to manhood they were initiated as warriors and lived together in age sets – the basis of defence of cattle from predators and raiders from other communities. These warriors also raided other communities for cattle.[77] By 700CE, Maa-Ongamo people (the ancestors of the Maasai and Samburu) had moved south from the increasingly arid Lake Turkana Basin into the plateau areas and plains east of the Rift Valley. Ehret says that one aspect of the development of the young warrior class among Maa and Kalenjin speakers was the emergence of rite-of-passage hunts for lions, during which initiates proved their valour.[78] This set lions apart from hyenas as predators worthy of respect and whose killing could confer prestige and wealth (in the form of cattle given to those who killed lions). Archaeological remains of Sirikwa encampments suggest that hyenas lived in close proximity to them. Mary Leakey excavated several Sirikwa burial pits and found that in most cases the skeletons were poorly preserved, with signs of disturbance and removal of body parts and bones by scavengers presumed to be hyenas.[79]

The Maa communities, especially the Maasai, started to push out the Sirikwa, or forcibly assimilated them, around the middle part of second millennium CE.[80] The Maasai are believed to have originally been agro-pastoralists, cultivating sorghum and millet and raising cattle and smallstock. Over time, a core group specialised in cattle-raising, developing a self-identity as pastoralists. Other Maa-speakers developed into communities with a variety of economic specialties – pastoralist, agro-pastoralist, farmers and hunter-gatherers.[81] The Maasai have gained a reputation as inveterate and courageous lion hunters, but beyond that of being relatively benign in their relationship with wildlife. Steinhart warns against mixing myth and reality when portraying the Maasai. He says that they don't routinely hunt wildlife for food but kill predators like hyenas and lions when they attack livestock.[82] Frank is "sceptical of the romantic notion that the Maasai have a great tolerance for wildlife" and in modern times have resorted to the spearing and poisoning of hyenas and lions, seen as a threat to their cattle, and killed species like rhino as a protest in response to loss of land in the 20th century to national parks and reserves.[83]

The Sirikwa and other Kalenjin groups pushed aside by the Maasai moved into the hillier country around Kerio to the west of Lake Turkana. They were called Suk by Europeans, but are now referred to as the Pokot. Most were pastoralist, but in suitable areas they also engaged in cultivation. The Pokot were active hunters. Hyenas were common and not universally hated. One Pokot clan has the hyena as its totem[84]. Most Kalenjin clans have an animal as a totem or symbol as part of their identity. Pokot had the custom that when a man died he was buried in his cattle or sheep enclosure. If he died without owning animals, he was buried in his hut. A goat would be killed and eaten at the grave and the remains of the goat left on the grave. The community would then move with their stock to another site. There can be little doubt that any hyenas resident in the area would have been attracted to the grave site initially by the goat remains and may have dug up and consumed the corpse. Open country disposal of bodies and shallow burials without stones or other obstructions was an invitation to hyenas to scavenge human remains. This, combined with the wide range of vocalisations of hyenas (especially the maniacal giggles and human-like laughter sounds), has helped to develop the association in many parts of Africa of hyenas with the night, death and the occult, including the belief that some hyenas are part wizard or witch. The latter belief was common among some peoples settled east of Lake Victoria. The Giryama of coastal Kenya, for example, have folktales of finding footprints that show a creature with one human leg and one hyena leg.[85]

At the southern end of Lake Victoria (Nyanza) and the western edge of the Serengeti region, in what is now the Western Corridor of Serengeti National Park, there was considerable human settlement in the first two millennia of the Common Era. Shetler's research shows the presence of the ancestors of the Ikoma, Nata, Ikizu, Ishenyi and Ngoreme peoples, dispelling myths deriving from 19th century European travellers that these areas were solely the domain of wildlife.[86] People coexisted with wildlife, though human settlements were never extensive and farming/pastoralism did not involve the use of fencing to keep wild ungulates or predators away from farmland.

Sudan and the Horn of Africa

The early history of northern Sudan and its role as a point from which pastoral peoples migrated south has been noted above. One cause of migration was climate change, another was the effect of warfare and regime change in Egypt and the areas of northern Sudan under Egyptian rule. In the eighth century BCE, Egypt still exercised hegemony from Lebanon down to Meroe, below the fifth cataract of the Nile in present day Sudan. Much of the area to the south of Egypt was part of the Kushite kingdom. The Kushites grew in power and by 727BCE had started to eat away at Egyptian territory, gaining control of the Nile as far as the Egyptian city of Thebes.[87] Kushite rule of large parts of Sudan and Egypt ended with the Assyrian invasion of Egypt in the seventh century BCE. The Kushites were pushed back into Sudan around 671BCE, and then further south in 591, establishing a new capital at Meroe.[88]

In southern Sudan, there was continuity of human habitation from the last millennium BCE and into the Common Era. In areas with adequate rainfall and fertile soils, communities mixed cultivation, pastoralism and, along the rivers and swamps, fishing. In the drier grassland areas people relied heavily on pastoralism.[89] In the agro-pastoral areas, cattle grazed on seasonal floodplains, then moved to fields used for cultivation after harvesting to feed on stubble and manure the fields. The main crops were sorghum, cowpeas and groundnuts.[90] Archaeological research indicated that pastoral peoples, the ancestors of the Dinka, established large cattle camps on the floodplains. Evidence focusing on bone assemblages from middens or other waste dumps at settlement sites in the region show cattle, smallstock and a wide variety of wild faunal remains, including wild ungulates, indicating that the pastoralist peoples still hunted to supplement food production. In areas with lakes, rivers or swamps, the assemblages include extensive fish remains and some crocodile bones. The middens used to dispose of animal remains would have been attractive to resident spotted and striped hyenas.

In their archaeological research in Lakes Province of South Sudan, Robertshaw and Siiriäinen found remains of settlements of Dinka ancestors. These were in areas of savannah/woodland, with evidence of large numbers of trees having been felled to clear land for grazing.[91] Where the terrain changed to open plains, these merged with the seasonal floodplains and swamps of the Sudd area of the Nile. Cattle would be moved across these areas, being gathered in camps situated near permanent rivers during the dry season to take advantage of remaining grazing and access to water.[92] Excavations of the layers of soil in the camps indicated that many camps on the floodplains were used year after year and contained middens and ash mounds. These showed evidence of the disposal of waste, made up of livestock, fish and wild ungulate bones, including kongoni, kob, bushbuck and other small to medium-sized antelope.[93] During the excavation, the research team saw giraffes and a variety of antelope in the vicinity of the site,[94] and found middens riddled with hyena burrows (not identified by specific species in the research paper).[95]

The Mandari of southern Sudan are a Nilotic community of the Karo group of peoples. For centuries they have inhabited central Equatoria province, north of the current South Sudanese capital, Juba. They combine cattle husbandry with cultivation of crops. The region theyoccupy is rich in wildlife. The role of hyenas as hunters, scavengers and consumers of the bodies of dead people have led the Mandari to associate them with death, nocturnal attacks on people and witchcraft.[96] Buxton says that each Mandari settlement has an area just beyond it known as a "social bush", where villagers defecate, throw rubbish and dispose of carcasses of animals used for food, including buffalo they hunted. This "bush" attracts hyenas, which dispose of much of the waste, but are especially attracted by the carcasses of animals killed for food or ritual purposes.[97] Predators are known to the Mandari as *gworoy* – animals that can kill. Lions and leopards were long present in the bush beyond the "social bush", but were not as common as hyenas. They have traditional rituals linked to the killing of leopards, which are clearly animals worthy of respect, but have no rituals relating to the killing of hyenas, which Buxton believes is "related to the Mandari image of the hyena as a low creature, typically mean and cowardly, which enters a homestead yard by night to snatch a baby or young animal, but slinks from an aggressive stand by an adult".[98]

To the south-east of Sudan, excavations by archaeologists in present-day Eritrea indicate early stone age settlements in the coastal lowlands. They revealed early to middle Holocene (8,000–5,000 years BP) settlements, with bones of spotted hyena, Soemmerring's and Dorcas gazelles, hamadryas baboon, dik-dik and wild ass present.[99] Many of the settlements showed that early communities lived on shellfish, meat from wild animals and harvested wild grains. They were an expanding population that would provide the basis for future, more developed societies with livestock keeping, cultivation, trade and some hunting for food, skins, horns and ivory. The most important of those, the Axumite Empire, rose to prominence in the region of northern Ethiopia and Eritrea in the fourth century BCE. It became the dominant regional power, with strong regional trading links.[100] To the south, in what is now Ethiopia, the population expanded from small stone age settlements to farming villages and then towns, which were the basis for more structured, hierarchical polities. Wildlife was plentiful in areas with sufficient rainfall and grazing/browse for ungulates, and there was a large and widely distributed population of carnivores, including striped and spotted hyenas. Strabo, the Greek historian and geographer who lived from 63 or 64 BCE to about 21 CE in Asia Minor and then Rome, wrote about Ethiopia and its wildlife. Much of his information must have come from long-held beliefs about other lands and observations made by traders and travellers with whom he came into contact. He referred to, with interesting and clearly none too accurate detail, a variety of what he said were carnivores, most of which were mythical creatures - including the sphinx, the cynocephali (dog-headed) hominid and carnivorous bulls. Strabo wrote of crocuttas, presumably one of the species of hyenas, which he claims to be the progeny of a wolf and a dog (a common misperception for centuries).[101]

In western Ethiopia, excavation of sites at Ajilak (in the Gambella region) indicated extensive hunting of wild fauna in the first millennium CE.[102] The region's long-grass plains provided ideal habitat for a variety of ungulates and predators like hyenas, lions, leopards, cheetah and wild dogs. Small communities there, often occupying what appear to be seasonal camps, hunted, fished and kept livestock. Finds there included faunal remains in middens, alongside remains of cattle which the researchers believed were slaughtered for rituals, weddings and funerals.[103] Between these settlements, in what is now South Sudan, nomadic pastoralists related to the Nuer of South Sudan, established huge cattle camps, again likely to have been a major draw for hyenas of both species.[104] González-Ruibal, who led the excavations, believed these regions along the Ethiopia-South Sudan border in the early centuries of the modern era contained a wealth of wildlife including hyenas, buffalo, giraffes, waterbuck, roan antelope, zebra, bushbuck, reedbuck, warthog, hartebeest, lions and elephants.[105]

Historically, spotted and striped hyenas were found throughout savannah; open woodland; and other suitable habitats in Ethiopia, Eritrea, Somalia and Djibouti.[106] Lions were also found in this habitat, but were exterminated in most of Djibouti and Eritrea in second half of 19th century,[107] leaving the spotted hyena as apex predator. The Zagwe dynasty ruled much of Ethiopia from 900 to 1270CE, when the Solomonic Dynasty (claiming descent from Solomon and the Queen of Sheba and their supposed son Menelik)[108] succeeded. At this time, hunting appears to have been a means of obtaining meat without slaughtering cattle, of protecting livestock and sport for the nobility. Early hunting was probably not vastly damaging to overall wildlife numbers, but that changed with the spread of firearms from the late 16th and early 17th centuries, when muskets were imported for the large imperial army.

One region with a curious and lasting relationship with spotted hyenas was the town and surrounding territory of Harar, near the border with Somaliland. It was said to have been formed from the merging of seven villages inhabited by ancestors of the Harari community, who had strong animist beliefs, which infused natural features of the landscape and animals with spiritual importance.[109] By the 10th century CE, the people of the region had been mostly converted to Islam, but traditional beliefs remained strong. A Yemeni Sufi called Abadir migrated to Harar after the conversion and established Islamic schools there. His influence was important and today many Hararis refer to themselves as sons of Abadir.[110] Baynes-Rock, the chronicler of Harar's hyenas, wrote that one of the stories of how Harar came to be tolerant of and have a symbiotic relationship with spotted hyenas says that Abadir was sitting on a rock, facing Mecca, when he saw a hyena running fast from east to west:

> He took from his pocket a toothbrush [made from a stick from a local tree] and threw it at the hyena. The toothbrush hit the hyena in the leg and, such was Abadir's formidable power, the hyena's leg broke instantly. Since that time hyenas in Harar were 'domesticated' or made peaceful and the evidence of that is the way they run. Nowadays, hyenas are known as 'Abadir Hokolo', meaning 'Abadir tamed' or 'Abadir disciplined'. To call a hyena 'Abadir' is confirmation of its status as a citizen of Harar.[111]

Through its trade with Yemen and the Arabian peninsula, and its development as a centre for Islamic teaching, Harar became the centre of an influential Muslim community.[112] A number of Muslim polities grew up in this part of Ethiopia. One known as Adal centred on Harar in the 14th century, formed as Muslims fought for their independence from the powerful Christian Solomonic rulers of much of Ethiopia.[113] Conflict over attempts at Christian hegemony and Muslim assertion of independence continued for two hundred years, but by the mid-16th century Muslim power based on Harar was weakened and Christian dominance less challenged by Muslims, but more so by the increasingly powerful Oromo people to the west.[114] In the latter part of the 16th century, the Oromo began to exercise hegemony over both Christian and Muslim territories. Around this time, the Muslim authorities in Harar built a large defensive wall around the town, with five gates. This prevented the Oromo from capturing the town, but they took over much of the surrounding countryside.[115]

Around this time, spotted hyenas were said by Baynes-Rock to have started attacking people, on the basis of his interviews with residents and local historians in Harar, perhaps because the building of the wall prevented them from foraging in the streets for edible refuse at night. Harari folktales talk almost of a war with the people killing hyenas with double-pointed spears made specifically for the purpose, and people being killed by hyenas.[116] The myths say Emir Nur, the Harari leader who built the wall, decided to go out to meet the hyenas. He went to a Muslim shrine on a mountain outside the town and called on Muslim saints to arrange a meeting with the leader of the hyenas. Nur then met the "white hyena king". The hyena king told Nur that they were fighting against the building of the wall which stopped them foraging for food in the streets.[117] The story goes on to say, as Baynes-Rock describes, that Nur agreed to make holes in the wall through which hyenas could enter the town to clean the streets, as long as they did not attack people.[118] The holes can still be found in the walls and are still used by some hyenas, though the modern town has expanded hugely and is not completely enclosed by a wall. This mythical agreement to end the conflict and create the hyena holes is said to be the origin of the relationship between the people of Harar and the hyenas – which led to the night time rituals (now a tourist spectacle) in which people feed the hyenas. Another ritual said to have developed from this "truce" is the annual feeding of porridge to hyenas on the 10th day of the Islamic month of Muharram at shrines outside Harar.[119] It is interesting that the Harar stories and feeding only involves spotted and not striped hyenas, which are also found in this part of Ethiopia and neighbouring Somaliland.

Elsewhere in Ethiopia there are also examples of surprising tolerance of hyenas. In Tigray province, folklore centuries old says that if a wolf (and this is interpreted to mean a striped hyena) kills a goat, men should drive it away throwing pebbles rather than stones. The myth says that if men injure the "wolf" it will dip its tail in its blood, flick it on to the men and they will die.[120] In parts of Ethiopia inhabited by people who have followed the Jewish religion since around 1270CE[121] – known as Beta Israel or Falasha – these communities were associated, in the beliefs of other peoples, with hyenas and witchcraft. Other Ethiopian communities referred to them as buda "and treated with a mixture of fear and repugnance", as the buda was considered to be both an evil spirit which possessed a person, who could turn themselves into a hyena and dig up graves to devour the corpses.[122] This is another version of the myths found across Africa and parts of West Asia linking hyenas with the occult. Interestingly, as will be seen in folktales from parts of West Africa, Ethiopia and particularly the Zulu of South Africa, there was also a link between hyenas, witchcraft and blacksmiths.[123] Because of their ability to turn ore into metal and forge weapons, blacksmiths were often seen as practising magic and using hyenas to get human body parts to be used in the transformation of ore into iron.

The Sahel and Central Africa

Across the Sahel belt from the Red Sea to the Atlantic coast of West Africa, there is evidence of the presence of spotted hyenas, some dating back several thousand years in the form of rock paintings in areas of the southern Sahara and Sahel.[124] They also show a variety of antelopes, ostriches, buffalo, giraffes, elephants and rhinoceroses.[125] There are no published sources this author could trace on the size or ranges of either species of hyena, striped or spotted, across the region during this period or references to the nature of the relationship between settled or nomadic communities and hyenas. The probable range of hyenas in this region was limited by the Sahara to the north and dense forests of West Africa and the Congo Basin, though spotted hyenas do survive in some heavily forested areas of sub-Saharan Africa. Striped hyenas are still found in the Sahel from Sudan, into Chad, northern Central African Republic and Cameroon and into the western Sahel belt of West Africa.[126] The IUCN Red List of species shows the current range of spotted hyenas spreading down from the Sahel into Central Africa, including the tip of north-eastern Democratic Republic of Congo (around Garamba National Park), the Republic of Congo, the Central African Republic, Cameroon, Rwanda and Burundi, but omits Gabon.[127] The striped hyena range is limited by habitat and climate and it is likely that it was only present in Chad, the northern tip of Central African Republic and the grassland and open woodland areas of Cameroon.[128]

An early Portuguese traveller in southern Congo and Angola in the late 16th century, Duarte Lopez (whose travels were recorded by Fillipo Pigafetta), has two possible references to hyenas following his account of lions there. Pigafetta says Duarte refers to tigers being present in Congo and being very much like the tame ones Lopez said he had seen in Italy. He specifically refers to them being like dogs, not least in the ability to domesticate ones brought up by humans from a young age. Duarte says these supposed tigers will enter people's houses and drag out victims to consume. The skin of the animal is referred to as spotted.[129] The colouring and ability to tame them suggests a spotted hyena.[130] A few pages later, there is reference to a type of wolf being found in the region. Again, there are no wolves in the Congo nor have there been, but in Central and Southern Africa wolf is used to refer to hyenas, especially brown hyenas. Pigafetta says that these wolves "are beyond measure fond of palm oil".[131] The consumption of fruit or vegetable matter is strongly suggestive of striped or brown hyenas (often called strandwolves), with the latter being the most likely contender in that part of Africa.

Southern Africa

Constructing an evidence-based account of the relationship between humans and hyenas in Southern Africa involves piecing together fragments to try to provide viable interpretations. The historian of Zimbabwe, David Beach, warns that sources on the development of communities and their relationship to their environment are thin and one should be careful not to adopt "an over-enthusiastic use of traditions to reconstruct the past".[132] Many communities in the region, like the Shona of Zimbabwe, had no strongly established system of traditional historians or custodians of the past.[133]

During the last millennium BCE and the opening centuries CE, southern and south-eastern Africa contained a range of habitats – forest, woodland savannah, dry savannah, montane forest, arid steppe and desert.[134] Ungulate species were present in large numbers in suitable habitat and included all the current species plus the now extinct bloubok and quagga, which were found in large numbers in South Africa's Cape. In grassland, savannah and semi-arid areas of South Africa there were huge herds of springbok, black wildebeest, bontebok and blesbok, which supported large numbers of spotted hyenas, lions, leopards, cheetah and wild dogs. Brown hyenas were found in suitable habitat across much of the region, scavenging the carcasses of ungulates hunted by predators and those which died of natural causes. These were found across the territory of South Africa, Namibia, southern Angola, Botswana, parts of Zimbabwe and Mozambique.

The San people and Khoikhoi peoples populated the region, with low population densities. The San are the original human inhabitants of the region. Around 12000–10000BCE, Khoikhoi peoples moved south into the region from near Lake Malawi through eastern Namibia and the Okavango. Others migrated from the northern end of Lake Malawi via the Zambezi and Limpopo. The San and Khoikhoi both had cultures of cooperative foraging and hunting, including scavenging from carnivore kills. Later contact with early Bantu migrants influenced Khoikhoi adoption of pastoralism, becoming their main form of economic activity, with hunting and foraging as supplementary methods of food acquisition.[135] Many of the hunter-gatherer communities of southern Africa were pushed into arid regions or were assimilated when Bantu communities, migrating southwards around 300–200 BCE, occupied the land for pastoralism and cultivation. The migrants brought iron tools, which replaced stone ones.[136] The San remained dependent on hunting and foraging for wild foods.[137]

The Bantu migration southwards continued in the first five centuries of the Common Era. Much of the area south of the Zambezi and east of the Kalahari was occupied by Bantu ironworking communities between 200 and 490 CE.[138] They grew sorghum and millet; established permanent homesteads; raised cattle, sheep and goats; and forged iron tools and weapons.[139] They settled areas rich with wildlife. The harder borders they created between the human world and wildlife began to exclude it from the best watered areas and those with good soils for farming. But much wildlife still remained across the region, with sufficient prey species to support a substantial population of predators, with spotted and brown hyenas still thriving across southern Africa, including on the fringes of human settlement.

As agriculture became more developed, settlements expanded and land was increasingly cleared of forest for cultivation or grazing – the forging of iron tools enabled more efficient forest clearance and the establishment of larger settlements in south east Africa by the third century CE.[140] The migrant communities grew in size and developed hierarchical societies ruled by chiefs or kings – developing eventually into large states like the Mapungubwe kingdom of northern South Africa and eastern Botswana, the Monomutapa of Zimbabwe and eventually the Zulu Kingdom, its Ndebele offshoot, the Xhosa chieftaincies of the Eastern Cape and the Tswana chieftaincies of South Africa and Botsawana. They hunted to supplement food production and, increasingly, for sport. But considerable numbers of wildlife and a diversity of species survived across southern Africa, in substantial numbers in some areas and with no evidence of extinctions. It was only the arrival of European settlers and hunters from the mid-17th century onwards that led to the extermination of game in much of the Cape, Natal, Free State and Transvaal (as the regions became known under Dutch and then British/Afrikaner rule).

By 500CE, pastoralism was expanding and in the Zoutpansberg area of northern Transvaal, communities were building homesteads of circular huts situated around central cattle pens to protect against large predators.[141] To the south-west, in Botswana, sufficient rainfall enabled the development of a pastoral culture known as the Toutswe tradition,[142] based on similar settlements around cattle pens – defence of the all-important cattle vital to food security. To the west of the Kalahari, nomadic pastoralism was dominant in the first millennium CE, in an area too dry for extensive cultivation.[143] By 700CE, Bantu-speaking communities had established settled communities in south-west Zimbabwe/north-eastern Botswana/northern South Africa in what became known as the Leopard's Kopje culture. They combined pastoralism and cultivation with ironworking. The Leopard's Kopje people hunted as a supplement to livestock and arable farming.

In eastern Botswana, Zimbabwe and Mozambique, the Mapungubwe and Great Zimbabwe states came and went with no substantial ecological effect, according to the sparse evidence available. Wildlife might have been pushed out of some areas by growth in cultivation, but there is no verifiable evidence of a major reduction in ungulate or predator numbers as a result of the growth of these states or of the rise of the powerful Monomutapa state. Despite a growth in archaeological/anthropological research into the kingdoms of Zimbabwe and neighbouring regions, as Pikirayi emphasises, "historical data on the Mutapa state is still inadequate".[144] The archaeological evidence and early accounts tend to be about the nature of the state itself and its customs and culture, with very little suggesting the nature of its relationship with environment or wildlife. At the time of Monomutapa rule, elephants, ungulates and lions were known to have been there in large numbers and one can presume that spotted hyenas were there in good numbers and, like East Africa and the Horn of Africa, adapted to human settlements, taking advantage of the scavenging opportunities they offered.

Woodland areas in southern Africa were only partially cleared at this time and those that remained were a source of wildlife products such as meat, skins and ivory– the hunting skills of the people were vividly described in the early Portuguese accounts of the region.[145] The ungulate species present would have provided prey for spotted hyenas and, in the right habitat, carcasses for brown hyenas. In the period leading up to the arrival of European traders, settlers and hunters, the southern African environment was a mixture of savannah in eastern and southern Angola, eastern Zambia and northern Namibia, often combined with dry woodlands, as in northern Botswana, southern Zambia and Zimbabwe, and semi-desert or desert in parts of Botswana, Namibia and South Africa.[146] In South Africa, the southern Kalahari and the Karoo were arid but supported a diversity of ungulates, including blaubok, blesbok, bontebok, gemsbok, hartebeest, quagga, springbok and zebra, which fed healthy populations of hyenas, lions, wild dogs, cheetah and leopards. San communities, like the !Kung, lived in areas with spotted and brown hyena, lion, leopard, cheetah and wild dogs and large ungulates like kudu, giraffe, gemsbok, wildebeest, eland and hartebeest. along with smaller antelopes like steenbok and duiker. The !Kung, according to contemporary research, are unafraid of the presence of predators and often sleep out in the open without fires as protection, perhaps because they don't have livestock to protect and seemed to be able to coexist with predators.[147] A San tracker and guide working in the Central Kalahari Game Reserve told me that among his community there was no oral history of conflict with predators such as hyenas or even lions, rather a tactic of respectful mutual avoidance – they did not hunt spotted hyenas or lions, and both predators avoided their camps or hunting parties, while brown hyenas presented no threat or real competition for food. Suzman says the Ju/'hoansi San whom he studied did not eat the meat of predators, and said that hyenas tasted "like shit" and could make you ill. The meat of the hyena would only be eaten in exceptional circumstances of need.[148]

The original San and Khoikhoi inhabitants of what is now Botswana, South Africa and Namibia followed lifestyles based on hunting, some scavenging of the carcasses of wild ungulates and foraging for wild plant foods. The San livelihoods and cultures there changed little over the last two millennia, until they were forced to migrate or adopt more "modern" lifestyles by social, political and economic pressures (whether by the kingdoms and chieftaincies established by the incoming Bantu or during apartheid rule in South Africa and Namibia). The Khoikhoi people adapted more to the cultures of the Bantu migrants, adopting pastoralism alongside hunting. By the 15th century, Southern Africa's population was a mixture of small groups of San hunter-gatherers, Khoikhoi hunter-pastoralists and Bantu agro-pastoralist communities.[149] Wildlife was still abundant, despite a gradual reduction in habitat and some losses through increased hunting by the growing Bantu communities. In 1487, the first European arrived in southern Africa, in the person of the Portuguese explorer Bartholomeu Dias. He sailed round the Cape, anchoring at Mosselbaai on the south coast. Ten years later, Vasco da Gama sailed into the Indian Ocean, opening the way for Portuguese trading expeditions. Commerce developed with East Asia and ports were established on the Mozambican coast, trading for gold, ivory and skins with the peoples of Mozambique, Malawi and Zimbabwe.[150]

West Africa

Grasslands, dry woodland and semi-arid savannah in West Africa provided the habitat and prey species necessary for spotted hyenas to hunt and scavenge and for striped hyenas to scavenge, forage for fruits and vegetables and hunt when the opportunity arose. They were found from northern Nigeria, the Lake Chad Basin, Niger, northern Benin, Burkina Faso, northern Ghana, Ivory Coast, Mali, small areas of Guinea and Guinea-Bissau and eastern Sierra Leone to Senegal and Mauritania. Both species were present in a variety of habitats. Striped hyenas were found mainly north of densely forested areas and stretching right up around the edges of the western Sahara to North Africa. Spotted hyenas had their northern limit in southern Mauritania but spread further south into forested areas; they were only absent from the dense forests inland from the Atlantic coast In the period from 4,000BCE to the Common Era, the grasslands of West Africa are known to have had a diversity of ungulates, which supported the full range of African carnivores.[151] The drying out of the region, starting around 2000BCE,[152] and the effects of human expansion and land degradation, combined with deforestation, progressively reduced wildlife ranges – hyena numbers (both species) may have declined as a result, but, being adaptable to living in proximity to humans, populations of both have survived into modern times.

Early stone age populations in West Africa had limited ability to clear thick forest, restricting the expansion of cultivated areas and limiting pastoralism and arable farming to existing areas of open woodland and grassland. Better tools and techniques emerged around 500BCE, as demonstrated by the Nok culture of Nigeria, according to Davidson.[153] This enabled the expansion of agriculture as well as improved weapons for hunting and protection against dangerous animals. Settled agriculture and more extensive production of yams, millet and rice developed in the river valleys and in the areas cleared of forest south of the Sahara. In areas with less reliable rainfall and poorer soils, shifting agriculture was practised alongside pastoralism.[154] Transhumant pastoralism was ubiquitous across the savannah and dry woodland/bush zone of West Africa. The population groups that coalesced to become the pastoralist Fulbe (called Peuhl in Senegal and Fulani in Nigeria) moved west and south as the Sahara dried out and grazing land was lost. They traversed the whole region of West Africa between desert and dense forest with their livestock.[155]

At the beginning of the second millennium CE, human populations increased and settled communities grew and became kingdoms or large polities like the states of Ghana, Kanem, Songhay and Takrur in the Sahel.[156] These states, exacting tribute from weaker surrounding communities, mixed cultivation in fertile valleys and floodplains with pastoralism and control of trade. In the better watered, more southerly regions extensive cultivation suggests the probable and gradual exclusion or diminution of wildlife. The states of the region had well developed ironworking industries, used horses or camels for herding and warfare, and had a greater ability to hunt from horseback with iron spears, killing game for food and predators to protect livestock. The Mali kingdom, established in the 13th century CE by the Mande leader, Sundiata Keita, stretched from the Atlantic Coast to present-day Mali. To emphasise his power, Sundiata had praise names such as Simbon Salata (Master Hunter) and Mari Diata (Lord Lion); needless to say, you do not find hyena used as a praise name.[157] Extensive crop cultivation and pastoralism in the Mali Empire pushed out wildlife, which survived mainly in regions unsuitable for cultivation and marginal for pastoralism.[158] A steady growth in the human population, the establishment of large towns, states and empires with extensive cultivation around river systems, large-scale pastoralism and progressive desertification along the northern Sahel all led to depletion of wildlife. The West African grasslands, open woodland and semi-arid regions fringing the desert did not retain the large herds of ungulates and numerous predators found in East and Southern Africa at the time of the arrival of Europeans. But the adaptable spotted and striped hyenas did survive across much of the region, though never in large numbers.

References to hyenas are few and far between in recorded sources for this period in West Africa and their presence has to be gleaned from folktales and myths. Glickman wrote that evidence of a history of hyena presence can be found among the Kujamaat Diola of Southern Senegal in West Africa, noting that in their folktales "the Hyena becomes a vehicle for a very definite character type which combines the perceived characteristics of greediness (especially with food), aggressiveness, trickiness of a particularly crude sort...and often plain stupidness".[159] Senegal still supports populations of both striped and spotted hyenas, which have adapted to living around humans as well as in protected wildlife areas such as national parks and reserves. In northern Ivory Coast and Senegal, hyenas were present during the whole of the period under review in this chapter. This is attested to by the folktales among the Beng of Ivory Coast and the Kujamaat Diola of Senegal about links between hyenas, lepers and blacksmiths.[160] While hyenas may have been objects of derision, they were also seen as ritualistically important by these two cultures.[161]

North Africa and Egypt

As is the case today, the range of the striped hyena, the only Hyaenidae species that survived outside sub-Saharan Africa, stretched from Morocco and Mauritania in the west, around the Sahel region to the west and south of the Sahara, into East Africa and the Horn, and then through Egypt into West Asia (encompassing the Middle East, Turkey and the Arabian Peninsula) into Central and South Asia. It was nowhere numerous but was widely distributed. Its nocturnal habits meant it was not often observed or recorded; it is hard to make any assessment of numbers or precise ranges. It was able, though, to develop a strategy of foraging around human settlements (including burial grounds), scavenging and hunting in suitable habitats at night, keeping out of sight during daylight. It was persecuted where it was blamed by pastoral and farming communities for attacks on stock (especially goats, sheep and poultry), supposed threats posed to children and the aura of the occult surrounding it because of its digging up and consumption of human corpses from graveyards.

Egypt's Nile Valley and Delta were key locations for the development of cultivation and animal husbandry in Africa. The year-round availability of water was important in the development of the increasingly sophisticated and hierarchical societies of the dynastic periods. By around 4000BCE, the increasing food production agricultural production and consequent increase in human populations enabled the emergence of towns, with populations as high as 1,300–2,000.[162] Towns were common in Upper Egypt from Aswan to the Delta by 3500BCE. Within 500 years, Egypt had been unified as a single state stretching about 1300km north to south.[163] Egypt's history from the third millennium BCE to 343BCE can be divided into four periods: the Old Kingdom (2,700–2,200 BCE); the Middle Kingdom (c2,050–1,800 BCE); and the New Kingdom (c1,550–1,100 BCE). In the pre- and early dynastic period (c3,100–2686BCE) and for many centuries after that, despite the expansion of towns and human population, the fauna of Egypt was far richer than it is today, with large populations of lions, hyenas, jackals, hippos, giraffes, wild cattle, diverse species of antelopes and pigs – though these were depleted by habitat loss and hunting and restricted to sparsely-populated regions before most of them disappeared, some smaller ungulates, striped hyenas, jackals and smaller predators survived.[164] But the creation of towns with waste dumps, often outside the boundaries, would have created opportunities for adaptable scavengers like striped hyenas and jackals. Evidence of the presence of hyenas is to be found in art, in representations of rituals involving the sacrifice or consumption of hyenas but also in images or descriptions of hunting.

Images that have survived from the Middle Kingdom dynasties (2134–1785 BCE) and the rule of Rameses II (1131 BCE) show hounds running alongside a hunter's chariot hunting hyenas.[165] There is an illustration from the time of King Sahure (Fifth Dynasty, c2475BCE) of a hyena being hunted and killed by archers.[166] From these times through to the present, farmers in Egypt are known to have hunted hyenas to avenge or prevent damage to crops and livestock.[167] Striped hyenas in Egypt have been recorded eating maize crops, dates, melons, tomatoes, bananas and plums, and excavations of a hyena den in a cave near the pyramid of Cheops revealed ancient bones of domestic stock and dogs.[168] There is also evidence, including from friezes and paintings in tombs and other buildings, of striped hyenas being captured or possibly bred in captivity and force fed before being killed as sacrifices in rituals or for food. Osborn and Helmy recount that in the tomb of Merekua (from the Sixth Dynasty, 2300BCE) there are depictions of hyenas being fattened up, by being fed meat from ducks and geese, and then slaughtered for food. They also record that there are accounts of hyenas being tamed and used to hunt for their owners. Despite being eaten and used for sacrifices in rituals there is no suggestion that striped hyenas were sacred to the Egyptians.[169]

Hyenas were often included in funerary art and on knife handles suggesting, according to Ikram, that the knives were used in ritual or food-related slaughtering of hyenas.[170] The funeral connection is enhanced by evidence that hyenas were sacrificed to accompany the deceased into the next world, perhaps because they were known to eat human corpses and so had a link with the dead.[171] Some illustrations of funeral processions showed hyenas being led to the funeral site along with other animals for sacrifice.[172] A carved and painted frieze from Saqqara dating to the Sixth Dynasty shows two men force feeding a striped hyena, which had its legs tied.[173] Legge lists 48 images of hyenas being led or carried, five references in lists of animals to be offered up to the gods, four references in royal place names in Egypt, one of a hyena involved as a tame animal in a hunt, four depictions in the wild and one of a hyena being hunted. Legge emphasises that there is clear evidence of hyenas having been captured and tamed, or bred in captivity.[174] Hyena bones found at excavation sites from this period show clear cut marks where the animal has been slaughtered for food.

There is no evidence from the Ptolemaic period (323–30BCE) of the continued use of hyenas for ritual purposes or their consumption. When the Romans overthrew the last Ptolemaic ruler, Cleopatra, they treated Egypt as a bread-basket for Rome and also a source of obtaining wild animals to be killed in the arena – either from Egypt or using Egypt and other North African possessions to trade with merchants from across the Sahara for wild animals. The 14th century Morccan scholar and traveller Ibn Battutah reported that in 1325–6 he travelled from Tangier to Egypt and then up the Nile. After passing through Luxor, he journeyed towards Sheikh Shazlyas, the shrine of the Sufi saint Imam Abu Hasan Shadhili, and wrote that he found the place "infested with hyenas". He said his party had to continually drive them off – one animal managing to rip open one of the camel sacks, which was full of dates, which it ate, suggesting that these were striped hyenas, though their presence in large groups around his caravan could equally indicate they were spotted hyenas.[175] Striped hyenas eat fruit and vegetable matter as part of their broad diet; spotted hyenas less so, but they have a reputation for destruction and eating anything edible they can get hold of in camps and settlements.

There is a gap in written and translated accounts of the survival of wildlife in North Africa from this period until the 16th century. In that century, the Moroccan diplomat and writer known as Leo Africanus commented in his three volume work on Africa that around the countryside near the town of Agmet, between the Atlantic Coast and the Atlas mountains in western Morocco, are found "wolves, foxes, deere, and such other wilde beasts". It is likely that the reference to wolves meant either striped hyenas or golden jackals, now called North African Golden Wolves.[176] The writer also noted that in this and neighbouring regions in the early decades of the 16th century, convicted thieves were executed and their bodies thrown outside the town walls to be devoured by "wilde beasts, and ravenous foules", which would have certainly included striped hyenas.[177] In districts of this part of Morocco the local rulers engage in great hunts, with the quarry including wolves[178] – again, presumably hyenas or jackals. While the hunts were for sport, they also served a purpose of removing predators from a region that relied heavily on livestock farming, particularly sheep for wool, cheese and meat.[179] On a plain leading up to the Atlas Mountains, which Leo Africanus calls Azgari Camaren, he not only noted the presence of large numbers of lions which, along with other predators, preyed on cattle, but also the presence of hyenas. In this area he reported finding four newborn "strange" beasts and seeing the mother, which he described as being "not much unlike to a shee-wolfe" and which the translator notes is a striped hyena. The hyena was not threatening and he was able to get safely out of the cave he was exploring when he found them.[180]

In his ninth book on his travels in Africa, Leo Africanus described the striped hyena, which he called *Dabuh*, a name he said was derived from usage in Arabia:

> in bignes and shape resembleth a woolfe, saving that his legges and feete are like to the legs and feete of a man. It is not hurtful unto any other beast, but will rake the carkeises of men out of their graves and will devour them, being otherwise an abject and a silly creature. The hunters being acquainted with his denne, come before it singing and playing upon a drum, by which medolie being allured foorth, his legs are intrapped in a strong rope, and so he is drawn out and slaine.[181]

Notes

1 Rosie Woodroffe, Simon Thirgood and Alan Rabinowitz (2005) The impact of human-wildlife conflict on natural systems, in Rosie Woodroffe, Simon Thirgood and

Alan Rabinowitz (eds) *People and Wildlife Conflict of Coexistence*, Cambridge: Cambridge University Press, 1–12, p. 1.
2 Woodroffe et al., 2005, p. 2.
3 Adrian Treves and Lisa Naughton-Treves (2005) Evaluating lethal control in the management of human-wildlife conflict, in Woodroffe et al., 86–106, p. 86.
4 Robin S. Reid (2012) *Savannas of Our Birth. People, Wildlife and Change in East Africa*, Berkeley, Calif.: University of California Press, 2012, p. 95.
5 Ibid., p. 63.
6 Christopher Ehret (2016) *The Civilizations of Africa. A History to 1800*, Charlottesville, Virginia: University of Virginia Press, p. 37.
7 Ibid., pp. 41–2.
8 Roberts, 2018, p. 203.
9 Mayas A. Qarqaz, Mohammad A. Abu Baker and Zuhair S. Amr (2004) Status and ecology of the Striped Hyaena, Hyaenahyaena, in Jordan, *Zoology in the Middle East*, 33, 1, 87–92, p. 91.
10 Nikunj Gajera, S. M. Dave and Dharaiya Nishith (2009) Feeding patterns and den ecology of striped hyena (Haeyena haeyena) in North Gujurat, India, *The Tiger Paper*, January–March 2009, 36, 1, 12–17, p. 12.
11 Florian J. Alberto et al. (2018) Convergent genomic signatures of domestication in sheep and goats, *Nature Communications*, 9, 813, https://doi.org/10.1038/s41467-018-03206-y accessed 9 May 2019.
12 Roberts, 2018, p. 208–9.
13 M. Orbach and R. Yeshurun (2019) The hunters or the hunters: Human and hyena prey choice divergence in the Late Pleistocene Levant, *Journal of Human Evolution*, published January 2018, https://doi.org/10.1016/j.jhevol.2019.01.005 accessed 25 May 2019.
14 Jason Bittel (2018) Striped Hyenas Don't Have Magical Powers, But Their Disappearing Act Is Real, *Ecowatch*, 25 January 2018, https://www.ecowatch.com/hyenas-in-india-25 28561632.html accessed 3 June 2019.
15 Ibid.
16 Madhav Gadgil and Ramachandra Guha (1992) *This Fissured Land, An Ecological History of India*, New Delhi: Oxford University Press, p. 63.
17 Ibid., pp. 64–5.
18 Roberts, 2018, p. 209.
19 J. F. Jarridge and R. H. Meadow (1980) The antecedents of civilization in the Indus Valley, *Scientific American*, 243, 2, 122–133.
20 See G. R. Sharma et al. (1980) *The Beginnings of Agriculture*, Allahabad: University of Allahabad Press.
21 K. A. Chowdhury (1977) *Ancient Agriculture and Forestry in Northern India*, Bombay: Asia Publication.
22 Roberts, 2018, p. 209.
23 Gadgil and Guha, 1992, p, 66.
24 Ibid.
25 John Iliffe (2007) *Africans. The history of a continent*, Cambridge: Cambridge University Press, 2nd edition, pp. 12–3.
26 For a more detailed examination of the origins of domesticated cattle, see Jared E. Decker et al. (2014) Worldwide Patterns of Ancestry, Divergence, and Admixture in Domesticated Cattle, *PLOS Genetics*, 27 March 2014.
27 Wilma Wetterstrom (1993) Foraging and farming in Egypt: the transition from hunting and gathering to horticulture in the Nile Valley, in Thurstan Shaw et al. (ed) *The Archaeology of Africa. Food, metals and towns*, London: Routledge, 165–226, p. 167.
28 Ibid., p. 201.
29 Gus Mills and Heribert Hofer (1998) *Hyaenas Status Survey and Conservation Action Plan*, Gland, Switzerland: IUCN/SSC Hyaena Specialist Group, p. 56.

30 Keith Somerville (2019) *Humans and Lions Conflict, Conservation and Coexistence*, London: Routledge/Earthscan, pp. 100–2.
31 M. AbiSaid and S. M. D. Dloniak (2015) *Hyaena hyaena. The IUCN Red List of Threatened Species*, https://dx.doi.org/10.2305/IUCN.UK.2015-2.RLTS.T10274A45195080.en accessed 29 January 2020.
32 Juliet Clutton-Brock (1993) The spread of domestic animals in Africa, in Shaw et al., 61–70, p. 66.
33 Roberts, 2018, p. 212.
34 Clutton-Brock, 1993, p. 67.
35 Brian Bertram (1978) *Pride of Lions*, London: J.M. Dent, p. 240.
36 Personal communication with Dr Amy Dickman, 10 June 2019.
37 Personal communication, 20 July 2019.
38 Personal communication from Laurence Frank, 22 May 2020.
39 Personal communication from Dr David Anderson of the University of Warwick.
40 Reid, 2012, p. 96.
41 Bassey W. Andha (1993) Identifying early farming traditions of west Africa, in Shaw et al., 241–254, p. 245.
42 Christopher Ehret (2016), *The Civilisations of Africa. A History to 1800*, Charlottesville, Virginia: University of Virginia Press, pp. 25–33.
43 Ibid., p. 13; see also International Atomic Energy Authority (no date) *Nubian Aquifer Project*https://web.archive.org/web/20071020163247/http://www-naweb.iaea.org/napc/ih/Nubian/IHS_nubian.html accessed 28 May 2020; M. Ramdani, N. Elkiahati and R. J. Flower (2009) Lakes of Africa: North of Sahara, in *Encyclopedia of Inland Waters*, Cambridge, Mass: Academic Press, 544–554; Matt McGrath (2012) Huge water resource exists under Africa, *BBC*, 20 April 2012, https://www.bbc.co.uk/news/science-environment-17775211 accessed 28 May 2020.
44 Christopher Ehret (1998), *An African Classical Age. Eastern and Southern Africa in World History 1000BC to A.D. 400*, Oxford: James Currey, p. 6
45 Ibid., pp. 73 and 75.
46 Iliffe, 2007, p. 17.
47 Ehret, 1998, pp. 14–5.
48 Iliffe, 2007, pp. 1–2.
49 Ehret, 1998, p. 41.
50 Ibid., pp. 79–80.
51 J. E. G. Sutton (1969) "Ancient Civilizations" and Modern Agricultural Systems in the Southern Highlands of Tanzania, AZANIA, *Journal of the British Institute in Eastern Africa*, 4, 1, 1–13, p. 10.
52 Mohamed Amin, Duncan Willetts and John Eames (1987) *The Last of the Maasai*. London: Camerapix Publishers International, p. 103.
53 Personal communication from Marcus Baynes-Rock.
54 Frank W. Marlowe (2010) *The Hadza Hunter-Gatherers of Tanzania*, Berkeley, Calif: University of California Press, p. 65.
55 Lane, 2004, p. 248.
56 Ibid., p. 252.
57 Kennedy K. Mutundu (2010) An ethnoarchaeological framework for the identification and distinction of Late Holocene archaeological sites in East Africa, *Azania: Archaeological Research in Africa*, 45, 1, 6–23, p. 7.
58 Ibid.
59 Peter Robertshaw (1993) The beginnings of food production in southwestern Kenya, in S. Thurstan Shaw et al., *The Archaeology of Africa. Food, metals and towns*, London: Routledge. 358–371, p. 365.
60 Richard H. Lamprey and Robin S. Reid (2004) Expansion of Human Settlement in Kenya's Maasai Mara: What Future for Pastoralism and Wildlife? *Journal of Biogeography*, 31, 6, 997–1032, p. 999.

61 Fiona Marshall (1990) Cattle herds and caprine flocks in Robertshaw (ed) *Early pastoralists of south-western Kenya*, Nairobi: British Institute in Eastern Africa, 205–260, p. 246.
62 Reid, 2012, p. 96.
63 Juliet Clutton-Brock (1994) The Legacy of Iron Age Dogs and Livestock in Southern Africa, *Azania: Archaeological Research in Africa*, 29–30, 1, 161–167, p. 164.
64 Marlowe, 2010, p. 18.
65 James T. Pokines and Julian C. Kerbis Peterhans (2007) Spotted hyena (Crocuta crocuta) den use and taphonomy in the Masai Mara National Reserve, Kenya, *Journal of Archaeological Science*, 34, 1914–1931, p. 1915.
66 See Ehret, 2016, p. 105, on the drying of the climate and its effects.
67 Ehret, 2016, p. 105.
68 Diane P. Gifford et al. (1980) Evidence for Predation and Pastoralism at Prolonged Drift: a Pastoral Neolithic Site in Kenya, *Azania: Archaeological Research in Africa*, 15, 1, 57–108, pp. 64–5.
69 Ibid., p. 79.
70 Ibid., p. 88.
71 David Wright (2003) Archaeological investigations of three Pastoral Neolithic sites in Tsavo National Park, Kenya, *Azania: Archaeological Research in Africa*, 38, 1, 183–188, p. 187.
72 Ibid., pp. 107–9.
73 Ehret, 1998, p. 31.
74 Ibid., p. 109.
75 Anonymous (1998) The later archaeology of the Central Rift Valley of Kenya, *AZANIA: Journal of the British Institute in Eastern Africa*, 33, 1, 73–112, pp. 77–9.
76 Ibid.
77 Ehret, 2016, p. 269.
78 Ibid., p. 270.
79 Leakey cited by Anonymous, 1998, pp. 80–2.
80 Ibid.
81 Dorothy L. Hodgson (2001) *Once Intrepid Warriors Gender, Ethnicity, and the Cultural Politics of Maasai Development*, Bloomington, Indiana: Indiana University Press, pp. 23–4.
82 Edward I. Steinhart (2006) *Black Poachers White Hunters. A Social History of Hunting in Colonial Africa*, Oxford: James Currey, p. 21.
83 Personal communication with Laurence Frank; see also, L. Hazzah et al. (2009) Lions and warriors: social factors underlying declining African lion populations and the effect of incentive-based management in Kenya. *Biological Conservation*, 142, 11, 2428–2437.
84 K. R. Dundas (1910) Notes on the Tribes Inhabiting the Baringo District, East Africa Protectorate. *The Journal of the Royal Anthropological Institute of Great Britain and Ireland*, 40, Jan–June, 1910, 49–72, p. 59.
85 Werner, no date, pp. 227–9
86 Jan Bender Shetler (2007) *Imagining Serengeti. A History of Landscape memory in Tanzania from Earliest Times to the Present*, Athens, Ohio: Ohio University Press, 2007, p. 3.
87 For more detail on the Kingdom of Kush, see László Török (1998) *The Kingdom of Kush: Handbook of the Napatan-Meriotic Civilization*. Leiden: Brill.
88 G. Mokhtar (ed) (1990) *General History of Africa. II Ancient Civilizations of Africa*, London: James Currey/UNESCO, – pp. 163–5.
89 Peter Robertshaw & Ari Siiriäinen (1985) Excavations in Lakes Province, Southern Sudan, *Azania: Archaeological Research in Africa*, 20, 1, 89–161, p. 152.
90 Ibid., p. 91
91 Ibid., p. 89.
92 Ibid., p. 91.
93 Ibid., p. 102.
94 Ibid., p. 92.
95 Ibid., p. 107–8.

96 Jean Buxton (1968) Animal Identity and Human Peril: Some Mandari Images, *Man*, 3, 1, 35–49, p. 36.
97 Ibid., pp. 36–17.
98 Ibid., p. 42.
99 Amanuel Beyin (2011) Early to Middle Holocene human adaptations on the Buri Peninsula and Gulf of Zula, coastal lowlands of Eritrea, *Azania: Archaeological Research in Africa*, 46, 2, 123–140, pp. 123–4.
100 M. Tefera (2004) Recent Evidence of Animal Exploitation in the Axumite Epoch, 1st–5th Centuries AD, *Tropical Animal Health and Production*, 36, p. 105.
101 The Geography of Strabo, published in Vol. VII of the Loeb Classical Library edition (1932), http://penelope.uchicago.edu/Thayer/E/Roman/Texts/Strabo/16d*.html?fbclid=IwAR0MqdK6vIFUBMAIFWqrmwu9zxFE20qOq7p9J1fdWV0lVjmn8DUtGPmVOBY accessed 3 June 2019.
102 Alfredo González-Ruibal et al. (2014) Late hunters of western Ethiopia: the sites of Ajilak (Gambela), c. AD 1000–1200, *Azania: Archaeological Research in Africa*, 49, 1, 64–101, p. 95.
103 Ibid.
104 Ibid., p. 96.
105 Ibid.
106 Siegbert Uhlig (ed) (2007) *Encyclopaedia Aethiopica, Volume 3*, Wiesbaden: Harrassowitz Verlag, p. 571.
107 Ibid.
108 Richard Pankhurst (ed) (1967) *The Ethiopian Royal Chronicles*, Addis Ababa: Oxford University Press, p. xi.
109 Marcus Baynes-Rock (2012) Hyenas like Us: Social Relations with an Urban Carnivore in Harar, Ethiopia, PhD thesis, Department of Anthropology Macquarie University, Sydney, pp. 38–9.
110 Ibid.
111 Ibid., p. 39.
112 I. Hribek (ed) (1992) *Africa from the Seventh to the Eleventh Century, UNESCO General History of Africa III*, London: James Currey/UNESCO, p. 45.
113 Philip Curtin, Steven Feierman, Leonard Thompson and Jan Vansina (1995) *African History. From Earliest Times to independence*. London: Longman, 2nd edition 1995, pp. 148–9.
114 Ibid., p. 150.
115 Baynes-Rock, 2012, pp. 39–40.
116 Ibid., p. 41.
117 Ibid.
118 Ibid.
119 Ibid.
120 Courlander, 1996, p. 561.
121 James Quirin (1979) The Process of Caste Formation in Ethiopia: A Study of the Beta Israel (Felasha), 1270–1868, *International Journal of African Historical Studies*, 12, 2, 235–258, p. 236.
122 Ibid., p. 249.
123 Ibid.
124 See Whitney Davis (1984) Representation and Knowledge in the Prehistoric Rock Art of Africa, *The African Archaeological Review*, 2, 7–35.
125 C. A. Guggisberg (1961) *Simba. The Life of the Lion*, London: Bailey Bros and Swinfen Ltd, p. 29.
126 AbiSaid and. Dloniak 2015.
127 T. Bohm and O.R. Höner (2015) *Crocuta crocuta*. The IUCN Red List of Threatened Species, https://www.iucnredlist.org/species/5674/45194782 accessed 12 February 2020.
128 AbiSaid and Dloniak, 2015.

129 Duarte Lopez and Filippo Pigafetta (translated by Maragerite Hutchinson) (1881 reprinted) *A Report of the Kingdom of Congo: And of the surrounding countries*, London: John Murray (reprinted by Leopold Classic Library), p. 49.
130 Ibid., pp. 49–50.
131 Ibid., p. 52.
132 D. N. Beach (1980) *The Shona and Zimbabwe 900–1850*, Gweru, Zimbabwe: Mambo Press, p. 53.
133 Ibid., p. 53.
134 Ehret, 1998, pp. 212–3.
135 Alan Barnard (1992) *Hunters and Herders of Southern Africa. A Comparative Ethnography of the Khoisan peoples*, Cambridge: Cambridge University Press, pp. 30–2.
136 Ehret, 2016, p. 165.
137 Ibid., pp. 170–2.
138 D. W. Phillipson (1975) The Chronology of the Iron Age in Bantu Africa Author, *Journal of African History*, 16, 3, 321–342, p. 332.
139 Ibid., p. 271.
140 Ki-Zerbo and Niane, 1997, p. 209.
141 Iliffe, 2007, pp. 100–101.
142 Ibid., p. 101.
143 John Kinahan (1994) A New Archaeological Perspective on Nomadic Pastoralist Expansion in South-western Africa, *Azania: Archaeological Research in Africa*, 29–30, 1, 211–226, p. 211.
144 Innocent Pikirayi (1991), *The Archaeological Identity of the Mutapa State. Towards an historical archaeology of northern Zimbabwe*, Uppsala: Societas Archaeologica Upsaliensis, p. 15.
145 Pikirayi, 1991, p. 29.
146 Curtin et al., 1995, p. 214.
147 Richard B. Lee (1979) *The Dobe !Kung*, New York: Holt, Rinehart and Winston, p. 23.
148 James Suzman (2017) *Affluence with Abundance. The Disappearing World of the Bushmen*, New York/London: Bloomsbury, p. 152.
149 Leonard Thompson (2000) *A History of South Africa*, New Haven: Yale University Press, pp. 10–11.
150 Ibid., p. 32–3.
151 Oliver Davies (1967) *West Africa Before the Europeans. Archaeology and Prehistory*, London: Methuen, pp. 148–9.
152 Basil Davidson (1998) *West Africa Before the Colonial Era. A History to 1850*, London: Longman, p. 10.
153 Davidson, 1998, pp. 12–3.
154 Davies, 1967, p. 149.
155 Curtin et al, 1995, pp. 84–5.
156 Ibid., p. 69.
157 Joseph Ki-Zerbo and Djibril Tamsir Niane (1997) *Africa from the twelfth to the Sixteenth Century. General History of Africa IV*, London: James Currey/UNESCO, p. 56.
158 Ibid., p. 65.
159 Stephen E. Glickman (1995) The Spotted Hyena from Aristotle to the Lion King: Reputation is Everything, *Social Research*, 62, 3, 501–537, p. 526; see also, J. David Sapir (1981) Leper, hyena, and blacksmith in Kujamaat Diola thought, *American Ethnologist,* 8, pp. 526–43.
160 Alma Gottlieb (1989) Hyenas and Heteroglossia: Myth and Ritual Among the Beng of Côte d'Ivoire, *American Ethnologist*, 16, 3, 487–501, p. 487.
161 Ibid.
162 Fekri A. Hassan (1993) Town and village in ancient Egypt: ecology, society and urbanization, in Shaw et al., 551–569, p. 551.
163 Ibid., p. 552.
164 Richard Carrington (1972) Animals in Egypt, in A. Houghton Brodrick *Animals in Archaeology*, London: Barrie and Jenkins, 69–89, p. 70.

165 Alan Mikhail (2014) *The Animal in Ottoman Egypt*, Oxford: Oxford University Press, p. 82.
166 Salima Ikram (2001) *The Iconography of the Hyena in Ancient Egyptian Art*, Mainz: Mitteilungen des Deutschen Arachaologischen Instituts Abteiling Kairo, Band 57, p. 130.
167 Dale J. Osborn and Ibrahim Helmy (1980). The Contemporary Land Mammals of Egypt (including Sinai). Field Museum of Natural History, digitized version, https://archive.org/details/contemporaryland05osbo/page/422 accessed 24 July 2019, p. 422.
168 Ibid., p. 431.
169 Ibid.
170 Ikram, 2001, p. 127.
171 Ibid., p. 130.
172 A. J. Legge (2011) The hyaena in dynastic Egypt: Fancy food or fantasy food? *Journal of Osteoarchaeology*, 21, 5, 613–621, p. 615.
173 Ibid.
174 Ibid., p. 620.
175 Ibn Battutah (edited by Tim Mackintosh-Smith) (2003) *The Travels of Ibn Battutah*, London: Picador, p. 23.
176 Leo Africanus (no date) *The History and Description of Africa, Volume II*, London: Hakluyt Society, republished by LightningSource, UK, p. 273.
177 Ibid., p. 282.
178 Ibid., p. 297.
179 Ibid., pp. 302–3.
180 Ibid., p. 556.
181 Leo Africanus (2018) *The History and Description of Africa and of the Notable Things Therein Contained, Volume 3*, London: Hakluyt Society, reprinted by Forgotten Books, p. 947.

4

HUMANS AND HYENAS IN WEST, CENTRAL AND SOUTH ASIA TO 1600CE

This chapter covers the period up to 1600 and concerns the striped hyena in West, Central and South Asia, and its interactions with man. There are huge gaps in the archaeological and written accounts, and the need to interpret what data is available in the light of what we currently know about the striped hyena. As Majumder notes in his 11 volume history of the Indian people, "prior to the thirteenth century A.D., we possess no historical text of any kind, much less a detailed narrative as we possess in the case of Greece, Rome or China".[1] The result, as Keay said, is that historians have to reconstruct the past from "reluctant materials", which has involved piecing together an historical account from "random inscriptions, titbits of oral tradition, literary compositions and religious texts" to add to what can be ascertained from archaeological research.[2] This applies even more strongly to environmental/conservation history, which is often missing from the record.

Europe, West and Central Asia

After the extinction of the Eurasian spotted or cave hyena (*Crocuta crocuta spelaea*) between 14–11,000 years ago,[3] hyenas in Eurasia and West, Central and South Asia were limited to the single species, the striped. It was found on the fringes of south-eastern Europe, the Balkans, southern Russia, Central Asia and Turkey, but eventually became extinct in Europe, Russia and the region of Central Asia that is now Kazakhstan. This limited their range to Turkey, the Caucasus, Iran, the Middle East, the Arabian Peninsula and parts of South Asia.[4]

Striped hyenas survived in Greece after the disappearance of the cave hyena, even though their range was progressively shrinking as a result of the pressure created by the growth in human settlements and agriculture. In the middle and at the end of the first millennium BCE, striped hyenas, as well as lions, still were present in parts of Greece and neighbouring regions, because of the presence of a strong prey base (deer species, boar, ibex and mouflon) and despite competition from humans and wolves. At the time of the historian Herodotus (484–430BCE), there were hyenas and lions in parts of Greece.[5] Herodotus described the hyena as a pig-like creature with a bristly mane running down its back – clearly identifying it as the striped rather than spotted hyena. He wrote primarily about its presence in North Africa, describing in about 425BCE the areas of Libya inland from the coast as "infested by wild animals".[6]

Aristotle's (384–322BCE) descriptions of hyenas are also clearly of the striped:

> The hyena in colour resembles the wolf, but is more shaggy, and is furnished with a mane running all along the spine. That is recounted concerning its genital organs, to the effect that every hyena is furnished with the organ both of the male and female is untrue.[7]

This makes it clear that he is talking about the striped, but it is also an attempt to counter the assumptions drawn by many at the time, and for centuries to come, from the female spotted hyena's curious genitals, that hyenas were hermaphrodites or could change sex. Glickman believes that Aristotle's correct rejection of the hermaphrodite label was made even though and perhaps because he had never seen a spotted hyena and so was not aware of the physical attributes that led to the assumptions of sexual ambiguity.[8] Aristotle reflected beliefs of the time when he wrote that:

> The animal called "glanus" by some and "hyaena"...will lie in wait for a man, and chase him, and will inveigle a dog within its reach by making a noise that resembles the retching noise of a man vomiting. It is exceedingly fond of putrified flesh and will burrow in a graveyard to gratify this propensity.[9]

He rather exaggerates the danger to man, describing the hyena in his *The History of Animals* as a predator that "secretly attacks men, and hunts them down".[10] A further description of the occult abilities of the hyena, ascribed to Aristotle, can be found in the works of the Roman teacher of rhetoric, Claudius Aelianus (aka Aelian). This account says that it:

> ...has magic in its left paw that can send someone into sleep with a mere touch. It will often steal into a stable and find an animal that is already sleeping, then put its paw on the creature's nose and keep it from breathing...Then the hyena will grab the animal's neck with its mouth and carry it away. It attacks dogs in this manner; when the moon is full, it will get in front of it so that the moonlight casts its shadow on the dogs. These, bewitched as if by a sorceress, are helpless.[11]

Hyenas in Greco-Roman writings are presented according to the mythical narrative of sinister, magical beasts frequenting graveyards.[12] There were only three species left in the parts of Eurasia and Africa that would have been known to the

Greeks and Romans through conquest, exploration and trade – the spotted and striped hyenas and the aardwolf.[13] Most descriptions or references from this period would appear to be of striped hyenas[14]. Accounts of spotted hyenas from travellers or traders may have become confused with what was known and believed about the striped hyena, leading to confusion and mixing up the two species. Funk says that linguistic experts point out that the nomenclature used in this period suggests that in Greco-Roman times two species were known, but that descriptions often "overlapped".[15] This is evident from pictorial representations such as the mosaic from Palestrina in Italy (c120BCE), which shows a hyena with horizontal rather than vertical stripes that is likely to have been a striped hyena but is not definitively so. The few images said to be spotted hyenas on close examination are clearly based on their striped relatives.[16] The Palestrina mosaic is one of the earliest known Roman representations of a hyena. The mosaic depicts the course of the Nile from the Blue Nile to the Mediterranean. Until the discovery by archaeologists of mosaics in the ruins of Pompeii, the Palestrina mosaic was the earliest Roman depiction of Nilotic scenes.[17]

In a manuscript known as the Venice Oppian there are depictions and descriptions of carnivores with speculation about certain spotted animals being the offspring of a wolf and a leopard. Some images, according to Meyboom, are spotted hyenas. He believes that Roman references to an animal called *thoantes* are to the spotted hyena, while the *krokotta* (now translated into Latin term *Crocuta* for the spotted hyena) refers to the striped hyena.[18] Pliny the Elder, in his somewhat fanciful account of the wildlife of Africa, says that:

> Aethiopia (Ethiopia) produces Lynxes in great numbers, and Sphinxes with brown hair and a pair of udders on the breast, and many other monstrosities. Winged Horses armed with horns, called Pegasi; Crocotas (Hyenas) like a cross between a dog and a wolf, that break everything with their teeth…the leucrocota, swiftest of wild beasts.[19]

The Crocotas could be spotted or striped hyenas, but the leucrocota sounds more like a cheetah. Pliny also noted in his *Natural History* that hyenas were scavengers and consumers of carrion, with a reputation for digging up graves;[20] a probable reference to the striped hyena, which was found within the bounds of the Roman Empire.

Some references to hyenas by Greek and Roman writers suggest they are talking about the spotted hyena. Ctesias, a Greek physician and writer who lived in the 5th century BCE, said that there was:

> …in Aithiopia an animal called krokottas, commonly called dog-wolf. It is marvellous in its strength. This animal is said to imitate the human voice and to call people by their names at night so that they come closer in response to the human voice. The animals then jointly fall upon the people and devour

> them. This beast has the courage of a lion, the swiftness of a horse and the power of a bull, but it backs away from iron.[21]

This description of the crocotta/krokotta is repeated by the historian Dalion in the first half of the third century, who says that the crocotta listens to human conversation to find out the names of children. Agatharchides, in the 3rd/2nd century BCE, says that in Ethiopia there was an animal called the crocotta which was:

> ...a kind of composite of wolf and dog, fiercer than both and with much larger head and paws. It is amazingly powerful, and its teeth and stomach are stronger than those of other animals, for they easily crush every kind of bone, and quickly consume the fragments.[22]

Funk cites the Greco-Sicilian historian Diodorus Siculus on the wolf-dog hybrid of Ethiopia which could crush and digest any bones easily.[23] The reference to vocalisations suggest the spotted hyena, but myths about striped hyenas calling to people and dogs persisted for centuries.

About 400 years after Aristotle's observations about the hyena, the Roman writer, naturalist and philosopher Pliny the Elder wrote about hyenas in his study of the natural world. His book was to be highly influential for nearly 1500 years as a text on animal species.[24] Much of what Pliny wrote was reliant on Aristotle, such as his refutation of hyena hermaphroditism, and reference to the magical powers of hyenas to freeze other animals in place.[25] He gives more detail than Aristotle on the vocal range of the hyena, claiming that it mimics human speech around shepherd's encampments "and picks up the name of one of them so as to call him to come out of doors and tear him in pieces".[26] As Glickman points out, after describing the animal and its supposed supernatural abilities, Pliny:

> ...devotes five pages (of the English translation) to describing the hyena as a walking pharmacopoeia, including the 'fact' that a person carrying anything made of hyaena leather is not attacked, while the skin of the hyena head, when tied on, cures headaches; their teeth (when touched to the corresponding human tooth) relieve toothaches, and the gall, if applied to the forehead, cures opthalmia.[27]

Pliny claimed that barrenness in women could be cured by consuming a hyena's eye mixed with liquorice and dill; he also suggested that hyena genitals eaten with honey aroused sexual desire in men who "hate" intercourse.[28] Despite Aristotle and Pliny rejecting the sex shifting/hermaphrodite narrative, the Roman poet Ovid (43BCE to 17 or 18 CE) wrote of the hyena changing its sex continually.[29] Another poet, Lucan, refers to hyenas frequenting battlefields and brandishing the skin of its still living victims flayed alive.[30]

One of the ways in which Romans became aware of exotic animals, beyond those studying natural history or serving as administrators or in the legions abroad, was through the Roman Games. Brutal animal hunts were staged at many of the games in Rome or in provinces around the empire. The games began as religious festivals, "but eventually became political as showcases of military achievement or celebrations of victory over a foreign power".[31] Rather like the safari hunting culture that developed in 19th and 20th century India and Africa under the European colonial powers, the animal hunts staged at the games, with hundreds of imported wild animals (the bigger or fiercer the better) killed, represented domination over foreign lands and nature.[32] The animal slaughter in the arena was "a ritualization of hunting, and the arenas basically showed animals as foes that had to be conquered".[33] Pliny the Elder wrote that the first recorded lion contests in the arena were staged under Lucius around 80BCE when, he says, 100 maned lions took part in fights. Under Pompey, games were held involving 600 lions.[34]

It is thought that the first hyenas imported to be killed in the arena could have been imported for Pompey's games in 55BCE, when chroniclers said a "chama" had been killed, and this is thought to have been a spotted hyena.[35] Hyenas referred to as crocuttas were killed at the Games of Antoninus Pius in 148CE and at the games held by Septimus Severus in 202CE.[36] Wild animals were traded to the Romans from Mauritania, Sudanic regions and across the Sahara, the latter using the routes once controlled by the Carthaginians. Wild animal hunts at the circuses across the empire often involved gladiators fighting a variety of carnivores, as well as the killing of criminals by animals – a practice believed to have been started by Augustus (27 BCE to 14 CE). Martial (c. 41 CE to 102 CE) wrote that during the rule of the unpopular Emperor Domitian, games were put on which lasted six days, at which 9,000 animals, including hyenas, were killed.[37] Striped hyenas were recorded as having been sent as gifts from Ethiopia in the reigns of Antoninus Pius and Septimus Severus (193–211 CE).[38]

There are few references to hyenas in Europe after the fall of the Roman Empire or even during the Crusades. Hyenas had gone from Europe itself by the Middle Ages, though it is possible some may have survived in south-eastern Russia, in regions and in regions bordering Turkey. Some were to be found in royal menageries or other private collections owned by the nobility in Europe. King Henry I of England had a menagerie at Woodstock in Oxfordshire around 1130 where exotic animals sent as gifts to the king were housed. This included lions, leopards, lynxes and several hyenas.[39]

West Asia

Striped hyenas were widely distributed and common in Lebanon, Syria, Palestine, the Arabian Peninsula and the area known as Fertile Crescent centring on the Tigris and Euphrates rivers in Iraq (Mesopotamia). They are still found across the same range today. Because of their nocturnal foraging and general wariness of humans, the numbers even today are hard to assess, and there are too few reliable accounts from earlier times to make any realistic assessment of precise range or numbers.

One of the first written references to hyenas in West Asia is in the epic of Gilgamesh, a long narrative poem from ancient Mesopotamia (Iraq), one of the earliest surviving works of epic poetry dating from 2100BCE. It contains the lines, "May the country-side weep for thee as though it were thy mother...May the bear, hyena, and panther weep for thee; May the tiger and the hart, the leopard and the lion, the ox, the deer, the ibex".[40] In Persia, where striped hyenas are still present, images on ancient friezes, seals and other forms of art or symbols of authority, show striped hyenas. One seal ornamented with a seated hyena dates from the Achaemenid Empire (550–330BCE).[41] Despite this use of hyena images, the animal was generally viewed as dangerous and associated with witchcraft and the devouring of children. Ivanow says that they were associated by Persians with demoniacal beings that attacked children and were called the *kaftar*, a name used in Persian for hyena.[42] *Kaftars* were believed to be evil and should be killed on sight. This resembles the Arab *ghul*, or Kurdish *murdor-muy*, both examples of witches or animal-like beings. The references are to be found in ancient Persian literature, but are not that common and there is little to explain why hyenas were viewed in this way.[43] Some of the stories about how to identify a person who was a *kaftar* are not unlike European myths of vampires and werewolves, usually involving fear of daylight.[44]

Israel, Palestine and Jordan all have populations of striped hyenas and there is evidence from archaeological investigations, folklore and oral history of a long history of residence by them. One artefact that shows a long-standing presence, and also suggests human-hyena conflict and hunting of hyenas, is the discovery by archaeologists of ancient stone carnivore traps in the Negev and Judaean deserts of Israel and neighbouring regions stretching down to Yemen, where they are called wolf-traps but are clearly used to trap any large predator.[45] Avner et al. discovered a number of these traps in the 'Ein Gedi Oasis in the Judaean desert and the 'Uvda Valley in the southern Negev desert. They are built of unworked stones and:

> ...The traps comprise an elongated cell built of two parallel rows of upright stones... The cell is roofed by large, flat stone slabs...A stone slab...was suspended by a rope stretched above the roof. One end of the rope was tied around the trap door and the other end passed through a narrow gap in the ceiling and was looped around a wooden peg to which the bait – a chunk of meat – was attached. When the animal, having entered the trap, pulled at the bait, it released the rope attached to the peg and the door slab fell down entrapping the animal.[46]

The exact age of the traps is not clear but lithic artefacts from 4500–3500BCE, iron age implements from the twelfth to eighth centuries BCE, and Roman lamp fragments (c30BCE–70AD) have been found in the traps.[47] There is also evidence of newer traps having been built by Bedouin pastoralists in the last few centuries. The likelihood is that the traps have been used for millennia to capture and then kill predators that could prey on livestock – striped hyenas, leopards, wolves and jackals.[48] With their excellent sense of smell, there is no doubt that hyenas would have sniffed out the baits and are very likely to have been caught and killed in the traps.

The presence and human awareness of striped hyenas in Palestine and neighbouring regions is evidenced by biblical references to them – notably in the book of Samuel (13:18), where there are four references to hyenas, one of which calls an area the Valley of Zevoim (a Hebrew word for hyena – the area was also known as Wadi Abu Diba in Arabic, *Diba* being an Arabic term for hyena)[49]. Another reference is a derogatory one in the book of Jeremiah (12:19), when the prophet laments, "Is my heritage unto me as the hyena's den?", while the book of Deuteronomy (16:7|) cautions against eating the flesh of the hyena.[50] In the book of Isaiah (31:21–2), the eighth-century BCE prophet talked of ruin that could come upon communities, saying that, "Wild animals will rest there, the ruined houses will be full of hyenas. Ostriches will live there, wild goats will skip among the ruin…hyenas will howl in the fortresses, and jackals, in the luxurious palaces".[51] Other biblical references also have negative connotations attached to hyenas. One epistle attributed to Barnabas (believed to have been killed for his faith in Cyprus in 61CE) repeats the myths about changing sex and relates this to sinful behaviour in humans: "And do not eat the hyena. Do not, he says, be an adulterer or a corrupter or make yourself like such people. Why? Because this creature yearly changes its nature and is sometimes male and sometimes female".[52]

South Asia

South Asia is still home to a widespread population of striped hyenas. From the start of the Holocene, striped hyenas were the only hyena species which survived there. The modern striped hyena (*Hyaena hyaena*) is thought to have evolved from the African *Hyaena makaponi* and it dispersed, according to Rohland et al.,[53] into West Asia, parts of Central Asia and India from Africa around or just after 130,000 years ago, with a further dispersal in the Neolithic period under 12,000 years ago. The arrival of the first dispersing populations of striped hyenas in India preceded the migration of lions there from Africa via the Middle East, with hyenas reaching India around 40,000 years ago.[54] The range of the striped hyena in South Asia spans Pakistan, India and Nepal and falls largely within the Indomalayan ecozone, which has a diversity of ecosystems of the types found in the South Asian, Malayan and African continents.[55] Striped hyenas are found in many of the ecosystems, excluding dense rainforest and high mountains.

As human societies developed in South Asia, agriculture largely replaced hunter-gatherer modes, with the cultivation of rice, pulses, vegetables and fruit and livestock husbandry, chiefly for milk rather than meat. While vegetarianism is closely connected with the Hindu and Jain religions, there is still substantial consumption of meat. In the past, meat was consumed by many communities and was largely obtained through hunting.[56] Hunting remained key for the surviving hunter-gatherer communities, who were increasingly pushed to the margins socially, economically and in terms of occupation of land. Hunting became the sport of rulers of Indian states in ancient times and right up to and through the period of colonial occupation, as well as a means of supplementing other food sources.[57] And, as Pabla argues:

> The interest in hunting as a means of recreation spread to the common man with the emergence of more egalitarian and democratic societies in recent times...most of us believe that hunting and consumption of wildlife is contrary to Indian cultural traditions. However, a closer look at the Indian culture, history and religious texts leads one to the conclusion that Indian culture is no different from other cultures of the world as far as the utilisation of natural resources is concerned.[58]

As adherence to religions like Hinduism and Jainism spread, the importance of hunting and meat consumption declined, as these religious beliefs excluded or discouraged consumption of meat. The killing of predators to protect stock and people or for sport did not disappear. When considering attitudes towards killing animals like hyenas, it should be borne in mind that communal action to kill predators like leopards, tigers and hyenas has always been present and is still today, including capturing hyenas and using them for brutal sports that involve fierce dogs being set on them. Film of this violent form of "entertainment" can be easily found on sites like YouTube today, showing the capture and beating to death or baiting with dogs of striped hyenas.[59] Franks told me that he had information that in North Africa, hyenas were sometimes trapped, blinded and then tormented to entertain people.[60]

Ironworking was introduced into India around 1000BCE,[61] enabling more efficient land clearance and more productive crop harvests as well as allowing the production of more advanced weapons for war and hunting. Livestock husbandry based on zebu cattle and increased cultivation of rice, gram, beans enabled populations to expand. Expanding communities would have coexisted alongside but also come into conflict with the predators indigenous to South Asia – tigers, leopards, wolves, wild dogs (dholes) and striped hyenas. Lions are believed to have dispersed from Africa, through West Asia and into South Asia at the time of the rise of the Indus Valley civilizations in India around 3300–1300BCE, equivalent to the Bronze Age cultures of Egypt and Mesopotamia. More advanced agricultural and societal forms spread with the peoples of this culture and reached Gujarat from the north.[62] A series of dynasties rose and fell in northern and western India, such as the Harappan (c3000–1700BCE), Yadavas, the Mauryan, the Kshatrap dynasty, the Gupta Empire and then the Pratiharas, from the second millennium BCE through to the ninth century CE.[63] The presence of hyenas is not as well recorded as that of lions or tigers in Indian, or other regional histories or in artistic or religious images of these dynasties. But carvings from Bharhut in Madhya Pradesh from the second century BCE depict hyenas and jackals as beasts of ill-omen and shows them in graveyards and on corpse-strewn battlefields[64] – not inaccurate, but something that would have helped instil prejudices against them as ghoulish scavengers and consumers of human flesh and bones. Indian history from these early dynastic periods is hard to piece together and relies heavily on tracing people and events, not to mention animal life at the time, from sections of literary and religious works such as the *Mahabharata* and the *Ramayana*. [65]

The Mauryan rulers in western India, who conquered the Saurashtra region of Gujarat in 323BCE, were avid hunters and established hunting grounds in and around the Gir Forest in the Kathiawar peninsula. The forest areas were well populated with a diversity of antelopes and wild boars, but also lions, leopards and hyenas. This is now the stronghold of the sole surviving Asiatic lions and is also important habitat for hyenas, partly as a result of the regular production of carcasses by the lions that provide meals for hyenas. The Mauryan rulers hunted for sport but also allowed farmers and hunting communities to kill wildlife that invaded farmland or killed stock.[66] This may have had the effect of limiting safe habitats available to wildlife and reducing the range and numbers of many species, though there is no suggestion of the total extermination of wildlife around inhabited areas. Striped hyenas would have been well adapted to survive the growth in population and regular hunting by people, being able to take advantage of carcasses from hunts and the edible waste to be found around settlements. They do take some livestock – especially smaller domesticated animals, poultry, dogs and sick animals – but this does not appear to be a large part of their diet.

As creatures of the night which scavenged carcasses, middens and other collections of waste from human settlements, hyenas were viewed as unclean and below contempt. From early on in their association with human populations in India, the striped hyena was also associated with death, sorcery and despoiling graves. This reinforced why "striped hyenas often get associated with the dead", according to Priya Singh, a wildlife biologist with the Researchers for Wildlife Conservation in Bangalore, India.[67] In some parts of India, hyenas became known as the "horses of witches", and legends of witches riding on the backs of the animals multiplied. The myths said that when a hyena found a carcass and ate the flesh, the witch then fed on the soul of the dead.[68]

Notes

1 E. C. Majumder (1950) The Vedic Age, in R. C. Majumder (ed) *The History and Culture of the Indian People, Vol 1*, Bombay: Bharatiya Vidya Bhavan, p. 47.
2 John Keay (2010) *India A History from the Earliest Civilisations to the Boom of the Twenty-First Century*, London: Harper Collins, updated edition, pp. xvii–xviii.
3 Mary C. Stiner (2004) Comparative ecology and taphonomy of spotted hyenas, humans, and wolves in Pleistocene Italy, *Revue de Paléobiologie, Genève*, 23, 2, 771–785, p. 771.
4 AbiSaid and S. M. D. Dloniak for IUCN, 2015.
5 Herodotus (trans by Robin Waterfield) (1998) *The Histories*, Oxford: Oxford World's Classics/OUP, p. 446.
6 Ibid., p. 107.
7 Aristotle (no date) *The History of Animals*, Seattle: Loki's Publishing, p. 149.
8 Stephen E. Glickman (1995) The Spotted Hyena from Aristotle to the Lion King: Reputation is Everything, *Social Research*, 62, 3, p. 510.
9 Aristotle, p. 172.

10 Aristotle (trans. Richard Cresswell) (1883) *The History of Animals*, London: George Bell and Sons http://www.vliz.be/imisdocs/publications/298626.pdf accessed 21 February 2020, pp. 204–5.
11 Gregory McNamee (2011) *Aelian's On the Nature of Animals*, San Antonio, Texas: Trinity University Press, p. 61.
12 Lloyd Llewellyn-Jones and Sian Lewis (2017) *The Culture of Animals in Antiquity A Sourcebook with Commentaries*, London: Routledge, https://books.google.co.uk/books?id=GvJFDwAAQBAJ&pg=PT446&lpg=PT446&dq=Hyenas+in+the+Iliad&source=bl&ots=Ekl0sM7wmf&sig=ACfU3U1UYbqeO7Jaf4VnP9YfBBNcJtUvCg&hl=en&sa=X&ved=2ahUKEwibivbjy8XiAhXqXhUIHatmA8gQ6AEwEHoECAcQAQ#v=onepage&q=Hyenas%20in%20the%20Iliad&f=false accessed 31 May 2019, no page numbers.
13 Holger Funk (2012) How the ancient "Krokottas" evolved into the modern spotted hyena "Crocuta Crocuta", *Quaderni Urbinati di Cultura Classica, New Series*, 101, 2, 145–166, p. 145.
14 Ibid., p. 147.
15 Ibid.
16 Ibid.
17 Paul G. P. Meyboom (1995) *The Nile Mosaic of Palestrina. Early Evidence of Egyptian Religion in Italy*, Leiden: Brill, pp. 115–6.
18 Ibid., p.116.
19 Pliny the Elder (trans. Rackham) (no date) *Natural History 8. 72 (Roman encyclopedia C1st A.D.): African Beasts*, https://www.theoi.com/Thaumasios/TheresAithiopikoi.html accessed 30 May 2019.
20 Stephen Jay Gould (1981) Hyena Myths and Realities Both male and female genitals are strikingly similar, *Natural History*, https://pdfs.semanticscholar.org/93b8/27819b950a097e9083bc34694449abfb6b5c.pdf accessed 24 January 2020.
21 Cited by Funk, ibid., p. 148.
22 Ibid.
23 Ibid.
24 Glickman, 1995.
25 H. Rackham (editor and translator) (1958) *Pliny's Natural History*. Cambridge, MA: Harvard University Press, Volume 3, Book 8, Paragraph 45, https://ia802309.us.archive.org/0/items/naturalhistory03plinuoft/naturalhistory03plinuoft.pdf accessed 3 June 2019, pp. 107–9.
26 Cited by Glickman, 1995; see also, Rachkam, ibid.
27 Glickman, 19995.
28 Rackham, 1958, Volume 8, Book 28, Paragraph 27, pp. 98–9.
29 Mikita Brottman (2012) *Hyena*, London: Reaktion Books, p. 38.
30 Lucan (2008) The Civil War, Oxford: Oxford University Press, Kindle Edition, loc. 2625–2626.
31 Ingvild Sælid Gilhus (2006) *Animals, Gods and Humans: Changing Attitudes to Animals in Greek, Roman and Early Christian Ideas*, New York: Routledge, 2006, p. 12.
32 Ibid.
33 Ibid., p. 36.
34 Pliny the Elder (2004) *Natural History. A Selection*, London: Penguin, revised edition 2004, p. 115.
35 Lloyd Llewellyn-Jones and Sian Lewis (2017) *The Culture of Animals in Antiquity A Sourcebook with Commentaries*, London: Routledge, https://books.google.co.uk/books?id=GvJFDwAAQBAJ&pg=PT446&lpg=PT446&dq=Hyenas+in+the+Iliad&source=bl&ots=Ekl0sM7wmf&sig=ACfU3U1UYbqeO7Jaf4VnP9YfBBNcJtUvCg&hl=en&sa=X&ved=2ahUKEwibivbjy8XiAhXqXhUIHatmA8gQ6AEwEHoECAcQAQ#v=onepage&q=Hyenas%20in%20the%20Iliad&f=false accessed 31 May 2019, no page numbers.
36 Ibid. See also, J. L. Cloudsely-Thompson (1967) *Animal Twilight Man and Game in Eastern Africa*, London: G. T. Foulis, p. 165.
37 Daniel P. Mannix, *Those About to Die*, St Alban's: Mayflower, 1973, p. 51.

38 Cloudsely-Thompson, 1967, p. 165.
39 Caroline Grigson (2016) *Menagerie. The History of Exotic Animals in England 1100–1837*, Oxford: Oxford University Press, p. 1.
40 Jasper Griffin (1992) Theocritus, the Iliad, and the East, *The American Journal of Philology*, 113, 189–211, p. 209.
41 Llewllyn-Jones and Lewis, 2017.
42 W. Ivanow (1926) Muhammadan Child-Killing Demons, *Royal Anthropological Institute of Great Britain and Ireland*, 26, p. 197.
43 Ibid.
44 Ibid.
45 Uri Avner et al. (2011) Carnivore traps in the Negev and Judaean deserts (Israel): function, location and chronology, *Prédateurs dans tous leurs états. Évolution, Biodiversité, Interactions, Mythes, Symboles. XXXIe rencontres internationales d'archéologie et d'histoire d'Antibes*, 253–268, p. 255.
46 Ibid., pp. 255–7.
47 Ibid., p. 259.
48 Ibid., p. 262.
49 Brottman, 2012, pp. 96–7.
50 Ibid.
51 Hyenas in the Bible, https://bible.knowing-jesus.com/words/Hyenas accessed 30 May 2019.
52 Mary Pendergraft (1992) "Thou Shalt Not Eat the Hyena". A Note on "Barnabas" Epistle 10.7, *Vigiliae Christianae*, 46, 1, 75–79, p. 75.
53 Nadin Rohland et al. (2005) The Population History of Extant and Extinct Hyenas, *Molecular Biology and Evolution*, 22, 12, December 2005, pp. 2435–2443, https://doi.org/10.1093/molbev/msi244 accessed 11 April 2019.
54 Sudipta Mitra (2005) *Gir Forest and the Saga of the Asiatic Lion*, New Delhi: Indus, p. 36.
55 A. Stephen, R. Suresh and C. Livingstone (2015) Indian Biodiversity: Past, Present and Future. *International Journal of Environment and Natural Sciences*, 7, 13–28 https://www.researchgate.net/publication/276410026_Indian_Biodiversity_Past_Present_and_Future accessed 8 January 2020, pp. 13–4.
56 H. S. Pabla (2015) *Wildlife Conservation in India – 1 The Road to Nowhere*, First E-Book Edition Publisher: H. S. Pabla, Kindle Edition, loc. 389.
57 Ibid., loc 394–402.
58 Ibid.
59 One example was posted by Arjun Dheer on Twitter, @wildpakistan tweet 1/12/2019. Yet another video emerges of an endangered striped hyena being forced to fight dogs in Baluchistan province of Pakistan. Shows striped hyena on a ling rope tied to a pole having a dog set on it while hundreds of men watch and cheer the dogs on.
60 Personal communication with Laurence Franks.
61 Madhav Gadgil and Eamachandra Guha (1992) *This Fissured Land. An Ecological History of India*, New Delhi: Oxford University Press, p. 67.
62 Mitra, 2005, pp. 36, 43.
63 Ibid.
64 Jeannine Auboyer (1972) Animals in India, in A. Houghton Brodrick, *Animals in Archaeology*, London: Barrie and Jenkins, 115–145, p. 127.
65 Keay, 2010, p. 2.
66 Mitra, 2005, p. 43.
67 Jason Bittel (2018) Striped Hyenas Don't Have Magical Powers, But Their Disappearing Act Is Real, *Ecowatch*, 25 January 2018, https://www.ecowatch.com/hyenas-in-india-2528561632.html accessed 3 June 2019.
68 Ibid.

5

HUMANS AND HYENAS FROM 1600CE TO THE END OF THE 19TH CENTURY

By the 17th century, the ranges of the spotted, brown and striped hyenas were well-established, with changes over the next few hundred years largely the result of continued human economic growth and the effects of European colonialism in Asia and Africa. Modern forms of cultivation with ever greater clearing of woodlands, fencing of arable land and the greater ability of pastoral communities to enclose and/or defend their livestock had begun to deplete populations of prey species and consequently affected the ranges of striped and spotted hyenas, though not substantially reducing them. Both species adapted surprisingly well to surviving in close proximity to people and didn't suffer the massive range depletion of predators like lions and tigers. Spotted hyenas were widely distributed across sub-Saharan Africa and some could still be found in southernmost Egypt and northern Sudan. Striped hyenas were resident in much of West Africa and round the Atlantic coast to Morocco and eastward to Egypt, from eastern Africa into the Horn, into the Middle East, Turkey, southern parts of Central Asia and South Asia. Brown hyenas were found only in southern Africa's more arid regions and were only to come under greater pressure from humans as more areas were settled and cattle, sheep and goats were introduced into even quite arid regions.

Africa in the centuries leading to colonial occupation

Despite the rise and fall of kingdoms and empires over the preceding centuries, in 1600 much of sub-Saharan Africa still had comparatively low human population densities. Most people lived in rural communities dependent on pastoralism, crop cultivation or supplementary hunting, or mixtures thereof, with the persistence in some areas of hunter-gatherer cultures.[1] Advances in livestock husbandry and settled forms of agriculture affected wildlife, chiefly through habitat loss and greater competition for grazing and water. Growth in pastoralism increased human-predator competition, with carnivores like hyenas, lions, leopards, wild dogs, cheetah and jackals killed in retaliation for stock-raiding and in some areas pre-emptively hunted to protect stock. The effects of this on prey species and predators was magnified by the arrival of European traders and hunters, who ventured into the interior for profit and adventure, and the resulting availability of firearms from the 16th century. European and Asian traders encouraged increased hunting by indigenous people for ivory, horn and skins to supply demand from Europe and Asia. Arab traders sold muskets and gunpowder in North Africa, parts of West Africa, along the Red Sea, and to kingdoms in the Horn of Africa and parts of East Africa, as later did the Portuguese, French and British.

East Africa – settlement, pastoralism and the arrival of Europeans

In East Africa, especially the savannahs and open woodlands of central-southern Kenya and northern Tanzania, human populations were in a state of flux in the 17th–19th centuries, with extensive movements of pastoralist and agro-pastoral communities and shifting use of land. It is hard to establish an exact timeline, but Robertshaw suggests that preceding the 17th century, which saw the arrival and establishment of dominance of the Maa-speakers (Maasai and Samburu), groups of Southern Nilotic-speaking peoples such as the Tatoga and southern Kalenjin occupied much of the land suitable for pastoralism in central and southern Kenya and what became later Tanzania's Maasailand. They mixed some cultivation with livestock husbandry.[2]

Maa-speaking communities moved into the area around 1600.[3] They were the ancestors of the Omotik and Okiek Ndorobo. They displaced many of the existing inhabitants but were then pushed out or assimilated by the Maasai/Samburu, who occupied grassland areas from Laikipia south through Kenya into Tanzania. The dominance of pastoralism at this time was based on the fertile grasslands and open woodland areas, which were the habitat of the huge herds of wildebeest, zebra, gazelles, buffalo and browsing antelopes like sable roan and kudu, and supported large populations of hyenas, lions, wild dogs, leopards and cheetahs. The growth in cattle numbers brought the pastoralists into conflict with predators. Hyenas would have been a particular problem, as they were more likely than other carnivores to frequent settlements or temporary cattle camps, given their wide diet and adaptability to the presence of humans and the opportunities they offered for foraging.

Maasai dominance was established in the 18th and 19th centuries from Laikipia south to Narok, the Mara and Tanzania's steppes.[4] Their expansion brought new factors into play in human-hyena coexistence, with their large herds of cattle grazing on savannahs alongside huge numbers of wild ungulates and the hyenas and other predators that preyed on them. This would have meant that Maasai who settled in the wider Serengeti-Mara ecosystem in the mid-19th century would have killed stock-raiding predators and very probably sought to drive them from areas where they grazed their cattle and other stock. For the Maasai and other pastoral communities, killing hyenas seems to be viewed as just vermin control. It didn't gain the cultural importance of lion hunting in communities like the Maasai, Nandi and Barabaig.[5] Maasai social organization is based on gender and age-sets. Young men from around 15 to 30 make up the warrior set, and among them bravery and protection of the community's livestock is a paramount duty. The image of the hyena as a scavenger and coward seemed to preclude any great prestige for killing it, according to Amy Dickman of the Ruaha Carnivore Project, who works closely with the Maasai and Barabaig of Tanzania to prevent the killing of predators by pastoralists.[6]

To the west of the Maasai, Nandi and other Kalenjin-speaking groups settled on the highlands of the Rift Valley. They combined livestock husbandry with crop production, supplemented with hunting when necessary. The Nandi developed male age-sets and through manhood rituals established them as warrior groups tasked with protecting cattle against predators and human raiders (chiefly Maasai). To the south-west, on what is now the western border of the Serengeti National Park, communities such as the Ishenyi and the Ikoma lived alongside the considerable concentrations of game animals and the hyenas, lions and other predators preying on them. They had strong hunting traditions and employed snares, pitfalls and other traps, as well as hunting on foot using dogs, to provide meat and hides for consumption and trade. Over centuries, snares took a regular toll of wildlife, including the unplanned-for killing of hyenas in snares – those caught by stepping into snares and those that scavenged dead animals caught in snares.[7]

The development of trade on the Indian Ocean littoral, controlled by Arab, Afro-Shirazi or Swahili traders, had an effect on the hunting practices of communities across the region and enhanced the ability of East African communities to kill hyenas and other predators, as it brought firearms into the region. By 1888, 100,000 guns a year were being traded through Zanzibar.[8] From mid-century, there was a growth in British expeditions seeking to obtain ivory, establish trade routes and map out the river systems to gain access to the interior. This involved the advent of hunting by Europeans with modern firearms to obtain wildlife products, and for sport, which was to deplete wildlife in many areas. The hunters documented their travels and repeated folktales that they collected from local communities, creating images for Europeans of African wildlife. These included stories and beliefs about hyenas from local peoples, which mainly stressed their scavenging their alleged cowardice and supposed links with the occult. This added to the existing European images of hyenas as despicable, skulking scavengers and grave robbers.

In July 1857, Richard Burton set off inland from the Tanzanian coast. He reported extensive grasslands, herds of zebra and large numbers of kudu. He said his guides warned of the presence of hyenas and lions.[9] He wrote that he heard hyenas near his camps at night. Burton repeated the inaccurate description of the hyena as a scavenger rather than hunter, noting that it rarely attacked people, except when it could attack someone sleeping, "and then it snatches a mouthful from the face, causing a ghastlier disfigurement even than the scalping of the bear".[10] Despite referring to it as a scavenger he said that during the expedition three of his donkeys were predated by hyenas when they raided his camp.[11] When he reached the trading and population centre of Unyanyembe (Tabora region in Tanzania), he reported that the forests and plains had numerous "cynhyaenas", lions, leopards, buffalo, elephants, rhinos, giraffes, kudu and zebra.[12] When he reached Lake Tanganyika at Ujiji, he commented on the wildlife in the bamboo forests along the lake, noting that around the town and villages, hyenas were "bold thieves", entering the villages to take what food they could.[13]

Burton was accompanied by an Indian army officer, John Hanning Speke. Speke planned to collect specimens to establish his own museum.[14] He corresponded with the Zoological Society of London (ZSL), reporting on the wildlife species and their habits. In one account for the ZSL, he said that the spotted hyena was "the common scavenger" of East Africa and was abundant across the whole of the region. He also referred to the striped hyena, called "jungle-dogs" by local communities, which were liable to run off when challenged.[15] Speke wrote that across Tanzania he found that, "Hyenas are numerous, and thievishly inclined".[16] His desire to collect specimens for his own natural history museum meant that while he traced the paths of the rivers and lakes, he hunted and skinned as many species he could. In the Usagara region (south of Lake Victoria) he recorded shooting "elephants, rhinoceros, giraffes, buffaloes, zebra, and many varieties of antelopes, besides lions and hyenas". He said local people believed hyenas were agents of sorcerers and traditional healers.[17]

Ludwig von Höhnel, who accompanied the Hungarian aristocrat Count Teleki on his expedition to Lake Turkana, makes only occasional references to hyenas in his narrative of the journey to and from the lake, even though that area of northern Kenya had and still has a substantial population of spotted hyenas, as well as striped hyenas. His first mention is of Teleki shooting a hyena in error. He said that "the Count would not have shot this one but, catching only a fleeting glance at a yellowish-brown body moving amongst the grass of the steppe, he mistook it for a leopard".[18] This shooting of hyenas by mistake crops up several times in accounts in Africa and India by hunters – indicating that hyenas were not viewed as "sporting" quarry. In the region of Lake Baringo, Höhnel noted that in soft rocks on the plateau by the lake there were "whole colonies of hyenas" living in caves that he thought they had hollowed out for themselves. The local Suk people (now referred to as Pokot) had a clan with the hyena as its totem and the related Kalenjin-speaking Kamasia also had a clan, the Tungaw, with the hyena as their clan totem.[19] The anthropologist, K.R. Dundas, who studied the Suk and other Kalenjin communities of the Baringo region, said that burial customs among them involved the burial of men but the "women, excepting perhaps very old ones, the mothers of many descendants, and children are thrown out to the hyenas".[20]

In Tanzania, spotted and striped hyenas were common at this time, though the latter were seen less often. The explorer, hunter, amateur naturalist and colonial administrator, Sir Harry Johnston, wrote a report for the ZSL in 1885 on the fauna of the Kilimanjaro region, noting that striped hyenas were not uncommon in the foothills of the mountain, while spotted hyenas were to be found mainly on the surrounding plains. The spotted hyenas were described by him as predatory and he recorded that they were known to take sheep and calves kept by local farmers, children and sick or injured adults.[21] The Wachaga (Chaga) farmers of the foothills of Kilimanjaro disposed of the bodies of dead children out in the bush. Lehmann wrote that while married adults were buried in the hut where they lived and a year later exhumed and their heads buried elsewhere with great ceremony, "dead children were left exposed beyond the heaps of dung in the banana groves, where the hyenas could find them".[22]

The Horn of Africa – the Harar story continues

Records of hyenas in Ethiopia and the Horn are sparse in this period, excepting the continuing story of the hyenas of Harar. Taking up from where we left the story in the last chapter, the narrative of the truce between the people of Harar and the hyenas, the ceremonial feeding of the hyenas with porridge continued, maintaining the appearance of a degree of human tolerance of hyenas and the ability of hyenas to scavenge, receive food directly from humans and coexist with limited reports of conflict.

The Muslim Hararis had at one stage exercised hegemony over the surrounding countryside.[23] But as the power of the Christian rulers of the central highlands of Ethiopia increased, there was growing conflict, with Christian control spreading and undermining Harari influence.[24] From the late 16th century, Harar was weakened further by conflict with Oromo groups.[25] The Oromo took control of the territory around Harar and its power dwindled, though its independence was clung to by the residents, the town wall as a symbol of this.[26] Centuries of conflict came to a head with the Egyptian seizure of the ports of Zeila and Berbera in 1875, through which trade from Harar passed. The Egyptians ended their occupation in 1885, but the restored Amirate was too weak to resist the expansion of control by the Christian ruler of Shoa, Menelik II. His army defeated a Harari force at the battle of Chelenko on 6 January 1887. Menelik occupied Harar, making it part of the empire.[27]

The Harari lost their political independence but retained their religious and cultural customs, which may have included the feeding of the hyenas. The defeat of the Hararis at Chelenko is bound up with another legend linking the Muslims of Harar with hyenas. It is said that there was a leading family in the town, known as the *waraba shure garach* (hyena porridge family), who claimed descent from Waraba Sheikh, the patron saint of hyenas. They were responsible for the ritual feeding of porridge to hyenas.[28] The male members of the family fought in the battle against Menelik and only a young boy survived. He fled into the hills and took refuge in a cave, where he was found and fed by hyenas until he was strong enough to travel. The hyenas then, the story goes, carried him on their backs all the way to Harar.[29]

The town's fortunes declined after the occupation, especially when cattle diseases and crop failures caused widespread malnutrition and death. A visiting British officer reported that the dead were left lying in the street and eaten by hyenas and dogs. A

French missionary built a home for orphans to save them from being eaten by the hyenas.[30] These later accounts suggest a breakdown in the supposed coexistence between people and hyenas. Other reports from the latter half of the 19th century also indicate a somewhat different story from the one of coexistence consistent with the dominant human-hyena narrative, with the British explorer Richard Burton suggesting that at times the people of Harar lured hyenas inside the walls, trapped them and speared them. The Austrian explorer Philipp Paulitschke said that he had joined local people in hunting hyenas in the countryside around Harar, where they were viewed as vermin.[31] These two accounts are in line with the details of conflict between hyenas and people before the "truce" between them.

One version of story about the origin of the feeding of porridge is linked with the periodic famines affecting Ethiopia. Baynes-Rock said that the oral history of the town is full of stories of severe famines, with people dying in the streets. Hyenas scavenged the bodies and were even said to have attacked starving people. To combat this, the town's Islamic preachers suggested giving the hyenas porridge to stop them attacking people, leading to the tradition of feeding the hyenas regularly.[32] In 2009, a child in Harar was eaten by hyenas, but the locals seemed to be able to accept this and not turn on the hyenas that regularly entered the town on the basis that they could convince themselves that the hyena was not a local one, according to Baynes-Rock.[33] He also told the author that the people have no real problem with hyenas eating the dead; they also have quite a hard-hearted attitude towards the abandoned children living on the streets which may fall victim to hyenas.

Both striped and spotted hyenas were found widely distributed across Ethiopia through the centuries under review in this chapter. Even with the rise of the Christian empire, with its expanding towns, growth in agricultural production and availability of firearms, wildlife remained abundant in more sparsely population areas and was not totally wiped out in those with greater human density. Both hyena species adapted, as the example of the Harar hyenas shows, to life on the fringes of towns and villages, using waste dumps, graveyards and other sources of refuse, human or animal remains to forage. By foraging at night, they avoided conflict with people and there seems to have been no concerted attempt to eradicate them from settled areas, even though they were viewed in some areas as vermin. The explorer and writer Mansfield Parkyns lived in Ethiopia from 1843 and 1846 and travelled widely throughout the region. He narrated how in the northern town of Ado, sacrifices were made during the Ethiopian Orthodox Christian Feast of the Cross: "One or two oxen or sheep are slaughtered, and their flesh is left on the spot till it is devoured by birds of prey, hyenas, and jackals".[34] Hyenas were resident close to towns and were to an extent tolerated as efficient street cleaners.

As in other areas of Africa and the Middle East, the hyenas of Ethiopia had the reputation of grave robbing. Parkyns wrote:

> The custom here is, after having washed the body, to wrap it in a sheet…the corpse is carried off at a trot to the cemetery, usually at some distance from the

> village; it is placed in the grave, which is generally very shallow, and a few stones are put down to mark the spot where it was laid, but not where it will lie forty-eight hours after the funeral, for there is no tranquillity here for the dead. The African insurrectionists, the laughing hyaenas, walk the body out of the tomb in no time.[35]

Similarly, he found that in parts of Ethiopia there were beliefs in the magical powers of blacksmiths, who could turn themselves into hyenas. He told a story he had received from local of people:

> I remember a story of some little girls who, having been out in the forest to gather sticks, came running back breathless with fright; and on being asked what was the cause, they answered that a blacksmith of the neighbourhood had met them ...they at length began to joke him about whether...he could really turn himself into a hyaena...

which he proceeded to do after sprinkling some ashes over his body.[36] Hyena skin was said to be used to make talismans for the sick by blacksmiths, who were often traditional healers.[37]

The regularity with which Parkyns referred to hyenas in his two volumes of memoirs of his period in Ethiopia suggests they were found across much of the country, and especially in settled areas. He wrote that, "the misshapen, disgusting, 'laughing hyaena,' called in Tigray 'zibby,' - in Amharic, 'jib'; he is to be found almost everywhere in the country, but, from his scavenger habits, chiefly in the most thickly peopled districts". He said they competed with feral dogs for offal left around settlements and would steal hides and anything leather to eat. Describing them as disgusting and ugly, he also repeated the accusation, widely used in Africa, that they were cowardly. The British soldier, hunter and naturalist Harald Swayne, noted for his accounts of hunting in Ethiopia and Somalia, reported that in the wake of the wars that led to and followed Menelik's accession to the imperial throne, hyenas (species not identified) hung around a Russian medical mission station near Addis Ababa that had been established to treat the wounded from the wars. He said that bandages and other medical waste were buried in a pit outside the mission and proved "attractive enough to the hyaenas to be dug up and scattered nightly".[38]

Spotted and striped hyenas were also common in Somalia (across the whole area divided up at the Congress of Berlin in 1884–5 by European powers into French, British and Italian Somaliland).[39] In October 1833, we get one of the first European accounts of hyenas in Somalia from a British naval officer, Frederick Forbes, who landed from his ship at Berbera. He reported that outside the port:

> The sand was completely covered with ye marks of cattle, hares, jackals, and some footprints of a large animal, but resembling those of a dog or jackal, probably ye hyena...I ascended ye high land to the right of the wells. It

> consisted of loose blocks & boulders of aluminous stone quite sonorous & as hard as flint. There were some caves in ye face of ye precipice near ye top which ye jackals or hyenas appear to frequent.[40]

The abundance and diversity of game, and particularly the large population of lions, made it a popular hunting ground for British, army officers, colonial officials and other sports hunters at the end of the 19th century. Prior to these incursions by European hunters, hyenas (along with other predators such as lions, leopards, cheetah and jackals) had only been killed by local communities to protect goats, sheep, cattle, camels and donkeys.[41]

Swayne hunted extensively in the region between 1885 and 1893.[42] He referred to the need of the nomadic pastoralists and trading caravans crossing the region to defend themselves against predators. They constructed camps called *zaribas*, with thorn fences to keep predators from their stock and to protect women and children, who he said were sometimes killed by hyenas and lions.[43] During his journeys in northern Somalia, he said he shot a spotted hyena at the request of local villagers, as it was blamed for killing several sheep from the village.[44] Further on in the journey, he referred to five hyenas prowling around his camp at dusk that had to be driven off with gunfire.[45] Swayne was told hyenas would eat the bodies of dead humans left outside settlements and the bodies of those killed in clan clashes. He noted on several occasions that he had to protect the bodies of animals he had shot, otherwise hyenas would come in the night "and spoil the skin". He also said on occasions when he tied out camels or other domestic stock as bait to attract lions, he would find that hyenas had killed and eaten them, spoiling his chances of shooting a lion.[46] As a result, he shot hyenas regularly, not for sport or their skins but to keep them from his camp. Drake-Brockman, in his survey of Somali mammals in 1910, said that even in the decades after European hunting started there and Somalis gained greater access to guns, "the spotted hyena [was] plentiful everywhere in Somaliland, from the maritime plain to the higher internal plateaux", while the striped hyena was present throughout Somaliland but was "considerably rare[r] than the spotted hyena".[47]

The Sahel and Central Africa

The arid Sahel belt, south of the Sahara but north of the Central and West African forests and stretching from the Red Sea to the Atlantic coast, remained the location of widely distributed populations of striped and spotted hyenas. Given that they are still to be found across these areas today, there is no reason to believe they were less widely distributed between 1600 and 1900. If anything, it is likely that they were more numerous and gradually retreated during the colonial period as human populations grew and cash crop agriculture developed alongside food production, taking up more land, water resources and leading to the exclusion of wildlife south of this belt. In Central Africa, the striped hyena was absent, except in parts of Cameroon and Central African Republic, while the spotted was found in savannah, open woodland and some forested areas.

At this time, the Sahel was not as dry or as bare of woodland as it is now, and was home to a wider range of wildlife. Cloudsley-Thompson cites an early 19th century French traveller, Linant de Bellefonds, as recording a diversity of ungulates and predators as far north as Dongola in Sudan in areas that were still forested.[48] In areas bordering the Sahara in northern and central Sudan, there was desert-adapted wildlife, including gazelles, striped and spotted hyenas, leopards, cheetah and lions. But wildlife was being progressively depleted in the more heavily cultivated and populated areas along the Nile, Niger and other river systems across this broad belt, and was increasingly restricted to areas unsuitable for cultivation. It is probable that hyenas were less affected than most carnivores by this process of habitat loss, given their adaptability in terms of diet and survival around human settlements.

The Scottish diplomat and explorer James Bruce, who wrote of his journey through Egypt, Sudan and Ethiopia in the 1770s to find the source of the Nile, said that at Atbara (120km north of Khartoum) the striped hyenas had seemed to abandon their consumption of vegetable matter and

> ...gone largely and undeniably into the slaughter of living creatures, especially that of man...There is another passion for which he is still more remarkable, that is, his liking for dog's flesh...there was not a journey I made that he did not kill several of my greyhounds, and once or twice robbed me of my whole stock.

He said hyenas would even enter tents where the dogs were tied up.[49] That there were still some spotted hyenas this far north is suggested by the account of a British traveller called Hoskins, who reported in 1835 that despite the attacks on livestock and camels, the local people rarely killed predators. In the region of the main north-south road near Atbara, he reported "numerous traces of lions, hyenas" and other wildlife.[50]

In the arid savannahs around Lake Chad, the German explorer Gustav Nachtigal was warned by his local guides on an expedition in 1869 of the presence of lions and other predators, and they were known also to be present in good numbers in the Wadai region, and the forested valleys of Butiha and Batha north of Fort Lamy (Ndjamena).[51] There were still appreciable herds of desert-adapted ungulates (such as bubal hartebeest, Soemmerring's, Dorcas and Dama gazelles and Addax).[52] These would have provided prey for spotted hyenas and scavenging opportunities for striped hyenas (it is not clear if spotted hyenas were found as far north as the Sahara Desert fringes, though striped ones were and still are).[53]

There are fewer accounts of sightings or the behaviour of hyenas in equatorial central Africa, though they can inhabit quite thick forest and are found in areas of Cameroon, the Congos and Gabon. The hunter and colonial administrator Harry Johnston travelled in the Congo in the 1890s and said that even in thick forest in some lowland areas there were striped hyenas, as well as lions, leopards, western gorillas, red river hogs, elephants and buffalo.[54] One might question whether he correctly identified the species of hyena as it is far more likely that they were spotted rather than striped ones, the latter not being otherwise reported this far south or in such dense, wet forests. Johnston wrote that local people said that hyenas were quite common in the forests around Stanley Pool, on the Congo River where Kinshasa and Brazzaville are now situated.[55]

Southern Africa – the European-led destruction of wildlife from Malawi to Namibia

By 1600, the region's population was a mixture of San hunter-gatherers, Khoikhoi hunter-pastoralists, and increasing numbers of Bantu communities descended from migrants from the north. The early narratives of Dutch settlers and Europeans travellers indicate that wildlife was very plentiful in many areas, with a broad range of ungulates and predators, including large populations of spotted hyenas, as well as the mainly scavenging brown hyenas. Humans had at this point only a minor effect on numbers and ranges of animals, with the small hunter-gatherer communities hardly denting wildlife numbers. There was gradually increasing conflict between the region's carnivores and both Khoikhoi pastoralists and the incoming Bantu farmers (who also kept livestock).

European penetration of the region started in 1483, when Portuguese ships landed at the mouth of the Congo river and established trading ties with the Bakongo kingdom (which straddled Angola and both Congo republics). The trade links and extent of Portuguese contact increased in 1488, when Bartholomeu Dias rounded the Cape. Ten years later, Vasco da Gama sailed into the Indian Ocean, paving the way for Portuguese trading expeditions. Portuguese ports were established on the Mozambican coast, trading for gold, ivory and skins with the peoples of Mozambique, Malawi and Zimbabwe.[56] By the end of the 16th century, Dutch, English and French ships had rounded the Cape to trade in Asia. In 1649, a Dutch expedition wintered in Table Bay and the Dutch East India Company decided to occupy the area to provision ships sailing to Asia.

In Angola, Portuguese colonisation expanded during the period under review, mainly due to the slave trade, which provided labour for their colony in Brazil. Plantation agriculture developed in northern and central Angola, where rainfall and soils were suited for growing cash crops for export, notably coffee and cotton. Portuguese settlement was low and the effects on wildlife negligible, limited to some clearance of forest to create plantations and a little hunting. In the south and south-east, wildlife was abundant in savannah, woodland and riverine habitats. The regions along the Kunene river and where the Kavango river flowed south, across the Caprivi Strip into the Okavango Delta, were rich in wildlife. The arid savannahs and dry woodlands supported a diverse range of wildlife with large herds of elephants, buffalo, zebra, and a diversity of antelope species, which supported large numbers of spotted hyenas, lions, leopards, cheetah and wild dogs, which preyed on them, and brown hyenas, which fed from the remains left by predators and natural mortality. This diversity of species stretched across south-eastern Angola, south-western Zambia, Namibia and northern Botswana in a large eco-system around the south-flowing rivers (Kavango and Cuito), the west-flowing Kunene and to the east the Linyanti, Chobe and then Zambezi – the area now designated as the Kavango-Zambezi Transfrontier Conservation Area (KAZA). Until the 1970s, there was little attempt by the Portuguese to survey the wildlife areas or establish reserves to protect the wildlife. It is clear from the surveys carried out then and the accounts of later conservationists, like Brian Huntley, that the region was rich in wildlife and had a large and thriving population of predators.[57]

One of the first extensive accounts of Angola's wildlife came from Joachim John Monteiro, a Portuguese colonial official and naturalist who was a corresponding member of the ZSL. He worked in Angola from 1858 to 1876 and was a keen collector of natural history specimens. Travelling widely across the territory, he noted that in Cuanza Norte, east of Luanda, extensive pastoralism and clearing of land for grazing meant that wildlife of all sorts was scarce.[58] Travelling about 300–400km inland and slightly south from the capital, Luanda, he recorded that while game was quite scarce, the spotted and striped hyena and the black-backed jackal were common – though there has to be some doubt about the presence of striped hyenas so far south. It is possible he was misinformed or mistook a brown hyena for a striped one. Monteiro said that hyenas:

> …used to visit us almost every night in Cambambe [Cuanza Norte province], and at one place, where my cook slept by himself in a small hut…they frightened him so by snorting under the door and trying to force their way in…[59]

The cook stopped sleeping in the kitchen hut, but the hyenas still were attracted by it and one night broke in, smashed crockery and dragged away a sheepskin used to cover a table.

Monteiro described the hyenas found in central Angola as voracious and bold but as rarely attacking humans, though they did enter villages to steal the hides of newly killed cattle. The local people stretched the hides on the roofs of their huts to dry, but the hyenas were able to jump and pull them down, dragging them away to consume.[60] Monteiro believed hyenas were found across the country and were plentiful around Benguela in Namibe province and around the port of Mocamedes. He recounted an episode where there was a smallpox outbreak at a place called Golungo Alto (150km inland from Luanda) and the hyenas consumed many of the victims, who were hastily buried outside the settlement. This, he said, led to a series of attacks on people by hyenas, but none were believed to have been fatal.[61] Monteiro wrongly believed that hyenas lived and hunted in pairs.[62] Boer settlers who trekked into southern and central Angola in the mid to late 19th hunted game for biltong and hides, and killed hyenas and other predators to protect their livestock. In central and eastern Angola, game numbers were reduced heavily by the Boers, many of whom worked as transport drivers when the Portuguese were building roads and later the Benguela Railway. In 1881 there were still large herds of game and plenty of predators in southern and south-east Angola. The hunters Axel Eriksson and William Chapman travelled across these areas in that year on a hunting expedition and reported that they still teemed with game including hyenas, lions, leopards, black rhinos, buffalo, elephants and a diversity of antelopes.[63]

European settlement in South Africa started in earnest in 1652, when Jan van Riebeeck arrived as commander of the settlement established to supply fresh food to Dutch East India Company trading fleets. The vegetation encountered by the first Europeans at the Cape was a mixture of fynbos (from the Dutch *fijnbosch*, meaning fine-leafed bush)[64], wooded mountains and grassland in the coastal plains and valleys, which supported a diversity of herbivores and predators, including spotted and brown hyenas, wild dogs, lions, leopards and cheetah. Company workers and settlers grew fruit and vegetables, kept cattle and sheep and hunted game for food and skins. They also began to shoot predators and scavengers to protect livestock and for sport. The possession of firearms and horses gave the Dutch a huge combat advantage over the resident Khoikhoi and San communities,[65] who were driven out or enslaved. Firearms also conferred advantages in hunting wildlife, something of which they made full and destructive use.[66]

Van Riebeeck kept a journal which provides periodic references to the presence of predators.[67] The journal, which commenced in July 1649, does not make mention of any predators until 25 December 1652, when Van Rieebeck noted the first case of stock predation, writing that seven or eight "wild beasts" killed a sheep, with no indication of whether the predators were hyenas, lions or wild dogs.[68] On 12 January 1653 he is more specific, saying that a lion killed a sheep and was driven off with musket fire.[69] On 21 and 22 August 1653, the journal reported raiding by predators of a kraal at the town, in which six sheep were taken and a number of geese and ducks killed.[70] The pattern of predation is suggestive of hyenas, given the way in which they rapidly to accustom themselves to hunting/foraging around human settlements.

Wildlife was a resource for food and commerce, and game around the settlement was soon depleted by hunting. Whether this led to predators raiding the livestock or whether hyenas, leopards and lions were attracted by enclosed, docile prey and the remains of animals slaughtered for their meat and hides is unclear. On 12th September, Van Riebeeck recorded that after a search on Lion Mountain at Cape Town (named for its shape), the body of a missing Dutch soldier was found: "the brain pan had already been bitten off, so that it is presumed that he must have been killed by a lion, from which animal our people sometimes suffer loss and attack".[71] That the skull had been broken open is not necessarily indicative of a lion attack, as they rarely break open skulls of their prey and instead suggests that either the man was killed by hyenas, or they ate from the carcass and broke open the skull after he had been killed by a lion. In his entry of 1 June 1655, Van Riebeeck referred to "leopards, lions and tigers" rarely killing cattle but more often smaller stock.[72] The reference to "tigers" clearly does not refer to the tiger species but is probably a misidentification or transfer of the description to spotted or brown hyenas. The only clear mention by Van Riebeeck of a hyena near Cape Town was in his journal entry of 30 July 16661, when he said that "A wolf was shot by the freemen during the night. The usual reward was paid to them".[73] In Afrikaans and among communities in southern and south-west Africa, spotted hyenas are often called wolves and brown hyenas strand-wolves.[74] That the "usual reward" was given to them suggests a persistent problem with "wolves" around the settlement and a bounty for their extermination.

The presence of predators threatening livestock, competing with hunters for wild ungulates and threatening human life, led to a hostile attitude from settlers and the administration. European settlers are widely perceived as having introduced an "extermination mentality toward predators and scavengers in Africa". As early as 1652, settlers in the Cape Colony targeted carnivores and crop-raiding animals because of perceived threats to humans, livestock and crops.[75] In 1656, "the Dutch East India Company began offering bounties of six Spanish reals for killing a lion, and four for a leopard or hyena, to the settlers in the Cape Colony".[76] One account of a hyena near Cape Town in the 1770s came from the Swedish naturalist Anders Sparrman, who reported that a drunken soldier was lying outdoors trying to sober up, when "a tiger-wolf" tried to drag him off, but gave up when the soldier startled it by blowing his trumpet;[77] the tiger-wolf presumably being a hyena. In 1814, the British authorities in Cape Colony "incentivized the destruction of carnivores" by poisoning, trapping and shooting them.[78]

Dutch and then, following Britain's seizure of Cape Town during the Napoleonic Wars, British occupation saw the growth of agriculture and the extension of the colonised area into the northern and eastern areas of the Cape. Following the British banning of slavery in the Cape in 1834, substantial numbers of the descendants of the Dutch settlers, known as Afrikaners or Boers, trekked out of British jurisdiction into what became the Orange Free State and Transvaal to evade British rule. Colonial wars were fought against the Xhosa and then the Zulu kingdom by the British, while the Boers came into conflict with the Xhosa, Sotho and Zulu kingdoms and later Mzilikazi's Ndebele. The British occupied Zululand after the 1879 war of conquest.

As the occupation of South Africa progressed, the growth of the livestock economy was rapid, particularly the introduction of merino sheep and angora goats in the Cape; the number of sheep and goats reached 23 million by the end of the 19th century. This led to conflict with predators – especially hyenas, leopards and jackals. It wasn't just a case of defence of livestock with fences and retaliatory killing, but concerted campaigns of extermination, using guns, traps and poison.[79] When a Game Protection Act was introduced in 1888, as it was perceived that wildlife numbers had fallen catastrophically and there needed to be controls on hunting, carnivores were excluded from it, as in livestock areas they still counted as vermin.[80] The Cape government established bounties to be paid for evidence of the killing of hyenas, leopards, caracals and jackals, and the farmers also established *poisoning clubs* to encourage the killing of predators.[81] From the mid-17th century to the mid-19th century, the settlers in the Cape and the trekboers moving north and east wiped out the huge herds of antelopes, zebra and quagga in the Cape and Karoo, removing the natural prey of hyenas and other predators (and the source of carcasses for scavengers like brown hyenas). The remaining predators were then more likely to attack cattle, sheep and goats, and were killed on sight. While it is possible to chart the denuding of the Cape of lions, whose presence and hunting was often noted,[82] the original range and subsequent extermination of hyenas was not worthy of mention in the historical accounts. Some accounts speak of the presence of hyenas in areas used for cattle or sheep farming and the use of trap-guns by the Boer farmers to kill "hyaenas that were too persistent" in their attempts to raid stock.[83]

British hunters and travellers provided some of the most interesting accounts of encounters with hyenas in southern Africa. Their observations must be treated with care, though they do give an idea of where hyenas occurred and human interactions with them. Sir John Barrow, private secretary to the first British governor of the Cape, wrote of his travels across the country at the end of 18th century and start of 19th, noting that human settlement, agriculture and hunting had reduced wildlife numbers dramatically.[84] He wrote:

> The peninsula of the Cape affords but a narrow field for the inquiries of the Zoologist. The wooded kloofs or clefts in the mountains still give shelter to the few remaining troops of wolves [probably spotted hyenas] and hyenas that not many years ago were very troublesome to the town. The latter, indeed, generally shuns the habitations of men; but the former, even yet, sometimes extends his nightly prowl to the very skirts of the town, enticed by the dead cattle and offal from slaughter-houses, that are shamefully suffered to be left or thrown even at the sides of the public roads.[85]

He added that "wolves", along with crows and eagles, were useful and to an extent tolerated as scavengers cleaning up human-created waste around inhabited areas.

Barrow was horrified by the practice of some indigenous communities of burying their dead in anthills or leaving them in the bush. He said the bodies were taken by what he called wolves (again, it is not clear whether he means brown or spotted hyenas, as he confuses the vocalisations and scavenging habits at times), noting:

> As these animals drag them away immediately into their dens, the relations of the deceased are in no danger of being shocked or disgusted with the sight of the mangled Carcase. A Kaffer, in consideration of this piece of service, holds the life of a wolf sacred, at least he never endeavours to destroy it; the consequence of which is, that the country swarms with them.[86]

Barrow seems to be going off on a flight of fancy, as hyenas of both species were treated as vermin and certainly not as sacred. In the area around the Great Fish River in what is now the Eastern Cape, Barrow recorded that "the lion, the leopard, wolves, and hyaenas, and other beasts of prey with which this wild part of the country abounds" were dangers to people and their livestock.[87]

More than a decade after Barrow, William Burchell went to South Africa to gather natural history specimens.[88] Throughout the narrative he mentioned species he saw, such as eland, kudu, hippos, lions and a diversity of birds. In the first volume he only once mentioned the hyena, when he said in the eastern Cape he came across the wild dog, which be believed was "a new and distinct species of Hyena".[89] In the second volume of his narrative of travelling in the Cape regions, he referred to the Hyena Mountains in the Karoo region, suggesting their presence then or in the past.[90] Later, while travelling in an area near the Gariep river, which he called Kosi Fountain, one of his Khoikhoi hunters shot a hartebeest and brought it back to the camp, "but during the night the hyenas or wolves as they are usually called by the Boors and Hottentots, devoured all the flesh, leaving us only the head and the hide".[91] He later compared vultures and hyenas and said that while the former were tolerated and even respected by local peoples, the latter were not because of their ability to do harm to humans.[92]

In the late 1840s, the hunter William Cornwallis Harris undertook a long expedition from Cape Town through the Eastern Cape into what became Transvaal, Bechuana (Tswana) territory and the lands of the Ndebele king, Mzilikazi. Although wildlife numbers had been reduced by the trekboers as they migrated from the Cape to Orange Free State and Transvaal, Harris reported seeing large herds of springbok and black wildebeest between Grahamstown and Graaff-Reinet and in the Sneeuwbergen.[93] He wrote that one night near the Maritsane River in Transvaal, "the hyaenas, attracted by the smell of our mutton, actually devoured a spring-buck within the limits of our camp".[94] As he progressed into Tswana territory south of Mzilikazi's kingdom, he noted that his hunting and butchering antelope for meat and skins meant that, "Our little band was also instinctively attended by a host of hungry vultures, who, little disturbed by the presence of man, divided the office of carrion scavengers with hyaenas and jackals".[95] In the Zululand, Cornwallis Harris said he learned that under Shaka the frequent executions of people meant hyenas were frequently found around major settlements. According to his account, when someone was executed, "The body was then dragged outside the kraal, and left to be devoured by hyaenas and carnivorous birds that were ever in attendance".[96] During Shaka's rule, great hunts would be organised in which huge numbers of beaters would drive wildlife into enclosed areas to be killed by the king and his hunters, including hyenas, wild dogs, leopards and lions.[97]

When a person died in Mzilikazi's kingdom, following a pattern found widely across eastern and southern Africa, Cornwallis Harris said that, "We could hear of no funeral ceremonies amongst them. High and low, their bodies are thrown forth upon the plain, soon after life departs, a prey to wild beasts…the howl of the hyaena being their only death-note".[98] This was also the fate of warriors who died in battle. The Matabele impis "left the bodies of their comrades, as usual, a prey to vultures and hyaenas".[99] The level of conflict within the region grew as the Zulu kingdom expanded. Mzilikazi, who was pushed northwards into what is now Zimbabwe by the Boer settlers in Transvaal, provided a periodic supply of corpses for the hyenas. Destitute survivors of the wars, Cornwallis Harris said, competed with hyenas and vultures for the remains of animals he shot.[100] Travelling through this region of north-western South Africa he noted the occasional forays by hyenas into the camp to try to kill the sheep he took with him:

> The moon was full on the night of the 23rd, and a spotted, or 'laughing' hyaena, superior in size to the largest mastiff, was shot through the head…as

> he was in the act of skulking under the sheep-pen. The great muscular power of this animal, which is called by the colonists 'the wolf', renders it exceedingly formidable.[101]

Another British hunter active in 19th century Africa was Roualeyn Gordon-Cumming. Gordon Cummings started hunting in 1843 in the Eastern Cape, where he found that in many areas hunting by Boer trekkers had depleted the game for miles around their settlements or trek routes.[102] When he reached Bechuana territory, he found that many communities "do not bury their dead, but unceremoniously carry them forth, and leave them lying exposed in the forest or on the plain, a prey to the lion and hyaena, or the jackal and vulture".[103] He feared that this accustomed predators to human flesh and encouraged man-eating. Later in his narrative, he wrote of shooting a wildebeest and a lion in country between Colesberg and Grahamstown and being desperate to find the body of the lion, because otherwise the skin would be spoiled by hyenas overnight, given their abundance in the area.[104] The author's regular references to hyenas raiding camps or trying to get hold of the carcasses of animals he has shot is an indication of the willingness of hyenas in the region to risk harm if there was a carcass there to be scavenged, and again suggests a rapid adaptation in the foraging behaviour of the hyenas in response to hunting expeditions by Europeans. This supposition is supported by Gordon-Cumming's account of shooting three black rhinos at a waterhole. He sat up over the carcasses the night after shooting them and counted 12–15 hyenas, six lions and 20–30 jackals surrounding the carcass.

Human-wildlife conflict was in an extreme phase in southern Africa and led to the depletion of wild ungulates and the predators/scavengers that relied on them over most of the Cape Colony and the newly-occupied territories to the north and east. Until the early-to-mid nineteenth century, the fauna of southern Africa had been "richly diverse, highly prolific, and widely dispersed", according to Mackenzie. The local communities had exploited it for food and defended their stock and crops against animal raiders. Game meat was a supplementary source of protein for pastoral or farming communities and vital for hunter-gatherers, but had not been exploited on a scale that wiped out herds, as the depredations of settlers armed with modern firearms did. The expansion of white settlement and hunting led to the fauna of southern Africa being "sent into rapid retreat".[105] Despite the rapid extermination in farming areas of most hyenas, lions and other predators, the remaining ones were viewed by settlers, commercial hunters and administrators as a threat to livestock and human life. Shooting, trapping and poisoning were used to exterminate hyenas, wild dogs, lions, leopards, jackals and caracals.[106]

In the 1870s, Frederick Selous started hunting in southern Africa to make money from ivory, horn and hides, and by supplying specimens for museums and collectors in Britain. His biographer, Etherington, says that Selous' and other contemporary accounts reveal that by 1870, the movement of settlers into the Orange Free State and Transvaal had "cleared the Highveld of big game all the way to the Limpopo".[107] In December 1871, in his first account of his hunting expeditions, *A*

Hunter's Wanderings in Africa, Selous said when he first travelled into the interior along the Orange River between Kimberley and Upington and reached Griqualand, he saw no game except for a few springbok and steenbok, and added "One might as well look for game in Hyde Park as in Griqualand".[108] In December 1877, hunting in south-western Zimbabwe, he said hyenas chased and killed two of his donkeys. He tried to shoot them as they retreated from the carcasses of the donkeys, but having failed to kill them, Selous decided to use poison. He had a supply of strychnine with him and laced the carcasses with it. He does not explain why he had taken the poison on a hunting expedition, but one can surmise that it was to kill pests.

When David Livingstone started his explorations of the Zambezi in Portuguese-ruled Mozambique 1858, wildlife was still plentiful even near the coast and Livingstone reported seeing large numbers of buffalo, warthog and antelopes on grassland by the river a few miles inland.[109] Livingstone rarely mentioned hyenas. When he did, it was in the context of a "troop of hungry, howling hyenas" which hung around his camp one night after his men had killed two buffalo and butchered them in the camp. The hyenas are described as "arrant cowards, and never attack either men or beasts, except when they can catch them asleep, sick, or at some other disadvantage".[110] He went on to suggest a way of dealing with them around camps:

> A piece of meat hung on a tree, high enough to make him jump to reach it, and a short spear, with its handle planted firmly in the ground beneath, are used as a device to induce the hyena to commit suicide by impalement.[111]

Like most of contemporaries, when Livingstone deigned to mention hyenas it was to propagate the myths that they were cowardly skulking, scavengers to be treated as worthless vermin. It is hardly surprising when men like Livingstone, viewed as intrepid explorers and colonial heroes, spread these views, they became embedded in the public mind.

European hunters and traders moved into Malawi towards the end of the 19th century, leading to the establishment of the Nyasaland and District Protectorate in 1891. A member of the protectorate administration, H. L. Duff, said that Malawi at the time had abundant wildlife, including the spotted hyena but probably not the striped hyena.[112] Reports had been made of the sightings of the latter, but Duff was not convinced.[113] Duff said that spotted hyenas were "extremely common" and a danger to people as they attacked people who were sleeping in inadequate shelters or in the open, biting the victim's face. They had also been reported, he wrote, ambushing people as they left their huts early in the morning. As detailed in later chapters, parts of southern Malawi and neighbouring Mozambique became notorious for man-eating hyenas.

In his Zoological Field Notes in his journal of hunting expeditions in Mozambique and Malawi, the hunter Frederick Vaughan Kirby referred to the spotted hyena as common and states that they "are to be met with in astonishing numbers". He then took up the common theme by describing them as "furtive and treacherous, bold in the entire of absence of cause for fear, and cowardly to a degree when the slightest danger, real or imaginary, threatens".[114] He also shared the long-held view that hyenas could not hunt, contending that they were "an animal ill-fitted by nature to attack and overcome a live prey".[115] The French explorer Edward Foa travelled through Mozambique into Malawi and Zambia in 1894–7. He frequently reported hyenas prowling round his camp at night.[116] He had no respect for the animal, finding it ugly and distasteful. He related how he used sulphur rockets to dislodge hyenas from their dens to shoot them. He also suggested that striped hyenas were found south of the Zambezi, presumably confusing it with the brown hyena.[117] He admitted poisoning hyenas that hung around his camp and gives the heartless account of wounding a hyena and finding it funny that the animal "turn[ed] on itself like a dog trying to catch its tail, uttering howls and grunts of mingled rage or pain, and [made] off, every now and then waltzing in the very comical manner described".[118]

In Namibia, spotted hyenas were found across much of the country, including the arid pre-Namib, Kaokoveld, Damaraland and Zambezi region, while brown hyenas had their strongholds in the semi-desert and desert regions such as Damaraland, the Namib and the coastal dunes – in the latter relying heavily on taking seal pups and carcasses from cape fur seal colonies. A diversity of antelopes (gemsbok, springbok, wildebeest, hartebeest, kudu and steenbok) were present, along with warthogs, giraffe, two species of zebra and ostriches. These were all potential prey for spotted hyenas. The substantial lion and cheetah populations provided carcasses to be scavenged by brown hyenas. Charles Andersson accompanied the British geographer Francis Galton on a Royal Geographical Society expedition to Namibia in 1850–54. He hunted throughout his travels in Namibia and recorded sightings of game and predators, but only occasionally referred to hyenas. Andersson recorded that the dry plains had little wildlife, but:

> …from the number of bleached bones of rhinoceros, giraffes and other wild beasts scattered about, it was evident that game had at one time been abundant in these parts…With the exception of hyaenas and jackals, beasts of any size were scarce.

He blamed this depletion on the introduction of firearms into the area, which had led to a scale of hunting that exterminated or drove out wildlife.[119] The density of the human population in Namibia at the time – a mixture of San, Nama, Herero, Damara, Ovambo and other communities, plus Boer farmers who had trekked north (known as the Thirstland Trekkers) – was low and the main forms of livelihood were pastoralism, hunting and some cultivation in Ovamboland and the Caprivi. The Nama, Damara and other peoples of Khoikhoi descent mixed livestock husbandry with hunting and foraging. The Nama and Herero rarely hunted for meat, except in times of need, but killed hyenas, lions, leopards, wild dogs and cheetah to protect their stock.[120]

Galton, in his account of the expedition, wrote that one of his animals was killed and he later found the "poor chestnut pack-mule half eaten, and a hyena devouring the remains of the carcase".[121] Galton referred to hyenas being a nuisance at night, regularly entering the camp to try to kill the sheep they had with them.[122] On one occasion, camping along the Swakop river, the hyenas around the camp were so annoying that Galton let loose the camp dogs. He wrote:

> We were perpetually teased by some hyenas they came most impudently in amongst us…but we never could catch them; at last the dogs overtook one on a bright moonlight night and held him at bay. and I put a bullet through his backbone.[123]

Galton also tried, on another occasion, to lure prowling hyenas to where he could shoot them by getting a Damara man to imitate their howls.[124]

Galton reflected the disdain which he and other travellers of the time had for hyenas when he wrote about setting gun-traps for them:

> Hyenas, perhaps, vex and trouble you night after night, and it is a horrid bore to sit up through the cold when sleep is in these tropical climes so peculiarly grateful, simply for the chance of shooting the worthless animal; it is far simpler to have a gun in his path, and let him pull the trigger himself, to his own destruction.[125]

A more disturbing part of Galton's narrative refers to an attack on a woman by a hyena:

> An old Bushwoman [San], who encamped under the lee of a few sticks and reeds that she had bent together, after the custom of those people…a hyena who was prowling about in the early morning, laid hold of her heel and pulled her bodily half out of the hut. Her howls alarmed the hyena, who quitted his hold; and she hobbled up the next morning to us for plaisters and bandages…I and one of Mr. Hahn's men sat up the next night to watch for the animal. I squatted in the shade of her house, my companion covered a side path, and the woman occupied her hut as a bait. It was a grand idea, that of baiting with an old woman. The hyena came along the side-path, and there received his quietus.[126]

The willingness to risk the woman's life to kill the hyena was typical of the attitude evident in journals like Galton's toward the indigenous populations they encountered.

At the site of a massacre of Damaras by Namas, Andersson wrote that they saw "the bleached bones of the victims scattered about, but we were unable to ascertain the exact number of people killed as the jackals and hyaenas had carried away and demolished many parts of the skeletons".[127] It is very likely that spotted and brown hyenas could both have been involved in consuming the corpses, with brown hyenas more likely to remove bones and other body parts, to feed their young or consume at dens, than spotted hyenas. Shortridge, in his survey of Namibian wildlife, recorded that in the 19th century brown hyenas were found "over practically the whole of South-West Africa" and were "plentiful" in most areas, except for the Namaqualand region in the south, where ungulates were scarce and there were fewer carcasses available or small mammals that could support a large brown hyena population, and in Caprivi, where it was very scarce and vastly outnumbered by spotted hyenas.[128] Shortridge said that brown hyenas were distributed from South-West Africa right down to the Cape and around Table Bay, but European settlement and the persecution of hyenas as suspected stock thieves had led to their virtual disappearance there. The survey detailed that in Namibia, the spotted hyena was less widely distributed even in the 19th century and more localised in its presence than the brown.[129] The spotted hyena, more dependent on hunting despite its well-established flexibility in diet and foraging methods, required the presence of sufficient numbers of prey species.

The relationship between people and wildlife in Namibia began to change in the 1880s when Germany laid claim to the region, with the first German outpost established at Luderitz in 1882. This enabled German Chancellor Bismarck to lay claim to what became known as German South-West Africa at the Berlin Conference of 1884, at which the major European powers divided up Africa between them. The influx of German administrators, traders, settlers and soldiers completely altered the nature of the region, with commercial farming, the search for valuable minerals and hunting for profit and to protect livestock taking its toll on wildlife in many areas.

West Africa

The rise and fall of empires in the grassland, riverine areas and semi-desert regions of Senegal, Mali, Niger and the northern Nigerian region of West Africa through the first 750 years of the second millennium CE were accompanied by growth in the population of the region, the rise of large settlements and the development of cultivation in river valleys and areas with sufficient rainfall. In more arid areas, nomadic pastoralism was the main form of subsistence. Wildlife suitable as prey for hyenas and a source of carcasses for striped hyenas – both species being common in suitable habitats in the region – was limited to the Sahel, grassland and more lightly-wooded areas between the dense forest and the Sahara. Wild ungulates in the arid savannah and dry woodland/bush north of the forests came to compete with the herds of livestock kept by the nomadic and settled farming communities across this belt of West Africa from the Atlantic through to Chad. The ability of humans to defend their stock against predators and to kill animals competing for grazing thinned out the wildlife as human populations and livestock populations grew. This reduced wild prey for predators and many of their species declined. Hyenas, as in other regions of Africa, proved more capable of adapting to a human-dominated landscape and both spotted and striped hyenas survived when the big cats and wild dogs disappeared or were limited to increasingly small and isolated ranges.[130] In some areas of West Africa, as the German explorer Henry

Barth wrote of what he witnessed on his expedition across the Sahara and into West Africa in the 1850s, communities followed the practice already described from eastern and southern Africa of leaving the bodies of their dead out in the bush to be cleaned up by hyenas and vultures. He said that the Kanuri of north-eastern Nigeria were "extremely negligent in burying their dead, leaving them without any sufficient protection against the wild beasts, so that most of them [were] devoured in a few days by the hyaenas".[131]

North Africa, the Middle East and Asia

By the period under review, the spotted hyena was gone from North Africa but the striped hyena was widely distributed across the region. Accounts of its range, numbers and interactions with people are few and far between, at least in English translations discovered by this author. Hasan Ibn Muhammad al-Wazzan (aka Leo Africanus), the Andalusian-born diplomat who served the Moroccan Wattasid Sultan Muhammad in the 16th century, in his *History and Description of Africa and of the Notable Things Therein Contained* described the wildlife of Morocco and neighbouring regions of Western North Africa. He said one notable animal was the "beast called dabuh", which is clearly the striped hyena, being described as wolf-like apart from its legs, to be "not hurtful to any other beast", "but will rake the carkeises of men out of their graves and devour them". He said that hunters are said to play music in front of its den and put a rope around its legs and drag it out to kill it.[132] Two hundred years later, a British diplomat, the Consul in Algiers James Bruce, said that striped hyenas were common in Algeria and lived mainly on succulent roots, fleshy vegetables and fruit, adding "and I have known large spaces of fields turned up to get at onions".[133]

The presence of striped hyenas in the region was noted in the mid-19th century by the British Consul-General in Tangier John Drummond Hay. He wrote of seeing hyena tracks in the region of Ksar elk-Kebir, in the mountainous region of northern Morocco. He said of his Moroccan companions that "their contempt of the cowardice and stupidity of the hyaena has no limit; indeed its Arabic name, 'dbaa', means addle-headed or stupid".[134] The local people said that hyaenas would hide their heads in holes believing their bodies could not be seen. They led him to a cave where a hyena was hiding inside with its head thrust into the corner of the cavern. One local man entered the cavern talking loudly about the stupidity of the hyena, tied a rope round its back legs and dragged it out of the cave to be killed with knives. His Moroccan companions said that hyenas would often grab a bone offered to them and be dragged out as they did not let go of the bone.[135] Osborn and Helmy, in their survey of the land mammals of Egypt for the Field Museum in Chicago, said that in the 19th century, striped hyenas were to be found in the Western Mediterranean coastal desert of Egypt, the Nile Delta and along the Nile Valley, particularly in the desert areas bordering the river. They were also found around oases in more remote desert areas and were believed to follow camel caravans traversing areas of desert in the hopes of feeding from the remains of camels that had died during the journey or other waste from the caravans.[136]

Striped hyenas were found east of Egypt across eastern Turkey, Palestine, Syria, the Arabian Peninsula, Iraq, Iran and into Central Asia. Accounts from the 17th to the end of the 19th century are few and far between, and we will continue their story in the following chapter, where accounts by European naturalists, archaeologists, diplomats and other travellers enable us to pick up the threads once more. In South Asia, there is more of a record and we can attempt to fit the story of the hyena's survival into the human history of the period.

South Asia

The Mughal Empire was established in 1526 in the plain of the Ganges River by a Muslim chieftain called Babur, who had been driven from his homeland near Samarkand by powerful Uzbek invaders. He led his army and other followers south from Central Asia and gradually took control over parts of Afghanistan, modern-day Pakistan and much of India, laying the foundations for an empire that eventually reached as far east as Assam and Bengal, and as far south as the Deccan Plateau.[137] Mughal power spread as Babur defeated powerful states, such as that ruled by the Sultan of Delhi.[138] The empire was to last in one form or another until the 1857 rebellion against British rule, but was at its most powerful between 1526 and 1720.[139] Mughal rule saw an expansion of agriculture, the taxing of agricultural produce and other economic activities by the Mughals and the gradual reduction in wildlife habitat and so the ranges of wildlife species, especially large predators and herbivores that were steadily excluded from arable land, much of which was created by clearing forests.[140] These changes were accelerated more rapidly as British penetration and control increased, and economic exploitation of the land and people along with it. The British East India Company (henceforth referred to as the Company), at first in competition with Dutch, Portuguese and French merchants and navies, was at the forefront of European exploitation and colonisation.

In 1612, James I of England instructed the Company's Sir Thomas Roe to visit the Mughal Emperor, Nur-ud-din Salim Jahangir, to negotiate a treaty. This treaty allowed the Company to establish a trading hub at Surat on the coast of Gujarat.[141] In 1634, Company traders gained access to trading stations in Bengal, through an agreement with the Mughal emperor Jahangir.[142] Trade was the centre of the Company's activities, rather than occupation of territory, for the first century or more of its existence. But over time, Company interests turned gradually from just trade to territory and hegemony over Indian rulers. This occurred amidst competition with France, Portugal and Holland for trade and as the Mughal empire declined in power, leading to its eventual disintegration into competing states with their own locally powerful rulers.[143] The progressive rise in trading dominance and gradual development of territorial control or effective hegemony over Indian rulers and states was made possible by the fast decline of the Mughals and the relative weakness of other rulers in the face of British pressure.

As late as 1739, the Mughals ruled a vast empire that stretched from Kabul to Madras. But in that year, the Persian warrior Nadir Shah came south through the

Khyber Pass with 150,000 of his cavalry and defeated a huge Mughal army.[144] The destruction of Mughal power by Nadir Shah increased its vulnerability to European penetration of India. The power of the Company increased as a result of British victories during the Seven Years War of 1756–63. During the war, the declining Mughal empire allied itself to the French, who were defeated by the British and their allies. The empire was weakened after the war and was in a terminal state. In 1771, a confederation of rulers known as the Maratha Confederation captured Delhi from Afghan forces, which had seized it from the Mughals, and brought the Mughal emperor under their control.[145] Maratha hegemony was brought to an end by the Third Anglo-Maratha War, in which the Company was victorious. Thereafter the British, through the Company, became overlords of the Mughal dynasty in Delhi.[146] As Dalrymple argues, "An international corporation was transforming itself into an aggressive colonial power".[147]

As the political shape of India was undergoing huge upheavals and colonial rule was being established through the Company, little of substance changed in the relationship between Indian farmers, peasants and the natural environment. In the 18th and early 19th centuries, north and central India had a diverse population of wildlife – large, medium and small ungulates that supported substantial numbers of tigers, lions, leopards, cheetah, wild dogs (dholes), wolves and hyenas, though their range was steadily shrinking in the face of human agricultural expansion. Large predators were being reduced in numbers through hunting by Indian rulers and by the ever-growing number of British soldiers, merchants and administrators. Much of the wild prey and carnivores were exterminated or had disappeared from substantial areas by the end of the 19th century, habitat loss, "strong human and wild animal interface" and "human-wildlife conflict" being the major causes.[148] While the growth in the human population and agricultural expansion forced lions, tigers, cheetah and dholes from much of their range, striped hyenas, with their ability to survive in close proximity to humans by scavenging human waste and carcasses of livestock and remaining wildlife, were able to survive in areas that no longer supported lions, tigers, cheetah and dholes. Extensive sports took a heavy toll of carnivores, with lions, tigers and leopards the favoured targets – hyenas were killed for sport and as vermin, but not with the single-mindedness and trophy-driven motivation that affected the cats.

While the colonial authorities – the Company and then the British Raj – restricted the relatively small-scale hunting by remaining hunter gatherer communities,[149] they positively encouraged hunting by British officials and soldiers. The devolution of local power to remaining Indian rulers allowed them to control hunting in their areas and continue extensive sports hunting. The British limited their interventions in the princely management of wildlife "to formal and informal recommendations and advice on wildlife management, good sportsmanship, scientific forestry, and related topics",[150] and did not place restrictions on them as they did on ordinary rural dwellers under British administration. Sports hunting was an obsession in British India, according to Mackenzie. He wrote that "Various forms of hunting were the standard recreation of officers of the civilian, military and forestry establishments".[151] They hunted deer, nilgai antelope, wild buffalo, gaur, tigers, leopards and lions.

There is little recorded in the late 18th and then 19th centuries by British colonial officials, hunters and naturalists about the presence, ranges and numbers of striped hyenas in British-ruled South Asia, only periodic references to hyenas being killed in hunts or used as alternative targets for those engaged in hunting wild boar from horseback, known as pig-sticking; striped hyenas would be flushed out and hunted in this fashion and killed with lances. There is an illustration of such a hunt in the *Illustrated London News* reproduced by Wikipedia, with no date on the piece – it shows British officers on horseback with one spearing a running striped hyena.[152] Hyenas were killed routinely as vermin and randomly as a sort of by-catch in large driven hunts, where literally every large animal or game bird flushed out by the beaters would be shot at.[153]

Daniel Johnson wrote about hunting in India in 1882, noting that many Indian hunters had great experience in hunting to protect crops, rid districts of man-eaters and assist the local Indian rulers in their hunts. He went on to extol the virtues of sports hunting there and wrote of paying local hunters to help him and fellow British hunters and to supply game meat for the table.[154] In his account of hunting in India, Johnson made few references to hyenas – only reporting that the servant of an acquaintance had a pet hyena and that wild ones killed domestic dogs, as well as scavenging tiger and leopard kills.[155] Henry Shakespear was a prolific hunter who wrote of his experiences hunting big game in India in the 1850s. He made no reference to hunting hyenas but noted with irritation that on one occasion hunting for a man-eating tiger near Dongargarh, Chhattisgar state, he tied out a calf which was killed by a tiger, but by the next day "had been torn to pieces by the hyenas".[156] He also wrote that good hunting dogs were often lost when hunting in thick scrub or forest, usually killed and eaten by hyenas or wolves.[157]

A British soldier and hunter, A. Leith Adams, reported to the Zoological Society of London in 1858 that the striped hyena was to be found in the Deccan, Sind, Punjab and the lower Himalayas, that it was almost totally nocturnal and had the habit of preying on poultry, sheep and dogs kept by local communities.[158] The hunter and naturalist William Hornaday in 1885 reported its presence in the Nilgiri Hills in southern India, where extensive forest cover provided habitat for "elephants, tigers, bears, hyaenas, elk [presumably a large species of deer is intended], and small deer".[159] He also wrote that a forestry official with whom he stayed in the Nilgiri Hills hunted keenly and on the floor of his bungalow in the forest "were spread, in the most indifferent way, skins of bear, hyaena, leopard, and deer, but of the half-dozen tigers killed by mine host only the skulls and claws remained".[160]

Lions were shot for sport and to protect livestock, and almost exterminated in South Asia. One British soldier and keen hunter, Walter Smee, wrote that in western India, there was extensive predation by lions on livestock.[161] Extensive hunting and habitat loss meant that lions had disappeared from most of India by 1880. The last lion outside Gujarat was killed in 1884. Lions survived in the Saurashtra district of Kathiawar peninsula, chiefly the Gir Forest, in Gujarat. Their survival is due largely to the Nawabs of Junagadh, who ruled from 1748. The Junagadh nobles hunted in Gir and invited British officials and Indian royalty to hunt there, killing many lions.[162]The decline was halted in the late 19th century by the Nawab of the time, in whose traditional royal hunting ground most remaining lions resided.[163] It should be noted that striped hyenas were resident in reasonable numbers in the Gir Forest and surrounding regions, and benefitted from the wild prey and livestock killed by the lions. The forest is today, as Menon points out in his *Field Guide to Indian Mammals*, one of the best places in India to see striped hyenas.[164] The protection of the lions and their forest habitat by the Nawabs of Junagadh, the British colonial authorities and then the Indian government after independence had the unintended consequence of protecting hyenas, their food sources and their habitat. The presence of large numbers of Maldhari pastoralists with their livestock in and around the Gir Forest would also have provided striped hyenas with the opportunity to scavenge carcasses of animals that died of disease or old age, and the chance to forage for other food on waste middens around the Maldhari settlements.

Notes

1 Keith Somerville (2019) *Ivory. Power and Poaching in Africa*, London: Hurst and Company, pp. 12–24.
2 Peter Robertshaw (1990) *Early Pastoralists of South-Western Kenya*, Nairobi: British Institute in Eastern Africa, p. 17
3 Ibid., p. 19.
4 Robertshaw, 1990, p. 19.
5 George B. Schaller (1972) *The Serengeti Lion. A Study of Predator-Prey Relations*, Chicago: University of Chicago Press, p. 7.
6 Personal communication.
7 The game wardens Myles Turner and George Adamson in their accounts of their work as wardens in Tanzania and Kenya made regular reference to the use of snares and the dangers these held for animals that weren't the original targets of hunters – see Myles Turner (1987) *My Serengeti Years*, London: Elm Tree Books, p. 27; and George Adamson (1968) *Bwana Game*, London: Collins and Harvill Press.
8 Robin S. Reid (2012) *Savannas of Our Birth. People, Wildlife and Change in East Africa*, Berkeley, Calif.: University of California Press, p. 105.
9 Sir Richard Burton (2001, originally published 1860) *The Lake Regions of Central Africa. From Zanzibar to Lake Tanganyika. Volume 1*, Santa Barbara, Calif.: The Narrative Press, pp. 61 and 65.
10 Ibid., p. 65.
11 Ibid.
12 Ibid., p. 12.
13 Ibid., p. 51.
14 John Hanning Speke (1864) *What Led to the Discovery of the Source of the Nile*, originally published by William Blackwood, London: Public Domain Book, Kindle Edition, loc. 1705.
15 Letter from Capt J. H. Speke commanding the East African Exploring Expedition, *Proceedings of the Zoological Society of London for the year 1863 (henceforth Proceedings)*, p. 4.
16 Speke, 1864, loc. 194.
17 Ibid., loc. 6788.
18 Ludwig von Höhnel (1894) *Discovery of Lakes Rudolf and Stefanie*, London: Longmans, Green and Co, reprinted 2015 by Forgotten Books, p. 261.

19 K. R. Dundas (1910) Notes on the Tribes Inhabiting the Baringo District, East Africa Protectorate. *The Journal of the Royal Anthropological Institute of Great Britain and Ireland*, 40, Jan–June, 49–72, p. 56.
20 Ibid., p. 59.
21 *Proceedings* for the year 1885, p. 217.
22 F. Rudolf Lehmann (1941) Some field-notes on the Chaga of Kilimanjaro, *Bantu Studies*, 15, 1, 385–396, p. 390.
23 Camilla Gibb (1997) Constructing Past and Present in Harar, in *Ethiopia In Broader Perspective: Papers of the XIII International Conference of Ethiopian Studies – Kyoto 12–17 December 1997 Volume II*, https://everythingharar.com/images/pdf/publication/Constructing%20Past%20and%20Present%20in%20Harar%20-%20Gibb.pdf accessed 30 April 2019, pp. 3–4.
24 Ibid.
25 Richard Caulk (1971) The Occupation of Harar: January 1887, *Journal of Ethiopian Studies*, 9, 2, 1–19, p.1.
26 Bahru Zewde (2001) *A Modern History of Ethiopia*, Oxford: James Currey, p. 19.
27 Zewde, 2001, pp. 61 and 63–4.
28 Marcus Baynes-Rock (2012) Hyenas like Us: Social Relations with an Urban Carnivore in Harar, Ethiopia, PhD thesis, Department of Anthropology Macquarie University, Sydney, p. 43.
29 Ibid.
30 Ibid.
31 Marcus Baynes-Rock (2015) *Among the Bone-Eaters. Encounters with Hyenas in Harar*, Pennsylvania: University of Pennsylvania Press, p. 6.
32 Ibid., pp. 58–9.
33 Personal communication from Marcus Baynes-Rock.
34 Mansfield Parkyns (1868) *Life in Abyssinia: Being Notes Collected During Three Years' Residence and Travels in that Country*, London: John Murray, Kindle Edition, loc. 4616–4619, loc. 4615.
35 Ibid., loc. 876–80.
36 Ibid., loc. 4834–43.
37 Ibid., loc. 4966.
38 Harald George Carlos Swayne (1895) *Seventeen Trips Through Somaliland: A Record of Exploration & Big Game Shooting, 1885 to 1893: Being the Narrative of Several Journeys in the Hinterland of the Somali Coast Protectorate, Dating from the Beginning of Its Administration by Great Britain Until the Present Time: with Descriptive Notes on the Wild Fauna of the Country*, London: Rowland Ward, Kindle Edition, loc. 4633.
39 Keith Somerville (2017) *Africa's Long Road Since Independence*, London: Penguin, pp. 6–7.
40 Roy Bridges (1986) The Visit of Frederick Forbes to the Somali Coast in 1833, *International Journal of African Historical Studies*, 19, 4, 679–691, p. 687.
41 Speke, 1864, loc. 159.
42 Swayne, 1895, loc. 104–109.
43 Ibid., loc. 72.
44 Ibid., loc. 813.
45 Ibid. loc. 1031.
46 Ibid., loc. 3179 and 3348.
47 Ralph Evelyn Drake-Brockman (1910) *The Mammals of Somaliland*, London: Hurst and Blackett, pp. 39–42.
48 Cited by J. L. Cloudsley-Thompson (1967) *Animal Twilight Man and Game in eastern Africa*, London: G.T. Foulis, p. 55.
49 Cited at length by Cloudsley-Thompson, pp. 168–9.
50 Cloudsley-Thompson, p. 60.
51 Annik E. Schnitzler (2011) Past and present distribution of the North African–Asian lion subgroup: a review, *Mammal Review*, 41, 3, 220–243, p. 231.

52 Henry Barth (1890) *Travels and Discoveries in North and Central Africa*, London: Ward, Lock and Co, reprinted 2015 by Forgotten Books.
53 M. AbiSaid and S. M. D. Dloniak (2015) *Hyaena hyaena. The IUCN Red List of Threatened Species*, https://www.iucnredlist.org/species/10274/45195080 accessed on 12 March 2020.
54 H. H. Johnston (1895) *The River Congo, From Its Mouth to Bolobo*, London: Sampson, Low, Marston and Co, reprinted 2018 by Forgotten Books, p. 175.
55 Ibid., p. 261.
56 Ibid., p. 32–3.
57 See Brian J. Huntley (2017) *Wildlife at War in Angola. The Rise and Fall of an African Eden*, Pretoria: Pretoria Book House.
58 Joachim John Monteiro (1875) *Angola and the River Congo, Volume II*, London: Macmillan, p. 79.
59 Ibid.
60 Ibid., p. 80.
61 Ibid., p. 71.
62 Ibid.
63 Huntley, p. 136.
64 http://www.fynboshub.co.za/fynbos-conservation/what-is-fynbos/ accessed 23 March 2018.
65 John Iliffe (2007) *Africans. The history of a continent*, Cambridge: Cambridge University Press, p. 127.
66 Noël Mostert (1992) *Frontiers. The Epic of South Africa's Creation and the Tragedy of the Xhosa People*, London: Jonathan Cape, p. 123.
67 Jan van Riebeeck (1897) *Precis of the Archives of the Cape of Good Hope, December 1651-December 1653, Riebeeck's Journal*, Cape Town: W.A. Richards and Sons, reprinted by LightningSource UK (no date), p. 9.
68 Ibid., p. 55.
69 Ibid., p 59.
70 Ibid., p. 79.
71 C. A. W. Guggisberg (1961) *Simba. The life of the lion*, London: Bailey Bros and Swinfen Ltd, p. 76.
72 Van Riebeeck, 1897, p. 223.
73 Jan van Riebeeck (1897) *Precis of the Archives of the Cape of Good Hope, Riebeeck's Journal, Volume 3*, Cape Town: W.A. Richards and Sons, reprinted by LightningSource UK (no date), p. 282.
74 This was confirmed to me by a number of South African based hunters and conservationists who have Afrikaans as their first languages.
75 Darcy Ogada (2014) The power of poison: pesticide poisoning of Africa's wildlife, *Annals of the New York Academy of Sciences*, 1–20, p. 2.
76 Vijaya Ramadas Mandala (2015) The Raj and the paradoxes of wildlife conservation: British attitudes and expediencies, *The Historical Journal*, 58, 1, 75–100, p. 80.
77 Cloudsley-Tompson, 1967, p. 125.
78 Ogada, 2014, p. 2; see also, Lance van Sittert (1998) "Keeping the Enemy at Bay": The Extermination of Wild Carnivora in the Cape Colony, 1889–1910, *Environmental History*, 3, 3, pp. 333–356.
79 Van Sittert, 1998, p. 333.
80 Ibid., p. 334.
81 Ibid., p. 336.
82 Keith Somerville (2020) *Humans and Lions. Conflict, Conservation and Coexistence*, London: Routledge, pp. 54–8.
83 Cloudsley-Tompson, 1967, p. 80.
84 Sir John Barrow (1802) *An Account of Travels Into the Interior of Southern Africa in the Years 1799 and 1798*, London: G.F. Hopkins, Kindle edition.
85 Ibid, loc. 362.

86 Ibid., loc. 2643.
87 Ibid., loc. 2158.
88 William John Burchell (1822, reprinted 2016) *Travels in the Interior of Southern Africa, Volume I*, London: Longman, Rees, Orme, and Brown, pp. 243 and 289.
89 Ibid., p. 456.
90 William John Burchell (1824) *Travels in the Interior of South Africa, Volume II*, republished by Cambridge: Cambridge University Press, 2015, pp. 27, 36 and 33.
91 Ibid., p. 277.
92 Ibid., p. 326.
93 Sir William Cornwallis Harris (1852) *The Wild Sports of Southern Africa*, Kindle Edition, (originally published in 1894), p.18. republished by H.G. Bohn, loc. 311.
94 Ibid., loc. 813.
95 Ibid., loc. 866.
96 Ibid., loc. 325.
97 Cloudsley-Thompson, 1967, p. 22.
98 Cornwallis Harris, 1852, loc. 2008.
99 Ibid., loc. 2138.
100 Ibid.
101 Ibid., loc. 2832.
102 Roualeyn Gordon-Cumming (2011) *Five Years' Adventures in the Far Interior of South Africa. With Notices of the Native Tribes and Savage Animals*, Milton Keynes: LightningSource, p. 4.
103 Ibid., p. 43.
104 Ibid., p. 96.
105 John M. Mackenzie (1988) *The Empire of Nature. Hunting Conservation and British Imperialism*, Manchester: University of Manchester Press, p. 86.
106 Ogada, 2014, pp. 2–3.
107 Norman Etherington (2016) *A Biography of Frederick Courteney Selous. Big Game Hunter*, Marlborough, Wilts.: Robert Hale, p. 31.
108 Frederick Courtney Selous (1881) *A Hunter's Wanderings in Africa*, republished in 2001, Alexander, N.C.: Alexander Books, p. 14.
109 David and Charles Livingstone (1865) *Narrative of an Expedition to the Zambesi and its Tributaries*, reprinted in 2005, Stroud, Gloucestershire: Nonsuch, p. 29.
110 Ibid., p. 173.
111 Ibid.
112 H. L. Duff (1903) *Nyasaland under the Foreign Office*, London: George Bell and Sons, p. 47.
113 Ibid., p. 106.
114 F. Vaughan Kirby (1899) *Sport in East Central Africa. Being an Account of Hunting Trips in Portuguese and Other Districts of East and Central Africa*, London: Rowland Ward, reprinted 2017 by Forgotten Books, p. 323.
115 Ibid., p. 324.
116 Edward Foa (1899) *After Big Game in Central Africa*, republished by Waxkeep Publishing, Kindle Edition, loc. 600.
117 Ibid., loc. 1435.
118 Ibid., loc. 1980.
119 Charles John Andersson (1856, reprinted 2012) *Lake Ngami*, Memphis, Tenn: General Book, p. 106.
120 Garth Owen-Smith (2010) *An Arid Eden. A Personal Account of Conservation in the Kaokoveld*, Jeppestown, SA: Jonathan Ball, p. 58.
121 Francis Galton (2012) *Francis Galton's Narrative of His Exploration of Namibia in 1851 (Annotated)*, Keith Irwin, Kindle Edition, loc. 1426.
122 Ibid., loc. 1628 and 2216.
123 Ibid., loc. 2430.
124 Ibid., loc. 3911.
125 Ibid., p. 4443.

126 Ibid., p. 1979.
127 Andersson, 1856, p. 36.
128 Captain G. C. Shortridge (1934) *The Mammals of South West Africa, Volume I*, London: William Heinemann, p. 157.
129 Ibid., p. 160.
130 Somerville, 2019, pp. 99–100.
131 Henry Barth (1857) *Travels and Discoveries in North and Central Africa. Being a Journal of an Expedition, Volume II*, New York: Harper and Brothers, reprinted 2018 by Forgotten Books, p. 364.
132 Leo Africanus (2018) *The History and Description of Africa and of the Notable Things Therein Contained*, Volume 3, London: Hakluyt Society, reprinted by Forgotten Books, p. 947.
133 Cited by Cloudsley-Thompson, 1967, p. 168.
134 John H. Drummond Hay (1861) *Morocco and the Moors. Western Barbary: Wild Tribes and Savage Animals*, London: John Murray, reprinted 2017 by Forgotten Books, pp. 78–9.
135 Ibid.
136 Dale J. Osborn and Ibrahim Helmy (1980). The Contemporary Land Mammals of Egypt (including Sinai), Field Museum of Natural History, digitized version, https://archive.org/details/contemporaryland05osbo/page/422 accessed 24 July 2019, p. 429.
137 Burton Stein (2010) *A History of India*, Oxford: Blackwell, pp. 158–160.
138 Marc Jason Gilbert (2017) *South Asia in World History*, Oxford: Oxford University Press, p. 75.
139 Stein, 2010, p. 159.
140 Madhav Gadgil and Eamachandra Guha (1992) *This Fissured Land. An Ecological History of India*, New Delhi: Oxford University Press, p. 94.
141 John Keay (1991) *The Honourable Company*, London: Harper Collins Publishers, Kindle Edition, loc. 1757–64.
142 William Dalrymple (2019) East India Company sent a diplomat to Jahangir & all the Mughal Emperor cared about was beer, *The Print*, 24 August 2019, https://theprint.in/pageturner/excerpt/east-india-company-sent-a-diplomat-to-jahangir-all-the-mughal-emperor-cared-about-was-beer/281255/ accessed 3 March 2019.
143 William Dalrymple (2015). The East India Company: The original corporate raiders, *The Guardian*, 4 March 2015, https://www.theguardian.com/world/2015/mar/04/east-india-company-original-corporate-raiders accessed 3 March 2020.
144 Dalrymple, 2015.
145 N. G. Rathod (1994) *The Great Maratha Mahadaji Scindia*. New Delhi: Sarup & Sons, p. 8.
146 Sugata Bose and Ayesha Jalal (2004). *Modern South Asia: History, Culture, Political Economy*, 2nd ed., London: Routledge. p. 41.
147 Dalrymple, 2015.
148 R. I. Meena, Sandeep Kumar and Shmshad Alam (2014) *Action plan for the Conservation of the Asiatic Lion (Pantheraleo persica Meyer, 1826)*, Junagadh: Gujarat Forest Department, p. xi.
149 Ibid., p. 129
150 Julie E. Hughes (2015) Royal Tigers and Ruling Princes: Wilderness and wildlife management in the Indian princely states, *Modern Asian Studies*, 49, 4, pp. 1210–1260, 1214–5.
151 Mackenzie, 1988, p.168.
152 https://en.wikipedia.org/wiki/File:Speared_hyena.JPG#file accessed 9 January 2020.
153 Divyabhanusinh (2008) *The Story of Asia's Lions*, 2nd ed., Mumbai: Marg Publications, p. 93.
154 Daniel Johnson (1822) *Sketches of Field Sports as followed by the 4 Natives of India*, London: Longman, https://archive.org/details/sketchesoffields00johnrich/page/78/mode/2up accessed 4 March 2020, pp. 39–40.
155 Ibid., p. 51.

156 Henry Shakespear (1860) *The wild sports of India*, Boston: Ticknor and Fields, Kindle Edition, 2017, loc. 878.
157 Ibid., loc. 1462.
158 *Proceedings*, Part XXVI, 1858, p. 514.
159 William Tempe Hornaday (1885) *Two Years in the Jungle: The Experiences of a Hunter and Naturalist in India, Ceylon, the Malay Peninsula and Borneo*, Kindle Edition, loc. 1940.
160 Ibid., loc. 2515.
161 Captain Walter Smee, Bombay Army, Fellow of Zoological Society (1833) Some Account of the Maneless Lion of Guzerat. *Proceedings*, 10 December, p. 171.
162 Sudipta Mitra (2005) *Gir Forest and the Saga of the Asiatic Lion*, New Delhi: Indus Publishing Company, p. 19.
163 Ibid., p. 16.
164 Vivck Menon (2003) *A Field Guide to Indian Mammals*, Delhi: Penguin/Dorling Kindersley, p. 77.

6

PERSECUTION INCREASES UNDER COLONIAL RULE

The end of the 19th and beginning of the 20th century saw the extension and entrenchment of colonial rule, magnifying the consequences for wildlife of European penetration and economic exploitation of Africa, and continued European hegemony or colonial rule in West and South Asia. The desire of colonising countries to make their *possessions* pay, through cash crops, settlement of Europeans, mining and other forms of extraction reduced wildlife habitat. Ungulates, seen as competitors with livestock or sources of disease, were excluded from farming areas. Predators were seen as threats to livestock and also, after the affair of the man-eating lions of Tsavo,[1] people.

But wildlife was also a source of meat, ivory, horn and hides to use or sell, tiding settlers over until cash crops could be marketed. Hunting provided sport for settlers, officials, soldiers and traders. Some wildlife species were exterminated completely by killing, to clear land for grazing or crops, for meat and for sport (for example, the quagga and blaubok in South Africa). Colonial rule entailed the partitioning of land with forced removals of indigenous people to make way for settlers or plantations, and later for game reserves, and restrictions on hunting by communities which had hunted wildlife for centuries without exterminating it.[2] In much of Africa, indigenous peoples were banned from hunting, and even from killing livestock predators without permission. European hunters had a free hand. In some colonial territories with few white settlers, hunting regulation involved the co-option of chiefs and other local leaders to control hunting under European tutelage.

Across the colonial empires, hunting by Europeans was viewed as ideal training for those running the imperial project. Mackenzie summed it up: "The importance of the Hunt can be identified at every level of the theory and practice of the imperial ethos".[3] Settlers in Kenya were encouraged to hunt to clear ungulates from farmland but also to provide food while they waited for the coffee or other crops to grow. Hunting of lions and other predators was encouraged as sport and for vermin extermination. The early-to-mid 20th century also saw the rise in the European application of early conservation measures in Africa and Asia. Game laws and game departments were established in colonies to regulate hunting and protect some species.

Sub-Saharan Africa

European occupation of Africa went into full swing after the Berlin Conference of 1884–5 divided the continent between European powers.[4] They worked to profit from their new territories through exploitation of land, natural resources and indigenous people. Settlement was encouraged in territories like Kenya, Southern Rhodesia, Angola and Mozambique. This brought a direct onslaught against wildlife as settlers, colonial officials, soldiers and traders hunted on a massive scale for food, commercial gain, to protect livestock or for sport. The designation of hyenas, lions, leopards, cheetah and wild dogs as vermin had a devastating effect on populations. They were killed in huge numbers with no licences necessary. Hunting lions was viewed as conferring prestige, but killing hyenas was done routinely with no kudos for the hunter. Predator extermination increased with the establishment of game reserves or national parks, through the extensive culling of carnivores to preserve large populations of ungulates, which continued up to the 1950s.[5]

Killing of wildlife was enabled on a huge scale by modern firearms. And in 1889, Europeans imported something as destructive as firearms – rinderpest, the cattle disease which wiped out millions of cattle and wild ungulates. It came from Italian imports of cattle from India to the Eritrean port of Massawa.[6] It spread rapidly south and west, affecting domestic cattle and many species of wild ungulates. The disease had a roller-coaster effect on predators. Initially, hyenas and lions increased in numbers, feasting on wildlife and livestock which died; and on the bodies of people who starved. When the epidemic receded, prey availability was reduced significantly, leading to livestock killing and increased man-eating. Many pastoralist communities across East and Southern Africa reverted to hunting to survive, reducing wild prey and increasing human-predator conflict.[7] Even then, African hunting was deemed poaching by the colonial authorities.

Hyenas, with their flexible feeding strategies were better able to survive increased conflict than lions. Keith Caldwell, head of Kenya's game department in the early 1920s, wrote that during the rinderpest outbreak in Kenya's Southern Reserve, thousands of cattle died along with wild game and hyenas thrived, increasing in numbers. When rinderpest declined, Caldwell inaccurately stated that the increased number of hyenas had "to give up their habits as scavengers and become hunters and killers of game",[8] adding that they had become such a pest that large groups were harassing lions and killing too many new-born wildebeest. He believed that "An intensive poisoning campaign [was] necessary to reduce the numbers of hyenas and restore the balance".[9]

When the epidemic ended, wildlife recovered but European hunting increased with the expansion of colonial settlement. The depletion of wildlife that resulted made some hunters and colonial administrators recognise that it, along with rinderpest, had "produced such a marked diminution of game that conservation measures seemed necessary".[10] This wasn't an altruistic measure to preserve wildlife, it involved fear that the decline deprived settlers and visiting hunters of sport and the colonies of income. The form preservation took "was shaped by the social and economic realities of empire...Access to animals was to be progressively restricted to the elite; animals were to be categorised...some were to be specially protected for their rarity, others shot indiscriminately as vermin".[11] The designation of protected areas did not stop the widespread killings of predators. Hyenas, lions, leopards, cheetah, wild dogs and jackals were still categorised as vermin.

Much of the impetus for the creation of reserves in British territories came from the formation of the Society for the Preservation of the Wild Fauna of the Empire in 1903 by a group of hunters and colonial administrators. This followed the meeting in London of colonial powers, which drew up the Convention for the Preservation of Wild Animals, Birds and Fish in Africa, signed on 19 May 1900.[12] It called for concerted efforts across colonial territories to conserve endangered species and habitats,[13] but schedule five of the convention called for the reduction in numbers of predators to preserve ungulates. At the heart of its ethos was that only Europeans could be trusted to manage wildlife. What was forgotten, out of self-interest and for imperial mythmaking, was that the inhabitants of colonies had long coexisted with wildlife without serious depletion, despite hunting for food and livestock protection.[14] As Reid argues, "Since the late 1800s, this ancient coexistence of people and wildlife in African savannas has often been ignored, often replaced by a modern practice of conservation which assumes that wildlife are best conserved in landscapes with no people".[15]

East Africa

There has long been a vision among Europeans of 19th century East Africa as an Eden or paradise, with huge areas of wilderness teeming with wildlife and largely untouched by humans. It is exemplified in books lamenting the decline in wilderness areas, shrinking of wildlife habitats and loss of species over the last 150 years, with titles like *Last days in Eden* or *Paradise Lost: A History of Game Preservation in East Africa.* [16] In the latter book, the author describes early 20th century travellers on the Uganda Railway from Mombasa seeing from the carriages "tens of thousands of animals of every species", and says that "in this now vanished world, the animals seemed inexhaustible".[17] While it is undeniable that early travellers would have been amazed (coming from the denuded landscapes of western Europe) at the profusion of wildlife, the image of an unspoilt paradise or Eden replicates the imperialist narrative that Europeans occupied a sparsely populated landscape with few indigenous human communities and no developed agriculture. It admits the destruction of habitats and wildlife under colonialism, but effectively denies the millennia of human coexistence and conflict with wildlife in Africa, which had gradually excluded wildlife from areas of settlement and agricultural expansion, but without the substantial destruction wrought by Europeans. In many areas, wildlife had survived and there was coexistence between wildlife populations and people.[18]

The rinderpest epidemic changed the situation somewhat, as European settlement accelerated. The disease's effects were felt across East and Southern Africa with livestock herds hugely diminished, significant human mortality through starvation, malnutrition and disease, notably smallpox epidemics. This reduced the human population in southern Kenya and northern Tanzania. The Ngorongoro region went from having a large human population with substantial cattle herds to an under-populated one with resurgent ungulate herds and predators.[19] Ngorongoro Crater was stripped of people and their livestock. Wildlife returned in large numbers after the departure of the Maasai and their herds.[20] The increase in ungulate herds, especially wildebeest, zebra and Thomson's gazelles, led to the re-establishment of a large concentration of spotted hyenas and lions.

Hyenas, despite their persecution as vermin by European settlers, were still common across East Africa, including around the growing European towns and farming communities. The Europeans, with their heritage of contempt for hyenas, picked up the prejudices of pastoralist communities like the Maasai and developed extreme distaste for the animal. This was reflected in a book by former US president Theodore Roosevelt, written after his hunting safari in East Africa, in which he and his co-author wrote that "the hyaena is a singular mixture of abject cowardice and the utmost ferocity", and normally feeds on carrion, often hesitating to attack even the weakest animal if it is unhurt.[21] But, typically contradictorily regarding European views of hyenas, in the next passage he talks of a group of hyenas attacking a half-grown rhinoceros which had returned to the body of its mother, killed by hunters. He added that they would also prey on sheep, goats, donkeys, mules, cattle and dogs, and would even enter large villages "killing men, women, and especially children".[22] He wrongly believed they were solitary and only gathered in large groups "around any big dead beast".[23] Roosevelt and Heller believed the striped hyena was not uncommon in East Africa, "but we never saw it abroad in the daytime, and only caught it in traps".[24]

The European impact on wildlife in Kenya increased at the turn of the century when the British High Commissioner, Sir Charles Eliot, charted a new approach to economic development. Where wildlife was a potential threat to large-scale crop production or livestock husbandry, it was to be exterminated. Eliot believed that extensive European livestock, food and cash crop production would enable the colony to pay its way and turn a profit.[25] Occupation and fencing of land by settlers and the creation of so-called tribal reserves to which pastoralist and other communities were sent were used to reserve the best grazing land for settlers, with extensive slaughter of herbivores and predators. The settler and hunter Lord Delamere obtained large areas of land to raise imported cattle and sheep[26]. To protect them, he became a determined exterminator of predators, zebra and other ungulates he saw as competition for his livestock.[27] This demand for land for settlers gave rise to the use of threat of force to expel communities like the Maasai. The effects of rinderpest and smallpox had destroyed much of their cattle wealth and reduced their population, rendering them too weak to resist the land-grabs by the colonisers. In some areas, the depletion of the populations of pastoralists and their cattle in areas not required for settlers enabled increases in wildlife and the reversion of grasslands to bush.[28] In 1913, the British compelled the il-Purko Maasai living in Laikipia to move to the Southern Maasai Reserve, which included rangelands and highland areas at Narok. Much of Laikipia was turned over to white settlers. The Maasai reserve areas in the south became increasingly crowded, with large numbers of cattle.[29] The white settlers who took the Laikipia land were principally livestock farmers.[30] They were intolerant of predators and shot them. They began to fence off their land to keep out both wild ungulates and the cattle of the remaining pastoralists. This reduced wild prey and predator populations, with lions and cheetah the worst affected as they were less adaptable than hyenas.

Hyenas, despite being killed by Maasai if they attacked livestock, were not exterminated in Maasai areas and continued to live alongside pastoralists, foraging from waste around Maasai manyattas and cattle camps. They also consumed the bodies of the dead that the Maasai left in the bush.[31]

Prior to WWI and the expansion of urban settlement, "hordes of hyenas" lived in the vicinity of Nairobi's slaughterhouses, consuming viscera, bones and any other leftovers from butchering carcasses.[32] The slaughter of cattle there during the war to feed the British and imperial troops fighting the Germans in Tanzania increased both the carcasses available and the numbers of hyenas. When the war ended and the supply of cattle remains was reduced, foraging in residential areas increased, with attacks on humans reported regularly. The hunter Peter Capstick wrote, "The hyenas, desperate for food, swarmed the place, eating everything that wasn't red-hot or nailed down, including clothes, stained cooking pots, brooms and finally several women working in the mealie patches".[33] Elspeth Huxley described how a Kikuyu cattle thief was badly injured by a gun-trap set for hyenas on a white farm close to Nairobi.[34] She also wrote that hyenas would enter hunting camps at night and grab leather saddles, even when the hunters were using them as pillows.[35] She recalled later in life that "Like everyone else in Africa, I was brought up to believe that hyenas were nasty, cringing, crafty scavengers living partly on scraps from kills made by others, such as lions, and partly on refuse and human corpses".[36] She added that hyenas had been known to carry off unattended babies. To call a man a *fisi*, the Swahili word for hyena, was a deadly insult in Kenya.[37] Karen Blixen wrote that the Indian carpenter living close to the mill on her farm regularly shot at hyenas at night when they tried to eat the strips of hide hanging near the farm ready for making leather straps.[38] On another occasion, three bulls had been castrated on her farm to make them into docile plough-oxen. They were shut up in a yard overnight. But hyenas smelt the blood, broke their way in and killed the animals.[39] Blixen added that the Kikuyu on her farm did not bury their dead but put them out in the bush for the hyenas and vultures to clear up – a practice she applauded rather than condemning.[40]

The biographical, narrative and fictional accounts of settlement and hunting in East Africa by Blixen, Huxley and others presented Kenya as a paradise with teeming wildlife and land for settlers, who could spend their spare time hunting. Hunting by visitors with money to spend was encouraged, as was hunting by settlers as a means of obtaining meat, hides and ivory to supplement farm income. A game department was formed, licences issued, and income generated from sport hunting and ivory. It was noticeable that in a territory where hunting became ubiquitous and attempts were made by the East African Protectorate authority to keep track of animals shot that the official returns for animals hunted on licence did not include hyenas, lions or leopards, which were considered vermin.[41] In regular reports on wildlife and the establishment of reserves in British East and Central Africa published by the Society for the Preservation of the Wild Fauna of the Empire's *Journal*, hyenas are scarcely ever mentioned. Hyenas, baboons and bush pigs were viewed as such serious vermin that the colonial administration carried out and encouraged settlers to engage in eradication campaigns using strychnine and other poisons, which would normally have been viewed by hunters as "unsporting".[42] While there was no official recording of the extermination of hyenas and other carnivores, you can get an idea of the scale given that in 1908, 150 lions were shot in Laikipia alone.[43]

West of Laikipia, near Lake Baringo and the main road route from the coast to the British Uganda Protectorate, the Tugen, Nandi and other Kalenjin-speaking groups had been forced into submission by British punitive expeditions. The communities were engaged in pastoralism, more so after floods destroyed fields used for cultivation. Following the floods there was a serious drought, which led to cattle deaths and widespread starvation. Anderson wrote that the heavy depletion of wildlife in the region by European hunters meant that resorting to hunting as a means of survival in times of food shortages was no longer available.[44] The Tugen and Nandi protected their stock against predators but had a far less antagonistic view of hyenas than of other predators, and didn't persecute them or kill in retaliation for stock losses as many pastoral peoples did.[45] As David Anderson told me:

> They leave them unmolested, on the grounds that the hyenas may be the conduits for the reincarnation of their kin. Their dead were not buried, but left out in the bush…The Kalenjin age-set system is cyclical, with recurring age-sets that are believed to be the reincarnation of the previous set of the same name. History thus only has a depth of one full cycle of the age-set calendar. Hyenas, in devouring the corpses are thus the conduits in this process. Kalenjin do not hunt or harm hyenas, although they will chase them away from homesteads, as their presence is thought to be a bad portent, and they keep them away from livestock as they are a pest.[46]

The Nandi of central Kenya and the Rift Valley also have strong beliefs about the hyena:

> …if not exactly sacred, is 'held in respect or fear' by all the clans alike, the hyena is, as has been said, 'the living mausoleum of their dead.' Only very old

> people and very young children are buried—other corpses are left for the hyenas to eat.[47]

Other legends about hyenas were common in Kenya, such as the belief that witches could turn themselves into hyenas to harm those hated by the witch or whom the witch had been paid to attack. Peter Matthiessen refers to beliefs that spirits who were hyenas in human form could be detected, according to folk stories, by finding they had a mouth on the back of their head.[48] Old women were often accused of being werewolf hyenas, while some witches were said to ride hyenas at night. The myths link in with a story which involved Baron Bror Blixen, the hunter and husband of Karen Blixen. Blixen was begged by villagers to kill a particularly voracious hyena which was raiding villages at night. The villagers wanted Blixen to kill it as they feared "reprisal from its witch". When Blixen shot it, it is said he found the hyena turned into a man.[49] During a revolt by the Giriama people of the Kenyan coast, the local colonial official, Arthur Champion, believed that it was caused by the elders of the Fisi (hyena) society; "Members of the Fisi or Hyena grade functioned as a secret society, enforcing their decisions by powerful oaths and spells".[50] The hyena society used its reputed supposed supernatural powers to overcome community elders and incite revolt.[51] The hunter Robert Foran wrote about beliefs in many parts of colonial Africa in were-animals, versions of the European myths about lycanthropy (people who turned into werewolves during the full moon). He blamed the continuation of such beliefs on traditional healers or witchdoctors, who:

> ...foster the idea in the minds of their gullible dupes that beasts are the natural receptacles of human souls and man is capable of transforming himself at will into the form of animate...the African continent happens to be particularly rich in stories concerning were-lions, were-leopards, were-jackals, and were-hyenas...[52]

He noted that accounts in Count de Prorok's *In Quest of Lost Worlds* indicated the continuation of such beliefs in Ethiopia, and that de Prorok "was an eyewitness of a remarkable instance of were-hyenas and the horrible orgies resulting therefrom when on an expedition of exploration in Ethiopia".[53] But Foran was broad-minded enough to conclude that these beliefs were not peculiarly African:

> We should not overlook the fact, I consider, that Europeans in the not too-distant past genuinely believed witches and wizards possessed the power to change themselves at will into the forms of various animals. Africans are equally positive about it...the subject is a matter of supreme and unshakable faith.[54]

The development of safari hunting as a sport for visiting hunters, settlers and officials took off in a big way in colonial Kenya in the early 20th century. Many hunters left accounts of their encounters with hyenas. Rainsford wrote of being plagued by hyenas at night on the Nzoia Plateau, and of setting up gun-traps to kill them. One hyena was killed by grabbing a bait tied to the trigger of the gun. But soon the clever animals:

> …knew that they must not come at the meat they wanted, and which they always carried away, no matter how securely it was tied on, from the front. It really looked as though they had placed a paw on the trigger and then proceeded to dine leisurely.[55]

Few hunters bothered shooting hyenas unless they tried to raid their camps. Abel Chapman wrote in 1908 of periodically seeing hyenas near Laikipia, but he paid them little attention as they were not a prized game species.[56] On one occasion he admitted shooting a spotted hyena by mistake, as he thought it was a lion.[57] When one of his retainers died on an expedition near Lake Elmenteita, they buried him near the camp, only to find a few days later that hyenas had dug up and eaten the body.[58] Most hunters and settlers accepted the traditional view that hyenas only ate carrion. But a few gave insights into their hunting ability.

Richard Tjader, who hunted in Kenya, Somalia and Tanzania, wrote:

> …it has been proven in a great many instances that the hyena often kills its own prey, which it devours with the most ravenous appetite. Very often horses, mules, and donkeys have been killed by these hyenas, although they have been tied up close to the campfire of the hunting party.

He added:

> …settlers and natives also complain of the continuous attacks of hyenas upon their cattle and sheep. One gentleman from German East Africa told me of how three spotted hyenas in one single night had killed and partly devoured not less than eleven cows from his herd.[59]

He also recounted the story of an English hunter sleeping in a hammock with his arm hanging over the side. During the night, a hyena crept up and grabbed his arm. It retreated when he shouted. The hunter grabbed his rifle and pretended to go back to sleep. When the hyena returned, he shot it in the head.[60] Tjader said he had many stories from African communities about hyenas dragging off old people and children. He blamed this liking for human prey on the fact that:

> many of the tribes do not bury their dead, but throw their bodies, as well as, in many cases, old, sick people, whom they think may be dying, out into the bush for the very purpose of having them eaten by hyenas and other carnivorous animals.[61]

Tjader wasn't interested in hunting hyenas, believing that:

> ...hyena killing can certainly not be classed among real 'sport'...[and] the hunter will sometimes have hard work to persuade the natives to touch it, as most of them will have nothing to do with a hyena...the hyena skin is hardly worth while preserving for the trophy room.[62]

Theodore Roosevelt and his son killed 512 mammals on their safari in 1909, including nine hyenas (their list does not specify species).[63] Many of the skins of these were preserved and sent to museums in America. Roosevelt shared the general disdain for hyenas. His first sighting was on the Mombasa-Nairobi train. He wrote that a hyena was nearly run over by the train, as it was not scared and was slow to move away. In his account of the safari, he said that "the hyena, too cowardly ever to be a source of danger to the hunter, is sometimes a dreadful curse to the weak and helpless".[64] But he went on to say that it was "a beast of unusual strength, and of enormous power in his jaws and teeth, and thrice over would he be dreaded were fang and sinew driven by a heart of the leopard's cruel courage".[65] Despite the supposed cowardice. he recounted that in East Africa:

> ...it is yet fraught with a terror all its own; for on occasion the hyena takes to man-eating after its own fashion...where famine or disease has worked havoc among a people, the hideous spotted beasts become bolder and prey on the survivors.[66]

He backed up his contention about man-eating with details of an outbreak of sleeping sickness in Uganda. Many sufferers were brought together in camps, Roosevelt said, where hyenas soon realised that the people there were sick and helpless. In 1908, they reportedly started:

> ...haunting these sleeping-sickness camps, and each night entering them, bursting into the huts and carrying off and eating the dying people. To guard against them each little group of huts was enclosed by a thick hedge; but after a while the hyenas learned to break through the hedges, and continued their ravages.[67]

Later in his narrative he referred to two striped hyenas being trapped by others on his expedition,[68] leaving one to presume that all his previous references were to spotted hyenas. Roosevelt recorded that he had been told in Kenya that a "native" hunter employed by the Kenyan settler and hunter Sir Alfred Pease was seized by a hyena as he slept beside the campfire, and part of his face torn off.[69] In 1904 Emperor Menelik II had sent a hyena as a present to President Theodore Roosevelt. Despite Roosevelt's recorded dislike of hyenas as cowardly, he is said to have grown fond of this one and at one time kept it in the White House, allowing it to beg for food from the table.[70] While it may seem excessive to have accorded so much space to Roosevelt, his safari was widely publicised in the media and even shot on film and shown in theatres. His propagation of contemptuous views of the hyena and his treatment of it as vermin will have reached a wider public audience than many accounts at the time.

Alfred Pease only devotes a few pages of his accounts of settler life in Kenya and his hunting exploits to hyenas. He adopted the contemptuous manner of typical of the time,[71] writing in an account of lion hunting that "hyaenas are utter cowards, but when they get together in packs of over a dozen, as they occasionally do, they may sometimes be dangerous".[72] He admitted that hyenas drove leopards from their kills and would attack and even kill lions, yet still called them "coward of the cowards".[73] Pease recorded that he saw striped hyenas in North Africa, Somalia and East Africa and believed the North African ones to be larger, though he said he "omitted to compare the skins and skulls of the specimens I killed".[74] He said the North African striped hyenas were more aggressive than their East African counterparts, "frequently killing donkeys, mules and occasionally horses, seizing them in the flank or belly and disembowelling them".[75] He rather fancifully says that the brown hyena ranged as far north as Mt Kilimanjaro in Tanzania – though no other hunter or naturalist of the time records this.

As Nairobi increased in population and settlers established farms in the surrounding countryside, hyenas were still common. They ventured on to farms and into the town at night. Mervyn Cowie, first chairman of Kenya's national parks board, said that in 1914 on his family's farm outside Nairobi, the kitchen was not secure at night and everything had to be packed away "to be out of reach of the hyenas".[76] He said the family's old Swahili cook believed that "hyenas haunt a graveyard as reincarnated spirits of the dead".[77] Cowie echoed accounts of hyenas thriving in the vicinity of abattoirs in towns, writing that when a slaughterhouse was established at Mbagathi, "there was such a generous supply of throw-outs" that the hyenas increased by "geometrical progression".[78] When the slaughterhouse was closed for health reasons, the hyenas had to find another source of food.

> They became the very embodiment of evil round our huts at night. Everything that had any semblance of food was carried out to be eaten or discarded.

They would attack cattle, sheep, goats and people. Women out cultivating their fields were on occasion attacked and eaten by the hyenas deprived of their accustomed source of food.[79] When he was first involved with conservation, Cowie said he could see good reasons to protect lions, elephants and rhino, but "could not see much sense in making a case to save ugly, useless animals like hyenas", adding that "Hyenas are generally despised by everybody. In appearance they are malformed and cowardly…cringing like a cur". But he gradually changed his mind and decided, "They are undoubtedly ridiculous animals, but I have learnt to have a warm feeling for them".[80] However, when the attacks on the women started in the late 1940s, he "set about reducing the hyenas by every means possible", including poisoning – and in one night he killed 78 in an area where they had killed many Maasai cattle.[81]

The regular game reports by Blayney-Percival, the first head of Kenya's Game Department, rarely mention hyenas, unless they posed a threat to livestock or people. In his book he made clear his dislike for them, yet how redolent of Africa they were, writing:

> I hate him…home leave comes seldom, but when, after a run home, I lie awake and listen to that weird, wild voice of his out in the darkness of the plains, I am softened towards the hyæna. His voice is unlovely, but it is the most typical of African sounds, and I feel that I am really back in Africa…[82]

Like Cowie, he also noted how common they were around Nairobi, where they could be heard every night, and that they thought nothing of entering store-houses and other buildings to forage for food, especially hides and leather[83]. He added that, in some districts farmed by settlers, hyenas had declined as ostrich and livestock farmers had shot or poisoned them in large numbers.

Often when livestock were killed or people attacked by hyenas, rumours started that occult forces were involved. Caroline Buxton wrote in 1917 that near her farm in Laikipia there was an outbreak of stock killing. Local communities blamed a mythical creature they called a *gadit*, which supposedly walked on two legs and cracked the skulls of animals to eat their brains. The killing of the stock ended when a particularly large spotted hyena was killed.[84] The chief game warden in the late 1920s, Capt Ritchie, wrote of the belief among some Kenyan communities of the existence of a *kerit*, also called a Nandi bear. This was a creature, reportedly having six toes on each foot, that killed children. Ritchie thought it was likely to be a large hyena with deformed feet.[85] In a district called Tuso, 12 cattle were killed and the *kerit* was blamed, but Ritchie said that other attacks in the region had been carried out by a large hyena that had been described as like the southern African brown hyena, which is not found in Kenya.[86]

By the mid-1930s, wildlife ranges and numbers had been progressively reduced by hunting and the clearing of land for cultivation or livestock. The Game Department report for 1934 noted that:

> When a district first became settled, heavy shooting of game of all kinds took place. New settlers had, many of them, in addition to the inclination, time and money to spend on shooting. Game herds became reduced and driven off while predatory animals, such as lions, leopards, and hyenas, were almost exterminated. In the course of time…shooting diminished…Herds began to regenerate…The carnivora, however, did not share in the process of reinstatement, for their attempts, where made, were met with rifle, poison, and trap.[87]

The imbalance created by the increase in wild ungulate numbers in many areas by taking out carnivores then led to severe problems during droughts with zebra, oryx, eland, wildebeest and Thomson's gazelles reducing grazing for livestock. At the start of WWII, the famed game warden George Adamson was tasked with killing over 1,000 oryx and zebra on farms on the border of Samburu country. He often killed 100 a day and for some inexplicable reason the Samburu were not permitted to use the meat. The carcasses were left to rot or be consumed by the remaining hyenas and other predators.[88]

Adamson, who became famous through the book and film *Born Free*, wrote in his work on locust control in the late 1920s and early 1930s, that at night in the region near Lake Turkana they had to build strong thorn bomas for their donkeys, "owing to the prevalence of hyena, and even lions, it was impossible to leave them out at night to pick up food".[89] He also tells of an Indian trader who had died at a waterhole in Turkana. His men surrounded the dead man's tent with a thorn fence, but the next morning they found that hyenas had broken through the fence and consumed the body entirely. Later, while camping on the Kerio River near Lake Turkana, six hyenas killed Adamson's donkeys after they had broken out of their boma.[90] During his travels in this region in the early 1960s, Hillaby came across hyenas on a regular basis. On one occasion a hyena entered his camp at night and ate one of his shoes.[91] He said that the Chalbi desert near the lake was known to local people as the home of the hyenas and that hyenas were sometimes called the sheep of the Turkana because they were so numerous. There were rumours, he said, that dead hyenas had been found with earrings in their ears identical to those of traditional healers and with necklaces. Hillaby thought them "a hideous-looking animal", but also a useful one because they cleared up the remains of diseased animals.[92]

Working as a game warden in Samburu country, Adamson said that he met Samburu who bore the scars of attacks by hyenas and that local people complained bitterly about the number of livestock lost to hyenas. Adamson became a notorious poisoner of hyenas – in the process killing many other carnivores that fed from the carcasses he poisoned.[93] He admitted to giving strychnine to the Samburu and instructing them how to kill hyenas with it. On one occasion, 20 hyenas, one leopard and four jackals were killed on one night as a result.[94] On another, he put out 30 poisoned baits to kill hyenas that had eaten a calf, but "to my dismay I found that two lions had eaten the baits".[95] Ian Parker, who worked with Adamson in the Game Department wrote, and later told the author, that, "As a poisoner George Adamson had no rival in the Game Department. He used it routinely on hyenas, for which he seemingly had a pathological dislike".[96] In his monthly report to the Game Department for February 1939, Adamson wrote that hyenas "are certainly far too many in many places, wherever I find them I always put down poison". Later, when he and his wife were raising the lioness Elsa, he shot wild hyenas that he deemed a threat to her, despite being a game warden.[97]

At the end of WWII, conservation policies led to the creation of fully protected national parks, in which hunting was prohibited. Nairobi NP, which contained hyenas and several prides of lions, and Tsavo NP, with a very large and diverse predator population, were the first.[98] Outside protected areas, there was a steady toll of predators through conflict with people. Parker wrote that by the 1950s large areas of Maasailand in Kenya and Tanzania were devoid of predators because of widespread poisoning by the Maasai with a pesticide used to keep their cattle free of ticks.[99] It was used to lace the carcasses of cattle to kill hyenas, lions, leopards and wild dogs blamed for livestock deaths.

In the former German colony of Tanganyika, taken over by the British after WWI, there had been little European settlement in comparison with Kenya and large areas of savannah and woodland were untouched, with less clearance for settler agriculture or cash crops. But these areas were not an Eden with no human activity. The Serengeti, famed in the European mind as a vast region populated only by wildlife, had long been utilized for seasonal grazing, with the Maasai present in the east, especially in what is now the Ngorongoro Conservation Area, and the Sukuma, Ikoma, Nata, Ishenyi, Tatoga and others in the western Serengeti, where they hunted. It was, as Shetler wrote, "a profoundly humanized landscape".[100] When the Serengeti became a national park, the territory was treated as though it was devoid of people and anyone using resources within the park boundaries were effectively poachers.[101] The huge wildebeest, zebra and gazelle migrations had occurred for millennia alongside humans grazing cattle and hunting; the latter had little long-term effect on ungulate numbers. Rinderpest had reduced cattle numbers and the pastoralist population. Zebra and wildebeest populations (unaffected to a great extent by the disease) increased.

The early 20th century had brought a massive increase in European hunting following German occupation and then British rule, which did reduce numbers of some species. This was somewhat reversed when the British expanded the number and size of game reserves, and wildlife repopulated areas in northern Tanzania denuded of cattle by rinderpest.[102] The Serengeti became a popular safari hunting destination in the 1920s, partly because it was accessible from Kenya. Two professional hunters, Simpson and White, led one group in 1925 which killed 50 lions – there is no record of whether or not they shot hyenas. There was no regulation by the Tanganyika Game Department, which allowed hunters to engage in what were considered unethical practices, such as shooting from vehicles and excessive killing.[103] In 1926, the colonial authorities introduced a system of hunting licences to control the level of hunting.[104] But the shooting of lions and other predators was seemingly unaffected. In 1928, a motorised safari in the Serengeti was credited with killing 60–65 lions.[105] This led the Kenya-based professional hunter Denys Finch Hatton to call for an end to the excessive killing of lions in the Serengeti.[106] He proposed the creation of closed districts and demanded that the administration in Tanganyika limit motorised safaris, referring to butchery by "tourist hunters".[107] He gained the support of the Prince of Wales, whom he had guided on a safari, and of Lord Onslow of the fauna preservation society.[108] Under pressure from such well-known figures, the government sent the scientist Julian Huxley to examine the situation in northern Tanganyika. The result was the declaration of a 900sq mile closed reserve in the Serengeti from Banagi Hill westwards between the Grumeti and Mbalagati Rivers.[109] In 1929, the Serengeti was declared a Complete Game Reserve, in which hunting was banned, much to the anger of the local Nata,

Ikoma, Sukuma and Musoma communities.[110] In 1959, it was expanded to include the area between Banagi and the Kenyan border, while the Ngorongoro Conservation Area was removed from the park.[111]

One of the first serious studies of Tanzania's spotted hyenas was carried out by L. Harrison Matthews in northern districts in 1935 – Matthews shot and poisoned 103 hyenas in the Ngorongoro area to carry out his studies of their reproductive tracts.[112] He noted that hyenas had "irregular movement patterns" during the wildebeest and zebra migration, with hyenas preying on the migrating herds and often following Maasai moving their cattle; but they didn't appear to engage in a full migration.[113] Hyenas were often to be found around concentrations of migrating ungulates. Matthews believed they did not abandon territories once the migration had passed a long distance away. Where game was plentiful, spotted hyenas were found in large numbers and would travel 10 or more miles at night to seek food.[114] The Maasai were often troubled by hyenas visiting their bomas at night to try to kill cattle, leading to the spearing of hyenas.[115] He also reported that:

> ...a big game safari that travelled through part of the Serengati area and killed over a hundred hyenas, in two or three weeks, by casual shooting when the animals came within range of the light of the camp fires at night, or when individuals happened to be seen in daylight. The hyaena has no friend, and usually gets a bullet sent after him if he shows himself within range of anyone with a rifle who is not stalking game. In settled districts farmers wage perpetual warfare against him, chiefly with poison.[116]

He added that hyenas were known to carry off children or old people and this may have been linked to the Maasai leaving their dead out in the bush.[117] There was another recorded episode of white hunters killing large numbers of hyenas. The American filmmaker and hunter Paul Hoefler bragged about the "uncountable number of hyenas his party shot and used to bait lions for photographing".[118]

In the Arusha and Manyara regions, the Iraqw people engaged in hunting to supplement crop and livestock farming. The hunting had little effect on wildlife numbers, though over several centuries land clearance for cultivation had reduced habitats.[119] There is little evidence of conflict between them and predators, and, like other East African communities, they left their dead in the bush. As Lawi explained:

> ...up till the 1920s homesteads usually had in their close proximity a *kiinta*, that is a patch of bush. This area served as a place for disposal of the dead, both people and livestock. Human bodies placed here were designated for collection by hyenas...It was common belief that although the hyena consumed the body as regular food, it would later deliver it to the ancestral land below...As embodiments of ancestral spirits, hyenas functioned as intermediaries between human life on earth and below. They brought dead persons to new life among

the ancestors and delivered ritual sacrifices offered by the living to their dead parents or close relatives.[120]

Families experiencing regular infant mortality would often name newborns "Baha or Hhawu, local names for the hyena, in hopes that on account of the shared name, the hyena (here symbolizing death) would spare the baby".[121] The practice of leaving the dead for hyenas gradually fell away under colonial pressure and had disappeared by the 1950s.[122]

In 1948, the Tanganyika administration passed an ordinance creating national parks. On 1 June 1951, Serengeti became a national park, with a total ban on hunting.[123] The continued hunting for meat by local communities, who had always used this to supplement farming, was deemed poaching.[124] Hyenas benefited from bushmeat hunting. Snares were set for antelope, buffalo and other ungulates. Frequently animals caught in the wire snares would be killed and completely eaten by hyenas before the hunters returned to check their snares.[125] On occasions hyenas would get caught in the snares, either by walking into them or while feeding on the ensnared ungulates. The creation of the park and protection accorded wildlife in the adjacent Ngorongoro Conservation Area, and the Crater, reducing the killing of game and of predators like hyenas and lions. The large number of hyenas in areas like Ngorongoro and Serengeti was seen by some as a threat to more charismatic species such as lions. The hunter Sydney Downey rather fancifully believed that hyenas killing lion cubs, along with Maasai ritual lion hunts, were to blame for falls in numbers of lions in some areas of East Africa, though he did not advocate the killing of hyenas in national parks, as he believed they had "an immensely important sanitary function in Nature's scheme".[126]

Hyenas were the most abundant large carnivore in the Serengeti ecosystem and the neighbouring Ngorongoro area, especially the Crater.[127] Despite the designation of the Serengeti as a fully protected national park, there were still Maasai with cattle, donkeys and smallstock within its boundaries, and they had seasonal rights to graze their animals in the Ngorongoro Conservation area. There were at least 20,000 cattle in the park and 100,000 in the Crater highlands.[128] This meant they both coexisted with wildlife and came into conflict with it, especially hyenas and lions. While wary of humans, the Serengeti/Ngorongoro hyenas foraged around Maasai bomas, the camps of game wardens and the growing number of tourists. George Adamson wrote that when he was trying to resettle Elsa's cubs around the Seronera region, hyenas used to raid his camp at night breaking into food boxes and stealing cheese, bacon and cans of Ideal milk (taking the latter away, puncturing the cans and lapping up the milk).[129]

In British Uganda, hyenas were listed among the animals deemed destructive to livestock and property. The report of the Uganda Game Department for 1925 recorded that "One would hardly associate the cowardly hyaena with man-killing, though this animal has been known to enter native huts and molest the sleepers, and there are cases of children having been seized and carried off".[130] In the Gulu region they had become regular killers of cattle.[131] Game ordinances did not include hyenas on hunting licence regulations, as they were deemed vermin and no permission was needed to kill them. As in Kenya, there were reports of mythical beasts attacking livestock, poultry and even children. One such creature was known as "*Ngagiya* or fabulous Mubende beast" was hunted down after raiding villages. It turned out to be a spotted hyena.[132] In his report to the Society for the Preservation of the Fauna of the Empire, a colonial Senior Commissioner, Col. Delme Radcliffe, said that hyenas were found across Uganda, but were not common in many areas except Karamoja in the north – there were regular reports of attacks on people there.[133]

The Horn of Africa

Ethiopia still had an unusual mix of coexistence and conflict between humans and spotted hyenas. In many areas of the country they were tolerated as disposers of waste in towns and villages, but in others were seen as a danger to livestock and people. Gade's study of the interactions noted that in Addis Ababa in the 1930s there were frequent reports of people sleeping out of doors being killed by hyenas.[134] He cites a German geographer, Philipp Paulitschke, as saying that by the end of the 19th century hyenas had killed "thousands of Harar's residents over recent decades" and that this "bold and rapacious scourge" descended from the neighbouring mountain to attack cattle and people in their coffee or banana groves outside the walls, or killed sick people inside the city walls;[135] giving a new angle to the Harar hyena story. Gade cited the 19th century British traveller Theodore Bent as saying that when he visited the northern Ethiopian town of Aksum in 1893, he was told to sleep within the walls and not set up camp outside them, due to the threat posed by hyenas.[136]

Frequent warfare, droughts and famines meant that human mortality rates were high and bodies were often not buried, with people fleeing man-made or natural disasters dying by the roadside. Wars and famines became feasts for hyenas, spotted and striped.[137] The rinderpest epidemic of the 1890s was another disaster that led to hyena consumption of the bodies of people who starved but also attacks on the living, as cattle herds were decimated along with wild ungulates.[138] There were also reports of people dying or being injured when competing with hyenas for the carcasses of animals that had died.[139]

During his extensive travels in the Danakil region of Ethiopia, the British explorer Wilfred Thesiger wrote that they regularly encountered hyenas. At Awash station he set up a camp on one trip and said immediately hyenas and jackals were loitering around them foraging for food.[140] He never specifies in his references whether the animals he sees are spotted or striped, though today striped hyena are found in the Awash NP. On a visit to the town of Bovu in October 1933, he saw the body of a criminal who had been hanged, still suspended on the rope, and noted that "the hyenas had got his legs".[141]

To the south, in what is today the self-proclaimed state of Somaliland (formerly British Somaliland), there was a diverse wildlife population and a substantial population of carnivores at the end of the 19th century and the beginning of the 20th century.

This made it a popular hunting ground for Europeans, particularly British hunters normally based in Kenya and officers in the colonial army in India. Swayne called it the best and most accessible hunting ground in the world.[142] In 1899, the British Consul-General in Somaliland, J. H. Sadler, listed the animals, skins and horns that sportsmen and collectors for museums had exported from the country: "lion, leopard, cheetah, zebra, wild ass, oryx, kudu, hartebeest, waterbuck, hyenas [presumably plural as it refers to spotted and striped], foxes, smaller mammals, large kudu, gazelle and other deer, rhinoceros, and elephants".[143] The main livelihood of the inhabitants was raising sheep, goats and camels. These pastoralist people came into regular conflict with carnivores that preyed on livestock; hyenas being common around settlements and cattle camps of nomadic herders.

The British authorities established hunting regulations and issued licences, recording what was shot on licence. The regular accounts of animals shot on licence published by the Society for the Preservation of the Wild Fauna of the Empire rarely mention hyenas – though in 1901 a striped hyena was recorded as having been shot.[144] The lack of references to hyenas is perhaps explained by the hunter and Fellow of the ZSL, Captain Mosse, who wrote that hyenas "hardly enter into the category of game…and a pest he is at times".[145] Mosse described the spotted hyena as a "black coward" and a "low-down character", but disagreed with the view that they were purely scavengers and said that "The Somali's herd of goats and sheep are continually being attacked by hyaenas in the daytime, and, if the prey is comparatively seldom carried away it is because the Hyaena's methods are clumsy and slow compared with the leopard's". Like many other writers, he also referred to hyenas attacking people if they slept outside at night. Those who survived the attacks usually had terribly disfigured faces.[146] Most of his references are to spotted hyenas and he noted that the striped hyaena is "much less abundant, though probably commoner than is often supposed, as it is shyer and less given to prowling round zaribas, and therefore is less in evidence". He said it was identical to the Indian species but more aggressive, and that Somalis accused the striped hyena of killing a dozen or more sheep at a time if they got among a flock.[147] The naturalist Drake-Brockman believed the striped hyena was present across Somalia but in smaller numbers than the spotted one, which was "plentiful".[148] He described the spotted hyena as a persistent stock-killer, adding that during a smallpox epidemic there were many reported hyena attacks on sick people – while noting the attacks on livestock and people and the regular marauding around his camp, he claims, as so many others do, that the spotted hyena is cowardly. He said that on one occasion he killed seven in one night as they ranged around his camp.[149]

Sudan

In the early accounts of the fauna of Egypt and Sudan sent to the Society for the Preservation of the Wild Fauna of the Empire by the colonial administrations there, there is no mention of any of the hyena species, though striped hyenas were certainly present in both and spotted hyenas across Sudan and possibly still in the southernmost areas of Egypt. In 1907 the Sudan game department reported that hyenas were common around the Shabluka hills near Omdurman, noting that they would have to be exterminated if the area was turned into a reserve for the preservation of wild sheep, and lamented that in the Dinder region, near the Ethiopian border, "the funds are not available to have teams of trappers on duty there to trap every hyena, lion, leopard and wild dog".[150] The report also highlighted the damage done to melon crops in Halfa province by striped hyenas.[151]

An expedition by Hubert Lynes and Willoughby Lowe to Darfur to collect animals for the British Museum in 1921 surveyed the area between El Fasher, Jebel Maidob and Jebel Marra. They collected 62 species, including the bones of a striped hyena from the foothills of the Jebel Marra and a spotted hyena from Wadi Aribo, reporting the spotted hyena as destructive of livestock, which "stole" sheep, goats and donkeys "out of our servants' huts".[152] The hunter Robert Henriques went to Darfur in 1937 and reported that his camps were visited most nights by hyenas. He hated the spotted hyena, describing it as "a vulgar and obscene brute".[153] In southern Darfur he said that the Taaisha people of the Baggara Arabs told him that they poisoned hyenas with strychnine to stop stock killing. At El Fasher, Henriques said that the British officials stationed there had assembled a pack of dogs with which they hunted hares, jackals and hyenas from horseback. He said only one hyena is recorded to have been killed by the hounds – "a large and bloated specimen of the striped variety". The hyena had been feeding on the carcass of a dead camel and had been discovered "slouching in the suburbs in the early morning".[154]

West Africa

Spotted and striped hyenas were both present in suitable habitats across West Africa, north of the dense forests found inland from the Atlantic Coast. They seem from accounts at the time to have been widespread but nowhere particularly numerous. One naturalist and colonial official, Col Haywood, recorded them as being present in Gambia and Ghana, though he failed to identify the species. He said they were found in the open woodland areas of central and northern Ghana.[155] In 1943, George Cansdale, a forestry office in Ghana later to become Superintendent of the ZS, surveyed the wildlife of the colony and said that spotted hyenas, wild dogs and lions were found "from just north of the forest edge in Eastern Ashanti to the northern frontier wherever it has suitable country".[156] To the north In Mali, hyenas of both species were present, and a British visitor to Timbuktu noted that cemeteries there were often surrounded by a high wall "to keep off hyenas, jackals and other night prowlers".[157]

Relatively little was recorded of the fauna of Nigeria by the British colonisers. Haywood wrote in 1937 that the striped hyena was fairly common in the scrub zone of northern Nigeria, but he made no mention of the spotted hyena.[158] In Adamawa district, at the junction of the Gongola and Benue rivers, an anthropologist studying the local people recorded that the area was "full of game of all sorts, ranging from occasional elephants, lions, leopards, hyena, bush-cow, to antelope of various kinds".[159] A British naturalist, Richard Oakley, detailing game found in Yola district, noted the presence of "innumerable large spotted hyaenas, and their smaller cousin, the striped hyaena", which were more numerous than lions.[160] Some references suggest that hyenas of both species were rare in much of Nigeria because of a bounty paid by the colonial authorities for their extermination.[161] The well-travelled Col. Haywood also surveyed the wildlife of Sierra Leone, reporting in 1933 that at least 32 mammal species were found there, including hyenas and leopards, but no lions.[162] An account of the beliefs of the inhabitants said many thought that people could be turned into or turn themselves into hyenas. The Temne people had superstitions about twins and it was not unusual for one twin to be taken out into the bush and left to die or be eaten by hyenas.[163]

Southern Africa: from Malawi to the Cape

Spotted hyenas were found across southern Africa from Malawi southwards into South Africa, though the spotted hyena was absent from most of the Cape and Orange Free State, having been exterminated by settlers. The brown hyena was only found in parts of western and southern Zimbabwe, southern Angola, Zambia and Mozambique, and across both Namibia and Botswana into South Africa – it was most abundant in semi-arid and arid areas.[164]

A British hunter, John Statham, undertook a hunting expedition across Angola in 1920. He referred regularly to lions, honey badgers, genets and other carnivores but only rarely to hyenas. He lamented the widespread killing of game by Afrikaner transport drivers working on projects like the Benguela Railway. He said their "massive slaughter of game" had denuded some areas of wildlife of all kinds.[165] This view is seconded by the conservationist Brian Huntley, who worked in Angola as a game warden. He believed that the Boers "shot for meat, especially to make biltong and cut huge swathes through the ungulate herds in the areas they worked and hunted – including killing large numbers of the endemic giant sable antelopes".[166] Statham did record that spotted hyenas were found across Angola, but thinly spread and nowhere particularly numerous. He added that they were feared because of their attacks on sleeping people and killing of children.[167]

In Botswana, the spotted and brown hyenas were distributed across the majority of the country. Spotted hyenas were numerous in the major game areas, such as the Okavango, Savuti and Chobe regions in the north, and were far from uncommon in the central and southern Kalahari. Brown hyenas were found in most areas, but never in large numbers in any one area, though counts were not carried out and the shyness of the animals meant they were not seen regularly, as they foraged at night. Unlike spotted or striped hyenas, they rarely foraged around human settlements. The British declared a protectorate over what is now Botswana in 1885. The territory was of no great economic value (diamonds only being discovered nearly a century later), with cattle raising the main economic activity. There were moves by the British to expand the cattle industry through settlement by Europeans, but also to control Tswana cattle husbandry practices to prevent the spread of disease, especially after rinderpest.[168] The territory was ruled directly from British South Africa but was not incorporated into the Union of South Africa in

1910. Settlers and indigenous farmers were hit by rinderpest and a serious drought in the 1890s. They complained regularly about the loss of cattle and other stock to predators, particularly hyenas and jackals.[169]

The response to rinderpest involved limiting areas in which Tswana herders could graze their cattle but also a new war on predators, with increased bounties paid for killing them, offering seven shillings for each jackal killed and 15 shillings for every wild dog; no record was available of the bounty for hyenas.[170] The rinderpest epidemic reduced some species of wild prey but also increased the availability of cattle carcasses for spotted and brown hyenas. In many areas, Ghanzi west of the Central Kalahari Game Reserve being one, settler and Tswana cattle competed with wildebeest, gemsbok, hartebeest and springbok in large numbers,[171] but there was no recorded mass shooting of wild ungulates as there had been in Kenya, though hunting was carried out by settlers, the Tswana and the San for meat. Game only began to be depleted by hunting when four-wheel drive vehicles became available and large numbers of ungulates could be harvested for meat and biltong.[172] Farmers killed lions regularly and waged relentless war on cheetah, leopards and wild dogs with guns, dogs and poison – accounts of predation of livestock make no mention of either species of hyena.[173] As in many other parts of Africa, belief in the occult role of hyenas was common in Tswana areas. Superstitions associated them with "night witches", who dug up newly buried bodies to get parts for use in witchcraft. These witches were linked to animal familiars, particularly owls and hyenas, riding the latter with one leg on the ground and the other on the animal's back.[174] The BaKgatla clan of the Tswana believed that witches made their own hyenas, moulding the body from porridge and bringing them to life by means of special medicines.[175]

In the British Central Africa Protectorate (Malawi), spotted hyenas were common in many areas, including on the outskirts of towns. The territory's commissioner, Sir Alfred Sharpe, himself a hunter and naturalist, said that a rough census had been carried out there of wildlife, showing a population of about 3,850 hyenas (he admitted there was a large margin for error); and it is unlikely there were striped hyenas.[176] Robin Wright, who shot animals for display in the natural history museums in Tring and Kensington, hunted in Malawi in 1900. He wrote that on one occasion near Blantyre, a hyena had come into the camp at night and dragged away and ripped apart a tent bag.[177] On another night, a hyena got into Wright's tent, where he had suspended some biltong from the roof pole. The hyena pulled down the biltong and escaped with it.[178] While hunting on the border of Zambia and the Belgian Congo, Wright was so irritated by marauding hyenas that he set up gun traps and put out bait to try to kill them. He managed to kill some using this method, but this did not stop others trying to raid his camp.[179]

The colonial authorities encouraged the killing of predators to protect livestock and to reduce the numbers of wild prey they killed, to ensure European hunters had enough sport and meat. In 1912, it was recorded that six hyenas were recorded as shot in the protectorate, almost certainly an underestimate, as in the same year 50 wild dogs and 20 leopards, out of a total of 3,676 mammals, were shot on licence.[180] After WWII, the main area of human-wildlife conflict was elephants, hippos, baboons, wild pigs and various other ungulates damaging crops.[181] But hyenas became a major problem around Mt Mulanje in the south in 1955.[182] In September 1955, a man with learning difficulties was killed and eaten on a path between two villages. It was believed that four hyenas had taken part in the kill.[183] A week later in the same area, an old woman was dragged from her hut by a hyena and lost an arm before the hyena was driven off. Later in the year a child was eaten by hyenas, after being taken from the porch of her house.[184] Kruuk estimated that 27 people had been killed by hyenas in the Mulanje area at this time.[185] Most of the killings took place between September and January, when hot, humid weather encouraged people to sleep on their porches or out in the open. In the late 1950s, Long reported that spotted hyenas were very common on the plains in the Shire Valley near Nsanje. He said there were several cases of hyenas attacking people, usually when they were sleeping outdoors on hot nights.[186]

Hyenas attacks on children, old people and those sleeping outside were a periodic problem in Mozambique, notably Benga, close to southern Malawi. Growing human populations and declining wildlife numbers led to hyenas and lions becoming regular predators of people there in the early to-mid 20th century. According to the hunter and ivory poacher John Taylor, who tracked down some of the man-eaters, they lived in dry thorn-bush country, which had little natural prey, and in areas where cultivated land attracted the remaining warthogs, bush pigs and baboons to raid fields at night.[187] He said relatively little about hyenas, but noted that after shooting two man-eating lions in Benga, hyenas came at night and consumed most of the carcass of one of the lions before he could skin it.[188] Like so many other European hunters, Taylor treats the hyenas, despite their depredations, as mere vermin whose killing is hardly worth mentioning. Another spate of man-eating developed among hyenas, leopards and lions in an area of western Mozambique near Mangoche in Malawi. Taylor said that there was little game in the area apart from warthog and bush pigs, which regularly raided farmers' fields, thereby bringing the predators into well-populated regions, where they regularly killed people for food.[189] It is likely that hyenas had taken up residence near human settlements, which were a regular source of food through foraging at waste dumps, as well as attacking livestock, dogs and sleeping or vulnerable people. Taylor believed that in many cases leopards had killed people, but most of the bodies had been eaten by hyenas.[190]

In Namibia (colonised by Germany and then governed by South Africa after WWI), brown hyenas were widely distributed and spotted hyenas common in areas with sufficient wild prey and opportunities to take livestock. From colonisation in 1884, one of the main issues for settlers was the arid environment in much of the county. For German settlers and then South Africans, the relationship with the environment and wildlife was one of "toil and battle to tame a hostile landscape".[191] Game animals were scarce in the southern and central regions by the end of the 19th century, and was found mainly in the north and in areas where indigenous peoples and settlers were thinly spread.[192] Ungulates competed with the cattle or smallstock of local pastoralists and settlers for the meagre grazing and water. Predators were seen by the settlers as a danger to their livestock, already under threat from rinderpest. As the Germans entrenched their rule, they defeated and decimated the populations of Herero and Nama pastoralists during the war of 1904–7.

In his thorough survey of the wildlife of Namibia, Shortridge said that spotted hyenas were relatively scarce and few were found south of Etosha Pan, with a small, scattered population in Damaraland. They were more numerous in the Caprivi Strip and along the Kunene and Okavango river valleys.[193] In the areas where they lived alongside livestock farmers there were "accounts of persistent destruction of stock by hyaenas…according to local report the Potted and not the far more numerous Brown Species is mainly responsible for the killing of adult domestic animals".[194] They were known to kill adult donkeys, mules and cattle, as well as sheep, goats and dogs. Shortridge said that brown hyenas were found across practically the whole of the territory and were only scarce in Namaqualand, where little game was to be found, and in the wetter areas like the Caprivi Strip, where there was also greater competition from spotted hyenas.[195] He noted its Afrikaans name of strandwolf, taken from his habit in coastal areas of patrolling beaches preying on fur seal pups and scavenging the remains of marine creatures. He said that they were not usually destructive of livestock but often were killed or trapped by mistake where farmers were trying to till other predators.[196] Oldfield Thomas reported accounts that brown hyenas had been suspected of taking young sheep and goats and of biting off the tails of cattle.[197]

After WWI and South African assumption of control over the territory, land policies progressively pushed local pastoralists and farmers from the best land for settlement by white South Africans. They established large farms but did not engage in transhumant pastoralism as the Nama, Herero and Himba had. To cope with the hostile environment for livestock, when compared to conditions in much of South Africa, the settlers and the colonial government adopted harsh measures of predator extermination, with white settlers allowed to eradicate predators, something which did not apply to African pastoralists.[198] In the mid-1920s, the South African governors of Namibia were issuing strychnine to settlers to kill predators and vermin clubs were set up among the settlers to encourage the killing of predators, as happened in South Africa.[199] In 1934 alone, the clubs reported that they had killed 10,221 predators – though no numbers were given for how many of those animals were hyenas.[200] When the Etosha NP was established in 1947, the area within the boundaries was about the only place where the killing of predators was not encouraged.

While South African settlers had brought with them the images and attitudes towards predators that had led to their extermination across much of South Africa, the Himba, for example, had their own views of the hyena. Even though they realised it took more of their stock than did jackals, they viewed it as stupid and gullible (like many peoples across Africa) as well as inordinately greedy.[201] They also believed it to be "a hermaphrodite and so abnormal and contemptuous".[202] Their folktales also have the cunning jackal endlessly outwitting the stupid but greedy hyena.[203]

By the beginning of the 20th century spotted and brown hyenas had been "practically exterminated in the settled districts" of South Africa. Spotted hyenas survived only in areas not strongly affected by "the advance of civilization", with the brown hyena "only being found in the north-west of the Cape province (Northern Cape), in small numbers in North-West Transvaal (Mpumulanga) and… arid regions of Rhodesia (Zimbabwe)".[204] In South Africa, the rapid expansion of the livestock economy in the Cape, especially the introduction of merino sheep, had led to concerted campaigns of predator extermination. This continued into the 20th century. The bounty system introduced at the end of the 19th century to encourage the killing of hyenas, jackals, caracals and leopards was retained after the formation of the Union of South Africa in 1910 and was only ended in 1956, with tens of thousands of predators in the Cape killed annually until the mid-1950s; jackals were the main victims.[205] The numbers of hyenas killed was rarely recorded but between 1889 and 1900, 1,934 hyenas and wild dogs were killed, according to the bounty payment records.[206] By the end of the bounty system in 1956, both species of hyena had disappeared from the Cape (excepting a few in what became the Kgalagadi Transfrontier Park bordering Botswana).

Across the rest of South Africa, excessive hunting for meat by settlers in the interior of South Africa; killing of large herds of herbivores to clear land for grazing; persecution of predators like hyenas, lions, wild dogs, leopards and jackals by livestock farmers; and hunting for sport or ivory, horn, hide, meat and biltong had cleared much of British-ruled South Africa of large mammals by the end of the 19th century. Even in areas not suitable for cultivation of large-scale livestock production, wildlife was scarce. When James Stevenson-Hamilton was appointed warden of the Sabi Game Reserve in the eastern Transvaal lowveld (later to become the world famous Kruger NP) in 1902, he found little wildlife there, something he blamed with some justification on hunting by members of a British army unit posted there during the South African War, and on local people, who had lost much of their stock to rinderpest and resorted to hunting to survive.[207] He saw a few zebra, waterbuck and impala and little other game.[208] He doesn't give an estimate of predator numbers in his early accounts of the numbers of the reserve's animals, but signals early on his intention to cull predators to encourage the recovery of game herds and allay fears of local white farmers that the reserve would be a breeding ground for carnivores that would take their stock.[209] Stevenson-Hamilton didn't give estimated numbers for hyenas in the reserve, but he believed that even after culling had started, they remained "very numerous" in most of the reserve.[210]

He didn't have much time for hyenas and in his book on Sabi and Kruger they rate just four mentions, compared with 260 mentions of lions. One of the early references is to his policy that "carnivora are hunted when possible, and twenty-two or twenty-three wild dogs and about twelve hyaenas have been accounted for".[211] This was the start of the 50 year policy of exterminating predators – especially hyenas, lions, leopards, wild dogs, jackals, crocodiles and even some raptors – based on the idea that the reserve was there to conserve herbivores, not carnivores. His report for 1904 revealed that from July 1903 to June 1904, 18 hyenas, 29 wild dogs, 51 jackals, one lion, one cheetah and five leopards had been killed by his rangers as part of the carnivore reduction. By the end of October 1904, he wrote that 230 carnivores had been killed in the reserve in the preceding 12 months, with more hyenas and wild dogs killed than any other predators.[212] In his report for 1911, he said that 50 hyenas were killed between January and December 1911, along with hundreds of other predators in a cull totalling "1,751 vermin".[213] Under Stevenson-Hamilton's leadership, predator culling continued, with 521 hyenas (both spotted and brown, though separate numbers are not given for each) killed between 1903 and 1927 – along with 1,272 lions, 1,142 wild dogs and 660 leopards.[214] The killing of carnivores in Kruger continued until at least 1958.[215] In 1956 alone, 55 hyenas, 55 wild dogs, 15 cheetah, 9 leopards and 50 crocodiles were killed in carnivore control programmes.[216] Despite this, by the mid-1960s there were again healthy populations of spotted hyenas, lions, leopards and cheetah in Kruger NP.[217]

In his 1917 book on wildlife in South Africa, Stevenson-Hamilton recorded that in areas around the Sabi and Shingwedzi reserves (which became Kruger NP in 1926), where the brown hyena occurred, there was evidence of it preying on young goats and sheep on an occasional basis.[218] He said the spotted hyena occurred in and around Sabi and Shingwedzi, in reserves in Zululand/Natal and in the region of the northern Cape that became the Kgalagadi Transfrontier Park. He wrongly believed that spotted hyenas lived mainly on carrion and obtained their food by following larger predators, and that they were solitary animals with no evidence of living together in families or groups.[219] He did note, however, that they had adapted to living near to human habitations in the game reserves and parks and would forage in and around the camps and even attack sleeping humans, usually biting at the face.[220]

In Transvaal as a whole, there was a policy of rewarding the killing of hyenas and other predators considered vermin. The Transvaal Game Preservation Ordinance of 1905 laid down that anyone applying for the reward for killing a hyena, wild dog or baboon needed to produce the head of the animal they had killed.[221] Records state that between January 1903 and December 1908, 118 spotted hyenas and four brown hyenas had been killed and a reward claimed.[222] The ordinance allowed predators to be killed by shooting, coursing with dogs, with nets, traps, snares, or by poison.[223] This resulted in the extermination of the brown hyena from the settled and farmed districts of the Cape, Transvaal, Free State and Natal.[224]

In the British colony of Northern Rhodesia (which became independent as Zambia in 1964), wildlife numbers were substantial, with a great diversity of ungulates and predators widely spread across the country (the diversity and numbers evidenced by the abundant wildlife in national parks, reserves, hunting concessions and unprotected areas after independence and before extensive poaching reduced numbers from the 1970s onwards).[225] Spotted hyenas were common and widely distributed. Striped hyenas were absent and brown hyenas thought to be absent,[226] but with a possible presence in the south-west corner bordering Angola and Namibia.

Under colonial rule, hyenas were labelled as vermin. The game regulations introduced by the British in 1905 listed hyenas (along with lions, leopards, hunting dogs, baboons and crocodiles) on Schedule 4, which was for vermin which could be shot by Europeans without a licence.[227] When new regulations were introduced in 1925, the schedule number had changed to 5 but hyenas were still classed as vermin.[228] Apart from the references in the game regulations, little was recorded by the colonial authorities about hyenas and it must be presumed that, as Northern Rhodesia did not have a substantial livestock sector or a large settler population, unlike South Africa, Namibia and Zimbabwe, whatever dangers hyenas posed to the livestock and lives of rural communities in the colony were not considered worthy of note in game department reports or accounts by colonial administrators and visiting hunters. There is little doubt that hyenas would have been killed where they took livestock or threatened human life, but this is not recorded. They would have benefitted, though, from the declaration of substantial areas of wildlife habitat in Kafue, the Luangwa Valley, Mweru Marsh, Sumba and Lukusuzi as protected reserves in 1942,[229] with Kafue being made a national park in 1950.[230]

In Southern Rhodesia (Zimbabwe from 1980), there was a long history of hunting among the Shona and Ndebele but also by visiting European hunters, like Selous. Colonisation by the British South Africa Company (BSAC) of Cecil Rhodes and then self-government as a settler-dominated British territory saw an influx of English and Afrikaner settlers, and the establishment of huge tobacco, wheat and livestock farms, which involved the extirpation of wildlife, including lions, in most farming areas outside what became reserves and national parks such as Hwange, Victoria Falls, Mana Pools, Matusadona and Gonarezhou. Wildlife policy under company rule and then self-government allowed for the mass slaughter of game for a variety of reasons, notably livestock protection, land clearance, rabies prevention and tsetse fly eradication. The Southern Rhodesia colony was where the effects of sleeping sickness on people and cattle was studied in detail in the early 1900s and game eradication programmes carried out. Scientists who studied the disease recommended the large-scale destruction of game in sleeping sickness areas; they found that hyenas could be infected and carry the disease.[231] Zimbabwe had a widespread population of spotted hyenas and brown hyenas in the south-west and south-east, in more arid regions. There was also a wildlife eradication campaign (focused on carnivores) to try to eradicate rabies. Between 1899 and 1912 thousands of predators, other wildlife and over 100,000 dogs were killed to prevent the spread of rabies.[232] This dovetailed with the sustained campaign against what the colonial authorities deemed vermin, namely hyenas, lions, leopards, cheetahs, wild dogs and baboons: "[these] species of animals were completely wiped out from European-owned land".[233] Over one million game animals were killed on and around European farms to create buffer zones to prevent tsetse fly.[234]

The first wildlife law to cover the whole country was the Game Preservation Ordinance, No. 6 of 1899, which set out categories of game for licensing purposes. It specifically excluded animals treated as vermin for whom no licence was required to shoot, poison or trap. As Mutwira recorded, the law offered rewards for the extermination of all animals classified as vermin and a threat to livestock or human life.[235] The BSAC Administration paid settlers who produced the skins of pest species; "£1/ 10sh was paid for each lion, leopard, cheetah, hyena and wild dog killed".[236] Records show that few farmers bothered to report their kills, as BSAC rewards were too low to warrant the trouble. But the programme was enthusiastically implemented, especially during the First World War, to increase beef production for the British war effort. Settlers and colonial officials enthusiastically set about killing predators like spotted and brown hyenas, which were viewed as substantial threats to livestock, as well as vectors for rabies. The tsetse fly eradication campaign had a huge effect on all wildlife, restricting most species to areas outside settler farming regions.

The wildlife eradication campaigns started in Hartley and Sebungwe in 1915; Victoria, Ndanga, Chibi, Gutu and Sebungwe in 1916; Wankie in 1919; Melsetter in 1920; and by 1924, Lomagundi, Chilimanzi and Darwin were included, with no licences needed to shoot wildlife.[237] The campaigns were organised by the Department of Agriculture. No record was kept of the numbers or species of animals killed but it is believed to have numbered in the tens of thousands. Most of the large game and the large carnivores were wiped out or reduced to small remnant populations in white farming areas. The African population of the colony had been moved to reserves, apart from those working for Europeans as servants or farm labourers. They had reduced access to meat from cattle and the land was of poorer quality for cultivation and grazing, resulting in increased hunting of game to provide protein. Larger game was soon depleted. Hunting by them on white-owned land was prohibited.[238]

The combined effect of rabies eradication campaigns, killing of wildlife to get rid of tsetse flies and the need of African communities to hunt to survive affected all farming areas and what were termed *native reserves,* depleting their wildlife. In such circumstances the range of the spotted and brown hyena shrank, despite the ability of the former to adapt to the presence of human settlements. The only areas unaffected were reserves or game sanctuaries, with the first established in Matopos in 1926. Over the next seven years, land in Wankie, Victoria Falls and Kazuma Pan was set aside as game reserves, with Matopos becoming a national park.[239] In these areas hunting was prohibited or severely limited and the tsetse eradication measures not applied. Both species of hyena may have survived in remoter areas outside the reserves and parks, but were constantly persecuted and only the protected areas contained safer populations. But even within those areas, there was shooting of predators to preserve ungulate populations. The warden's report for Wankie (Hwange) reserve said that spotted hyenas were uncommon and brown hyenas "exceedingly rare"; nevertheless, two spotted hyenas were shot in the reserve as part of predator control programmes in 1934.[240] Two were shot in 1935. The warden's report said they were:

> …endeavouring to do damage to domestic stock at the Homestead and had this not been so they would not have been destroyed, as these animals are

becoming rather scarce, and no steps are being taken to destroy them as has been done in the past.[241]

The numbers and distribution of protected areas increased in the 1930s, 1940s and into the 1950s, allowing some increase in the habitat available for predators like hyenas and their prey species. A national parks board was set up in 1949 and there was an increase in safari tourism involving white settlers, visitors from South Africa and some from Europe and North America. At this time, there was a shift towards conservation, with the government in Salisbury realising that if some species were to survive there would have to be habitat preservation and regulation of hunting. Safari tourism began to be seen as a source of income. Hunting by African communities was increasingly restricted.

Antipathy towards predators was still strong among livestock farmers, as the Fauna Society's *Journal* recorded at the end of the war.[242] The general tenor of the views was exhibited at a meeting of the Trypanosomiasis Committee of Southern Rhodesia on 4 January 1945.[243] The committee stated that "Southern Rhodesia is primarily an agricultural country, and there was unanimity of opinion that game preservation and farming do not go well together", adding that while some farmers like to have game animals present on their farms if they don't compete with livestock for grazing or browse, but that "the presence of more than a few herbivorous game animals is always associated with that of their natural predators; and Lions, Leopards, Hyaenas, Wild Dogs, Jackals, etc., can only be described as unmitigated nuisances in any stock country".[244] This antipathy, combined with continuation of the tsetse eradication campaign, meant that wildlife continued to be exterminated in white farming areas. Between 1948 and 1951, 102,025 animals were killed, and that overall around 750,000 were killed in the tsetse eradication campaigns, denuding huge areas of wildlife.[245] It was reported in 1956 that "a record total" of 36,910 animals had been killed in the last year. That total included 3,219 baboons, 61 wild dogs, 35 hyenas, 19 leopards, 4 lions, 55 elephant, 8 rhinos, 313 zebra, 950 bushpigs, 4,503 warthogs, 377 buffalo, 1,788 bushbuck and 2,259 impala.[246] As a result, by the 1960s, most game, and certainly most predators, were restricted to reserves and national parks.

North Africa, West, Central and South Asia

The striped hyena continued to be widely distributed over a belt stretching from Algeria to the Middle East, Turkey, the Caucasus, parts of Central Asia and into southern Asia from Pakistan across much of northern India and down into areas of central and southern India, where there was suitable habitat. Its mainly nocturnal habits, avoidance of contact with humans and use of dens and caves during the day meant that it was often overlooked in accounts of wildlife and estimates of its exact range and numbers are lacking. It was, though, persecuted in many areas as a stock killer, grave-robber, raider of food crops like melons and dates, and even a threat to human life. A bounty system for hyenas had operated in French-ruled Algeria from the 1880s and in 1881 and 1882 a total of 196 were killed, contributing along with habitat and prey loss to a decline in numbers.[247] In the neighbouring French colony of Tunisia, the striped hyena was distributed over much of the territory, often in close proximity to human settlements, but was described by the British hunter and colonial administrator Sir Harry Johnston as sparse in its numbers.[248] But he had no qualms about shooting one on a hunting trip there. He noted in his account of it that a hyena had been shot in the suburbs of Tunis, the main town, in 1880.

Across the North African coastal regions, striped hyenas were found in scattered populations in woodlands and mountainous regions. They were also found in oasis regions of the Sahara and around the fringes of the desert. The current wide distribution but low densities are likely to be similar to those during the colonial period. In Libya, hyenas were present but not common and there were few accounts of Libya's wildlife until occupation by the Italians in 1911 led to greater investigation by naturalists of the species found there.[249] A British traveller noted in 1912 that while jackals were common along the Libyan-Egyptian border and often entered towns, hyenas were much more rare and kept mainly to the hills around towns,[250] presumably only venturing out at night. They were also to be found in small numbers at the Siwa Oasis, which straddled the border. Masseti has noted that wildlife numbers were in decline with gazelle, addax and other ungulate populations falling as a result of hunting and the loss of habitat to pastoralism. His account does not specify the effect on hyenas, but does confirm that carnivore numbers had been reduced substantially over the last 100 years.[251]

As detailed in previous chapters, striped hyena were found in much of Egypt from the western desert to the Nile, eastern desert and across the Sinai peninsula to the border with Palestine.[252] Osborn and Helmy said they were widely distributed, with regular sightings recorded between 1902 and 1956 in the Jebel Musa; Jebel Umm Rijlein; Wadi Gazzah; Wadi areas of Sinai; particularly around St Catherine's Monastery in Ismaila around what were called the Hyena Quarries east of Lake Timsah; on the western edge of the Nile Delta; from Alexandria west along the coast to Matruh; and in the Nile Valley south of Cairo, where they were most numerous, had the main concentrations.[253] They reported that the numbers declined during the 20th century, but even then hyenas were still reported entering towns at night foraging for waste.[254] Away from population centres and cultivated land, they depended on the carcasses of domesticated camels and they would frequent the caravan routes. This source of food declined as motor vehicles began to replace camels and roads were built between major towns and farming areas.[255] Hostility towards them was widespread among pastoralists and farmers who grew melons, dates, bananas, plums, maize and other crops raided by the hyenas.[256] This had led, after the advent of motor vehicles, to hunting by using headlights to illuminate hyenas so they could be shot.[257] Along the banks of the Nile, striped hyenas scavenged dead fish and fed on fish offal produced by fishing communities.

In Palestine and what is now Israel, the striped hyena population was widely but thinly spread. As with the North African population it relied heavily on scavenging carcasses, raiding fields and digging up graves. It was greatly reduced in numbers during the British period of control under the League of Nations mandate after WWI. The British conducted an anti-rabies campaign aimed mainly at jackals. The British poisoned donkey carcasses and both hyenas and jackals died after eating the poison.[258] Poison continued be used there after much of Palestine became Israel, after the 1948 war between Jewish settlers and the neighbouring Arab states. This continued into the 1970s. Hyenas also fell victim to cars, as road traffic increased in the region in the second half of the 20th century, with an estimated 20–30 dying annually.[259] They often use the roads at night for foraging over long distances from their dens and are attracted by other roadkill. The attrition from poisoning has meant that numbers have fallen, and the population is relatively young with few reaching old age. As Osborn and Helmy wrote, "a shift in age distribution over the past 50 years comes from the observation that in the 1940s old hyaenas with worn teeth were found, whereas today no hyaenas are found that are older than 5–6 years".[260] Across the whole area that was known as Palestine until 1948, one naturalist noted in 1947:

> Most of the great animals mentioned in the Bible, like the Lion, have already been exterminated from that part of the Middle East and few of the country's big game other than the Red Gazelle and the Hyena have survived the war.[261]

They also survived to the north in Lebanon, despite a relatively high density of human population. The hyenas there may have hunted small prey animals but were primarily scavengers.[262]

In Turkey, records from the 19th and 20th centuries show that the striped hyena was present in several regions but always scarce.[263] They were rarely mentioned by travellers in Turkey in the late 19th and early 20th centuries and do not figure much in accounts of hunting in Turkey.[264] The Turkish hyena range stretches from southern Marmara in the west, along the Aegean and Mediterranean cost to south-east Anatolia. Kasparek et al. believe that the eastern Anatolian population has not been linked with the population of hyenas to be found in the Caucasus and there have been no records of them across the Dardanelles in Greece.[265] They were sometimes hunted, when suspected of killing livestock, and traps were also used to catch them.[266] The hyenas of the Caucasus, until 1991 part of the Soviet Union, were found in Armenia, Azerbaijan and Georgia, with other populations to the north and east in the Central Asian Republics of the Soviet Union, and before that the Russian Empire.[267] They were most common in Caspian Sea coastal regions and in reasonable numbers in Turkmenistan, Tajikistan and Uzbekistan.[268] They inhabited the foothills of the mountains, areas with rocky outcrops and ravines and the steppe regions of the Central Asia Republics.[269]

In the Caucasus and Central Asia, over the past 100 years persecution by humans has been a major source of mortality, as striped hyenas were believed to attack livestock and children.[270] In places persecution reduced numbers, but there was a contiguous population stretching into northern Iran, from where animals were believed to disperse into the Caucasus to take over vacant territory.[271] Hofer wrote that in the 1880s hyenas were blamed for the deaths of 25 children and three adults around the Armenia capital, Yerevan, as a result of which a bounty of 100 roubles was paid by the government for every hyena killed.[272] Attacks continued over the following years and even in the 1930s and 1940s there were reports from Armenia and Azerbaijan of children being taken at night by hyenas, and of children being killed and eaten near Turkmenia's Bathyz Nature Reserve around the same time.[273] The Armenian population was mainly to be found in semi-desert plains and scrubland, the foothills of the mountains and in river valleys, especially in the Ararat Valley, the Meghri district and Shamshadin and Ijevan districts in the north-east.[274]

In Turkmenistan and other Central Asian regions, hyenas were often trapped or poisoned by hunters intending to kill leopards and wolves for their fur. If they were killed, their fur was labelled in Tsarist Russia and the Soviet Union as "minor quality wolf" and "fox".[275] In the Caucasus and Central Asia in the Soviet era, 200 hyena skins were bought by the Soviet government in the 1930s and 100 in the 1950s. In Turkmenistan 130 hyena skins were sold annually between 1931 and 1937.[276] Hyenas, given their heavy dependence on scavenging carcasses, were very easy to trap and poison, and would also fall victim to poisoned bait intended for other carnivores, especially wolves, leopards and jackals. The combination of hunting, human population growth, the development of more intensive animal husbandry by livestock farmers and the progressive disappearance of wild ungulates – notably goitered gazelles, mountain sheep and onager (wild ass) on whose carcasses hyenas fed – served to reduce the overall population of hyenas in the Soviet republics during the 20th century.[277]

South from the Caucasus, striped hyenas were still to be found in Iran, Iraq, Syria and Jordan, although numbers were falling and, in some regions, it was totally absent.[278] Qumsiyeh et al. recorded that the hyaena was common in every part of Palestine and Jordan in the 19th century and early 20th century, and they had a great range of colour variations. Hyenas frequently strayed near human villages and agricultural areas in search for food and water. In this region, they inhabited caves and cracks in rocks, as well as ruins and tombs, foraging for carcasses of almost any medium sized to large mammals.[279]

Monchot and Mashkour said the striped hyena was most common at the end of the 19th century and beginning of the 20th century in southern Iraq, southern, western, and northern regions of Afghanistan and Baluchistan (Pakistan and neighbouring regions of Iran). In Iran, the striped hyena was said to be quite common throughout the 19th century and was even found near big cities such as Tehran, "but apparently was not greatly feared, as it does not attack humans".[280]

The southern boundary of the striped hyena range in West Asia was the Arabian Peninsula coast on the Gulf of Aden and Arabian Sea. It is widely distributed in small populations around the fringes of the Arabian desert, especially in the better watered mountainous areas of Dhofar in Oman and adjoining areas of Yemen. The central desert areas were avoided.[281] Hyenas were often blamed, sometimes with good reason, for predating young goats, sheep, dogs and poultry. As a result, they were hunted and their bodies often hung from trees and signposts, supposedly to warn off other hyenas.[282] Numbers fell progressively during the 20th century from 1900, when Oldfield Thomas reported to the ZSL that striped hyenas were not rare in south-western Arabia. He said they were elusive, shy and very adept at avoiding their traps and poisoned baits, which ended up killing feral dogs.[283] Reports from southern Arabia said that wolves and hyenas were still found there in 1952, but the latter were nowhere common, being found in scattered groups along the Red Sea coast of Yemen and Arabia, where they scavenged along beaches.[284] The growth of hunting in southern Arabia and Jordan was reported as a problem for carnivore conservation. Ungulates like oryx and gazelle were hunted extensively for meat and sport. Talbot wrote in 1960 that oryx were once threatened by hunters but had become scarce and shy and were no longer predated much by carnivores. He said "Predators used to abound in Arabia, cheetah, leopard, wolf, lion, hyena, jackal, fox, and wild cat. Motorized hunting and the modern rifle have exterminated or so depleted these animals that they probably present no danger to the oryx"[285].

India

By the start of the 20th century, India's rich and diverse wildlife had been reduced in numbers, diversity and range of habitat by expanding human populations, the growth in cash crop and subsistence agriculture and by extensive hunting by British colonial officials, soldiers and visiting hunters, as well as Indian nobles. Some local communities still hunted for meat, and for horns and hides to sell. During much of the 19th century and well into the 20th century, huge hunts were held, in which several hunters would wait on machans (seats built in trees) as huge teams of beaters drove quarry towards the guns; or the hunters proceeded on elephant back through areas well-stocked with game and shot whatever (mammals of all shapes and size and birds). The hunter turned conservationist Jim Corbett records one such hunt in which a line of 17 elephants carrying nine hunters and some spectators advanced in a line hundreds of yards long. They killed a variety of game birds and mammals. The favoured targets being leopards and tigers,[286] but hyenas would be shot if encountered.

Corbett was not alone in encouraging conservation. In 1929, Lord Irwin, the British Viceroy from 1926–1931, wrote that there was an obvious need for preservation of habitats and wildlife species. He was especially concerned about the conservation of forests.[287] Voices were raised by Indian naturalists, too. Wabis Ameer wrote in 1930 that the massive increase in population, which had reached 319 million, and demand for land for cultivation had hugely reduced grazing land for livestock and wildlife.[288] He lamented the attitude of many of the British and:

> ...the ordinary people of India those who take the trouble to think consider that the game, whether four-footed or feathered, is their own with which to do what they like as a source of immediate food or profit...The result of this spirit is the remorseless harrying of any bird or animal, in season and out of

season, which is eatable or reputed to be of medicinal value according to Eastern ideas.[289]

Hunting for pleasure, profit, crop or livestock protection or body parts for medicinal uses could lead to high rates of extermination. In 1924–5 in a small forested area of the United Provinces (Uttar Pradesh and Uttarakhand) 23 hyenas, 93 tigers and 36 leopards were shot, and in 1931–2 28 hyenas, 112 tigers and 110 leopards – the average annual shooting during the 1924 to 1932 period was 24 hyenas, 87 tigers and 98 leopards.[290] In most areas of British India, even once regulations were put in place to protect some habitats and species, there was rarely any protection for hyenas, and they were often persecuted.[291]

One area in which regulation, but not complete cessation, of hunting helped preserve hyena numbers was the Gir Forest in Gujarat, controlled, until the partition of India in 1947, by the Nawabs of Junagadh. There, the gradual moves towards conservation of the last population of Asian lions in the forest and surrounding regions worked to preserve dry forest habitat and the wild prey species that inhabited and provided food for lions and hyenas. By 1900, the Gir lions had increased in numbers from 12, at the end of the 19th century, to 31, but a drought in the early 1900s stopped the recovery and led to increased human-wildlife conflict.[292] But over the next two decades, conservation efforts bore fruit and by 1920 it was estimated that there could be 50 spread across the whole range on the Kathiawar peninsula.[293] Increasing efforts to protect the forest bore fruit for the lions but also their prey and hyenas, which scavenged carcasses of both wild prey and the cattle of the Maldharis people who still inhabited the forest.[294] The Maldharis did not kill the lions or hyenas, despite some livestock losses. But Jim Corbett wrote in 1944 that in many areas of western, central and southern India, hyenas were "blamed often with little proof for killing people", as well as livestock,[295] and killed as a result.

Notes

1 Keith Somerville (2019a) *Humans and Lions: Conflict, Conservation and Coexistence*, London: Routledge/Earthscan, pp. 81–2.
2 See Keith Somerville (2019b) *Ivory. Power and Poaching in Africa*, London: Hurst and Co, updated paperback edition, pp. 57–79; Somerville 2019a, pp. 77–107.
3 John M. Mackenzie (1987) Chivalry, social darwinism and ritualised killing: the hunting ethos in Central Africa up to 1914, in David Anderson and Richard Grove, *Conservation in Africa people, policies and practice*, Cambridge: Cambridge University Press, 41–61, p. 41.
4 Keith Somerville (2017) *Africa's Long Road Since Independence. The many histories of a continent*, London: Penguin, pp. 6–7.
5 Ibid.
6 Elspeth Huxley (1935) *White Man's Country. Lord Delamere and the Making of Kenya, Volume One: 1870–1914*, London: Chatto and Windus, reprinted new edition 1980, p. ix.
7 Ibid.

8 Keith Caldwell (1924) Game preservation, its aims and objects. Lecture delivered before the Kenya and Uganda Natural History Society, 14 March 1924, *Journal of the Society for the Preservation of the Fauna of the Empire* [henceforth *Journal*], 4, p. 47.
9 Ibid.
10 John M. Mackenzie (1988) *The Empire of Nature. Hunting, Conservation and British Imperialism*, Manchester: University of Manchester Press, p. 201.
11 Ibid.
12 Ibid, pp. 208–9.
13 IUCN Environmental Law programme, *An Introduction to the African Convention on the Conservation of Nature and Natural Resources*, http://www.sprep.org/attachments/Legal/IUCNApia.pdf accessed 29 May 2018.
14 Somerville, 2019b, p. 60.
15 Robin S. Reid (2012) *Savannas of Our Birth. People, Wildlife and Change in East Africa*, Berkeley, Calif.: University of California Press, p. 105..
16 Elspeth Huxley and Hugo van Lawick (1984) *Last Days in Eden*, London: Harvill Press; Thomas B. Ocansky (2002) *Paradise Lost: A History of Game Preservation in East Africa*, Morganstown, West Virginia: University of West Virginia.
17 Ofcansky, 2002, p. xi.
18 Helge Kjekshus (1996) *Ecology Control and Economic Development in East African History*, 2nd ed., London: James Currey, p. 70.
19 Ibid., pp. 74.
20 Ibid., pp. 74–5.
21 Theodore Roosevelt and Edmund Heller (1914) *Life histories of African game animals*, 2 vols, vol 1, New York: Scribner Sons, reprinted by Franklin Classics (no date), p. 259.
22 Ibid.
23 Ibid., p. 258.
24 Ibid., p. 253.
25 Elspeth Huxley (1935) *White Man's Country. Lord Delamere and the Making of Kenya, Volume One: 1870–1914*, London: Chatto and Windus, new edition 1980, p. 79; see also, Joanna Lewis (2000) *Empire State-Building. War and Welfare in Kenya 1925–42*, Oxford: James Currey, p. 29.
26 Ibid. p. 85.
27 Ibid, pp. 149–50.
28 Richard H. Lamprey and Robin S. Reid (2004) Expansion of Human Settlement in Kenya's Maasai Mara: What Future for Pastoralism and Wildlife? *Journal of Biogeography*, 31, 6, 997–1032, p. 1006.
29 Ibid.
30 David Collett (1987) Pastoralists and wildlife: image and reality in Kenya Maasailand, in David Anderson and Richard Grove (eds) *Conservation in Africa people, policies and practice*, Cambridge: Cambridge University Press, 129–48, pp. 138–9.
31 Donna Hart and Robert W. Sussman (2009) *Man the Hunted. Primates, Predators, and Human Evoloution*, Boulder, Col.: Westview Press, pp. 96–7.
32 Ibid.
33 Peter Hathaway Capstick (1977) *Death in the Long Grass*, London: Cassell, p. 276.
34 Huxley, 1935, p. 200.
35 Elspeth Huxley (1935) *White Man's Country. Lord Delamere and the Making of Kenya. Volume Two: 1914–1931*, London: Chatto and Windus, new edition 1980, p. 11.
36 Elspeth Huxley and Hugo van Lawick (1984) *Last Days in Eden*, London: Harvill Press, p. 43.
37 Ibid.
38 Karen Blixen (1937) *Out of Africa*, London: Century, illustrated edition 1985, p. 73
39 Ibid., p. 104.
40 Ibid., p. 254.
41 See for example, East African Protectorate 1904–5, Return of Game killed on 98 Sportsmen's Licences, *Journal*, 1907, pp. 85–7.

42 Edward I. Steinhart (2006) *Black Poachers, White Hunters. A Social History of Hunting in Colonial Kenya*, Oxford: James Currey, p. 162.
43 Laurence G. Frank et al. (2005) People and predators in Laikipia District Kenya, in Rosie Woodroffe, Simon Thirgood and Alan Rabinowitz (eds) *People and Wildlife. Conflict or Coexistence*, Cambridge: Cambridge University Press, 286–304, p. 303.
44 David Anderson (2002) *Eroding the Commons. The Politics of Ecology in Baringo, Kenya, 1890–1963*, Oxford: James Currey, p. 75.
45 Personal communication from Dr Mordecai Ogada.
46 Personal communication with Prof. David Anderson, University of Warwick.
47 A. C. Hollis (1911) The Nandi, their Language and Folk-lore [a summary of sections of the book], *African Affairs*, X, XXXVIII, 237–240, p. 240.
48 Peter Matthiessen (2000) *An African Triology. The Tree where Man Was Born, African Silences, Sand Rivers*, London: Harvill Press, p. 82.
49 Ibid., p. 83.
50 K. David Patterson (1970) The Giriama Risings of 1913–1914, *African Historical Studies*, 3, 1, 89–99, p. 90.
51 Ibid., p. 98.
52 W. Robert Foran (1956) Lycanthropy in Africa, *African Affairs*, 55, 219, 124–134, p. 126.
53 Ibid., p. 127.
54 Ibid., p. 128.
55 Ibid., pp. 3448–51.
56 Abel Chapman (1908) On Safari: Big Game Hunting in British East Africa (1908), Longmans, Green, Kindle Edition, loc. 285.
57 Ibid., loc. 1331.
58 Ibid., loc. 2065.
59 Richard Tjader (1910) *The Big Game of Africa*, New York: D. Appleton and Co, Kindle Edition, loc. 2901–2925.
60 Ibid.
61 Ibid., loc. 2935–58.
62 Ibid., loc. 2958–64.
63 Theodore Roosevelt (1910) *African Game Trails*, New York: Charles Scribner's Sons, Kindle Edition, loc. 5361.
64 Roosevelt, 1910, loc. 714.
65 Ibid., loc. 714–745.
66 Ibid.
67 Ibid.
68 Ibid., loc. 2284.
69 Ibid. loc. 745.
70 Theodore Roosevelt's Hyena (no date) *Presidential pet museum*, http://www.presidentia lpetmuseum.com/theodore-roosevelts-hyena/ accessed 5 June 2020.
71 Sir Alfred Pease (1913) *The Book of the Lion*, London: John Murray, reprinted by Forgotten Books, pp. 40–1.
72 Ibid., p. 63.
73 Ibid., p. 64.
74 Pease, 1913, p. 65.
75 Ibid., p. 66.
76 Mervyn Cowie (1961) *Fly, Vulture*, London: George G. Harrap and Co, p. 16.
77 Ibid., p. 21.
78 Ibid., p. 51.
79 Ibid.
80 Ibid., p. 50.
81 Ibid., p. 52.
82 A. Blayney Percival (1924) *A Game Ranger's Note Book*, Read Books Ltd, Kindle Edition, loc. 2197.
83 Ibid., loc. 2260.

84 Caroline Buxton (1921) African Notes, *Journal*, 1, pp. 49–50.
85 Capt. T. A. Ritchie, Game Warden (1928) Extracts from Game Department Annual Report, 1926, *Journal*, 8, pp. 77–8.
86 Ibid.
87 Kenya Colony Game Department (1936) Game Department Annual Report for 1932, 1933 and 1934, *Journal*, , 27, pp. 47–8.
88 George Adamson (1968) *Bwana Game*, London: Collins and Harvill Press, pp. 97–8.
89 Ibid, p. 25.
90 Ibid., p. 27.
91 John Hillaby (1973) *Journey to the Jade Sea*, London: Paladin Books, p. 65.
92 Ibid., pp. 157–8.
93 Personal communication from former Kenya game warden, Ian Parker, who worked with Adamson and said he was an avid user of strychnine with a blind hatred of hyenas.
94 Adamson, 1968, p. 75.
95 Ibid., p. 103.
96 Ian Parker (2004) *What I tell You Three Times is True*, Kinloss, Moray: Librario, p. 53.
97 Joy Adamson (2011) *Born Free*, London: Pan Macmillan, Kindle Edition, loc. 816.
98 Steinhart, 2006, p. 184.
99 Parker, 2004, p. 45.
100 Jan Bender Shetler (2007) *Imagining Serengeti. A History of Landscape memory in Tanzania from Earliest Times to the Present*, Athens, Ohio: Ohio University Press, 2007, p. 1.
101 Ibid., p. 2.
102 Kjekshus, 1996, pp. 176–77.
103 Anthony R. E. Sinclair (2012) *Serengeti Story. Life and science in the world's greatest wildlife region*, Oxford: Oxford University Press, Kindle Edition, loc. 120–3; George Schaller (1972) *The Serengeti lion. A Study of Predator-Prey Relations*, Chicago: University of Chicago Press, p. 6.
104 C. F. M. Swynnerton (1926) The working of the game ordinance Tanganyika Territory, *Journal*, 6, pp. 31–2.
105 Guggusberg, 1961, p. 190.
106 Sara Wheeler (2007), *Too Close to the Sun. The Life and Times of Denys Finch Hatton*, London: Vintage, p. 188.
107 Sinclair, 2012, p. 120.
108 Wheeler, 2007, pp. 203–4.
109 Ibid., p. 123; Guggisberg, 1961, p. 191.
110 Shetler, 2007, p. 181.
111 A. R. E. Sinclair (1995) Serengeti past and Present in A. R. E. Sinclair and Peter Arcese, *Serengeti II. Dynamics, management, and Conservation of an Ecosystem*, Chicago: University of Chicago Press, 3–30, p. 4.
112 Personal communication with Laurence Frank.
113 L. Harrison Matthews (1939) The Bionomics of the Spotted Hyena (Crocuta crocuta), *Proceedings of the Zoological Society of London* [henceforth *Proceedings*], 109, 1, 43–56, pp. 43–5.
114 Ibid.
115 Ibid.
116 Ibid.
117 Ibid.
118 Paul L. Hoefler (1931) *Africa Speaks: A Story of Adventure*, Chicago: John C. Winston and Co, pp. 139–40.
119 Yusuf Q. Lawi (1999) Where Physical and Ideological Landscapes Meet: Landscape Use and Ecological Knowledge in Iraqw, Northern Tanzania, 1920s–1950s *International Journal of African Historical Studies*, 32, 2/3, 281–310, p. 281–2.
120 Ibid., p. 288.
121 Ibid.
122 Ibid., p. 309.

123 Alan Root, History of the Serengeti, http://www.serengeti.org/footprints_mh_hi.htm l accessed 6 June 2018.
124 Myles Turner (1987) *My Serengeti Years*, London: Elm Tree Books, p. 27.
125 Anthony Cullen and Sydney Downey (1960) *Saving the Game*, London: Jarrolds Publishers, p. 72.
126 Ibid., pp. 86 and 131.
127 Brian Bertram (1978) *Pride of Lions*, London: J. M. Dent, p. 224.
128 W. H. Pearsall (1957) Report on an Ecological Survey of the Serengeti National Park Tanganyika, November and December 1956, *Oryx*, 2, pp. 71–136, p. 83.
129 Adamson, 1968, p. 225.
130 Uganda Protectorate (1925) Extracts from the Annual Report of the Game Warden, *Journal*, 1927, 7, p. 46.
131 Ibid.
132 Uganda Protectorate (1927) Extract from Annual Report of Game Department, *Journal*, 1929, 9, p. 89.
133 Uganda Game Department (1949) From the 1949 Report of the Game Department, *Oryx*, 1950, 1, p. 96.
134 Daniel W. Gade (2006) Hyenas and Humans in the Horn of Africa, *Geographical Review*, 96, 4, 609–632, p. 615.
135 Ibid.
136 Ibid.
137 Ibid., pp. 625–6.
138 Waktole Tiki and Gufu Oba (2009) Ciinna – the Borana Oromo narration of the 1890s Great Rinderpest epizootic in North Eastern Africa, *Journal of Eastern African Studies*, 3, 3, 479–508, p. 486.
139 Ibid., p. 487.
140 Wilfred Thesiger (1998) *The Danakil Diary. Journeys through Abyssinia, 1930–34*, London: Flamingo, p. 15.
141 Ibid., p. 43.
142 H. G. C. Swayne (1903) *Seventeen Trips Through Somaliland and a Visit to Abyssinia.* London: Rowland Ward, p. xi.
143 Cited by Mohamed Jama (2004) The Political Ecology of colonial Somaliland, *Africa*, 74, 4, 534–565, p. 536.
144 Extract from the Report of H. E. S. Cordeaux, Commissioner for the Somaliland Protectorate, 1901 and 1902, *Journal*, 1904, Volume 1, p. 71.
145 Captain A. H. E. Mosse (1913) *My Somali Book. A Record of Two Shooting Trips*, London: Sampson Low, Marston and Company, reprinted 2015 by Forgotten Books, pp. 10–11.
146 Ibid.
147 Ibid.
148 Ralph Evelyn Drake-Brockman (1910) *The Mammals of Somaliland*, London: Hurst and Blackett Ltd., pp. 40–1.
149 Ibid., pp. 41–2.
150 Soudan. Report of Game Preservation Department (1907) *Journal*, 1909, p. 118.
151 Ibid.
152 Oldfield Thomas (1923) On the Mammals obtained in Darfur by f.he Lynes- Lowe Expedition *Proceedings*, 92, 2, 247–271, p. 254.
153 Robert D. Q. Henriques (1938) *Death by Moonlight. An account of a Darfur journey*, London: Collins Publishers, pp. 90.
154 Ibid., p. 333.
155 Colonel A.H. Haywood (1932) The Gambia. The Preservation of Wildlife, *Journal*, 1932, 19, p. 34; Colonel A. H. W. Haywood (1932) Gold Coast. Preservation of Wildlife, *Journal*, 1933, 18, p. 33.
156 G. S. Cansdale (1943) A Regional Account of the Mammals of the Gold Coast, *Journal*, 1943, 47, p. 12.

157 Duncan Johnstone (1932) Gordon Laing memorial plaque: Account of unveiling at Timbuctoo, *African Affairs*, XXXI, CXXIV, 282–292, p. 290.
158 A. H. W. Haywood (1937) Game Animals of West Africa, *African Affairs*, XXXVI, CXLV, October, 421–432, p. 423.
159 Ethel Fagan (1930) Some notes on the Bachama tribe, Adamawa Province, Northern provinces, Nigeria, *African Affairs*, XXIX, CXV, 269–79, pp. 271–2.
160 Richard Oakley (1931) Game Preservation in Nigeria, *Journal*, 14, pp. 34–5.
161 Personal communication with Andrew Dunn of the Wildlife Conservation Society in Nigeria.
162 Colonel A. H. Haywood (1933) Sierra Leone. The Preservation of Wildlife, *Journal*, 1933, 19, pp. 21–22.
163 For a detailed early account of Temne customs, see: Rev. C. F. Schlenker (1861) *A Collection of Temne. Traditions, Fables and Proverbs, with an English Translation; as also some Specimens of the Author's own Temne Compositions and Translations; to which is appended A Temne-English Vocabulary. Church Missionary Society*, http://www.sierra-leone.org/Books/A_Collection_of_Temne_Traditions_Fables.pdf accessed 17 April 2020.
164 R. I. Pocock (1934) The Races of the Striped and Brown Hyaena, *Proceedings*, 104, 4, 799–825, p. 820.
165 John Charles Baron Statham (1922) *Through Angola A Coming Colony*, Edinburgh: William Blackwood and Sons, p. 157.
166 Brian J. Huntley (2017) *Wildlife at War: The Rise and Fall of an African Eden*, Pretoria: Protea Book House, pp. 40–1.
167 Statham, 1922, pp. 272–3.
168 Gary Marquardt (2005) Water, Wood and Wild Animal Populations: Seeing the Spread of Rinderpest through the Physical Environment in Bechuanaland, 1896, *South African Historical Journal*, 53, 73–98, p. 74.
169 Ibid., pp. 94–5.
170 Ibid., p. 105.
171 Vivien Tempest Kent (2011) The Status and Conservation Potential of Carnivores in Semi-Arid Rangelands, Botswana The Ghanzi Farmlands: A Case Study, Durham theses, Durham University, available at Durham E-Theses Online, http://etheses.dur.ac.uk/728/ pp. 34–5.
172 Ibid., p. 35.
173 Ibid., p. 36.
174 Isaac Schapera (1952) Sorcery and Witchcraft in Bechuanaland, *African Affairs*, 51, 202, 41–52, p. 44.
175 Ibid.
176 Sir Alfred Sharpe (1904) Game Census of the British Central Africa Protectorate, *Journal*, 1, p. 74.
177 Robert Wright (1904) *Bwana Mkubwa: Big Game Hunting and Trading in Central Africa 1894 to 1904*, copyright Keith Chiazzari 2012, Kindle Edition, loc. 1046–51.
178 Ibid., loc. 1056–60.
179 Ibid., loc. 2792.
180 Extracts from Blue Book, Summary of Game killed in various parts of the Nyasaland protectorate under Licences during the year ended 31 March 1912, *Journal*, 1913, 6, p. 18.
181 Forest and Game Reserves Commission, Nyasaland (1946) Nyasaland. Protection of Native Crops, *Journal*, 1946, 54, pp. 23–4.
182 F. A. Belstra (1962) The man-eating hyenas of Mlanje, *African Wildlife*, 16, pp. 25–27.
183 C. K. Brain (1981) *The Hunters or the Hunted. An Introduction to Africa Cave Taphonomy*, Chicago: University of Chicago Press, p. 65.
184 Ibid.
185 Hans Kruuk (1972) *The Spotted Hyena A Study of Predation and Social Behaviour*, Chicago: University of Chicago Press, pp. 64–5.
186 R. Charles Long (1973) A List with notes of the mammals of the Nsanje (Port Herald) district, Malawi, *The Society of Malawi Journal*, 26, 1, pp. 60–78, p. 66.

187 John Taylor (1959) *Maneaters and Marauders*, London: Frederick Muller Ltd, p. 16.
188 Ibid., pp. 28–9.
189 Ibid, p. 131.
190 Ibid., p. 168.
191 Christo Botha (2005) People and the Environment in Colonial Namibia, *South African Historical Journal*, 52, 1, 170–190, p. 170.
192 Ibid.
193 G. C. Shortridge (1934) *The Mammals of South West Africa*, London: Heinemann, p. 160.
194 Ibid., p. 164.
195 Ibid., p. 153.
196 Ibid., p. 156.
197 Oldfield Thomas (1926) On Mammals from the Gobabis district, Eastern Damaraland, South-West Africa, obtained during Captain Shortridge's fourth Percy Sladen and Kaffrarian Museum Expedition., *Proceedings*, 97, 2, 371–398, pp. 378–8.
198 John Heydinger (2020) 'Vermin': Predator Eradication as an Expression of White Supremacy in Colonial Namibia, 1921–1952, *Journal of Southern African Studies*, 46, 1, 91–108, p. 92.
199 Ibid., p. 101.
200 Ibid.
201 D. Crandall (2002) Himba Animal Classification and the Strange Case of the Hyena, *Africa*, 72, 2, pp. 293–311, p. 293.
202 Ibid., p. 302.
203 Ibid., p. 307.
204 J. Stevenson-Hamilton (1917) *Animal Life in Africa, Book I, Carnivora*, London: Heinemann, reprinted by Leopold Classic Library, pp. 85 and 88.
205 L van Stittert (1998) "Keeping the Enemy at Bay": the extermination of wild carnivora in the Cape colony, 1889–1910, Environ. Hist, 3, 333–356, p. 334.
206 Ibid.
207 James Stevenson-Hamilton (2012) *South African Eden. From Sabi Game Reserve to Kruger National Park*, London: Penguin (originally printed in 1937), Kindle Edition, loc. 124–8.
208 Stevenson-Hamilton, 2012, loc. 446–455.
209 J. Stevenson-Hamilton (1905a) Game reserves, *Journal*, 2, 1905, p. 25.
210 Stevenson-Hamilton, 2012, loc. 545, 541 and 895.
211 Stevenson-Hamilton, 1905a, p. 30.
212 Ibid.
213 James Stevenson Hamilton (1913) Government Game Reserves, Sabi and Singwitsi, South Africa, Part of Annual Report 1911, *Journal*, 6, p. 15.
214 G.L. Smuts (1982) *Lion*, Johannesburg: Macmillan, p. 174.
215 Laurence G. Frank and Rosie Woodroffe (2001) Behaviour of carnivores in exploited and controlled populations, in John L. Gittleman, Stephan M. Funk, David W. Macdonald and Robert Wayne (eds) *Carnivore Conservation*, Cambridge: Cambridge University Press, 419–442, p. 424.
216 Smuts, 1982, p. 183.
217 Ibid., p. 21.
218 Stevenson-Hamilton, 1917, p. 86.
219 Ibid., p. 89.
220 Ibid., p. 96.
221 Transvaal Game Preservation Ordinance (1909) *Journal*, p. 85.
222 Smuts, 1982, 102.
223 *Journal*, 1909, p. 84.
224 Stevenson-Hamilton, 1917, p. 85.
225 Somerville, 2019b, pp. 163–8.
226 I. Wiesel (2015) *Parahyaena brunnea. The IUCN Red List of Threatened Species*, https://dx.doi.org/10.2305/IUCN.UK.2015-4.RLTS.T10276A82344448.en accessed 30 June 2020.

227 Game Ordinance, Northern Rhodesia (1905) *Journal*, 1905, p. 74.
228 Game Ordinance, Northern Rhodesia (1925) *Journal*, 1927, 7, p. 9.5
229 Somerville, 2019a, p. 95.
230 A National Park in Northern Rhodesia, *Oryx (*formerly *Journal)*, 1950, 1, p. 15.
231 Ernest E. Austen (1913) The present position of the problem of big game tstse flies, and sleeping sickness, *Journal*, 6, pp. 57 and 67.
232 Roben Mutwira (1989) Southern Rhodesian Wildlife Policy (1890–1953): A Question of Condoning Game Slaughter? *Journal of Southern African Studies*, 15, 2, Special Issue on The Politics of Conservation in Southern Africa, 250–62, p. 250.
233 Ibid.
234 Ibid.
235 Ibid., pp. 252–3.
236 Ibid., p. 254.
237 Ibid., p. 257.
238 Ibid.
239 Ibid., p. 258.
240 Wankie Game Reserve, Warden's Report (1934) *Journal,*1936, 26, p 39.
241 Wankie Game Reserve (1935) Annual Report, 1935, *Journal*, 1936, 29, pp. 42–3.
242 H.E. Hornby (1945), A note from Southern Rhodesia, *Journal*, 52, pp. 23–5.
243 Ibid., p. 23.
244 Ibid., pp. 23–4.
245 Mackenzie, 1988, p. 240.
246 Game and tsetse fly in Southern Rhodesia (1956) *Oryx*, 5, p. 264.
247 Mills and Hofer, 1998, p. 25.
248 Sir Harry Johnston (1898) On the Larger Mammals of Tunisia, *Proceedings of the Zoological Society of London*, 3 May, p. 351.
249 M. Masseti (2010) Holocene mammals of Libya: A biogeographical, historical and archaeozoological, *Journal of Arid Environments*, 74, 794–805, pp. 794–5.
250 C. V. B. Stanley (1912) The Oasis of Siwa, *African Affairs*, XI, XLIII, 290–324, p. 310.
251 Ibid.
252 Dale J. Osborn and Ibrahim Helmy (1980). The Contemporary Land Mammals of Egypt (including Sinai), Field Museum of Natural History, digitized version, p. 422–3.
253 Ibid., pp. 428–9.
254 Ibid.
255 Ibid.
256 Major Stanley Smyth Flower (1932) Notes on the Recent Mammals of Egypt with a list of the Species recorded from That Kingdom, *Proceedings*, 102, p. 369.
257 Osborn and Helmy, 1980, p. 429.
258 Mills and Hofer, 1998, p. 25.
259 Osborn and Helmy, 1980, pp. 428–9.
260 Ibid.
261 Captain Eric Hardy (1947) The Palestine Leopard, *Journal*, 1947, 55, pp. 16–17.
262 Lee Merriam Talbot (1960) A Look at Threatened Species. A report on some animals of the Middle East and southern Asia which are threatened with extermination, *Oryx*, 4, p. 155–293, p. 279.
263 Max Kasparek et al. (2004) On the status and distribution of the Striped Hyaena, Hyaena hyaena, in Turkey, *Zoology in the Middle East*, 33, 93–108, p. 93.
264 Ibid., p. 95
265 Ibid., p. 93.
266 Ibid., p. 96.
267 See V. G. Heptner and A. A.Sludskii (1992) *Mammals of the Soviet Union: Carnivora (hyaenas and cats)*, Volume 2, Smithsonian Institution Libraries and National Science Foundation, https://archive.org/details/mammalsofsov221992gept/page/10 accessed 24 July 2019, for a very detailed account of the regions in which they were found.
268 Mills and Hofer, 1998, p. 22.

269 Heptner and Sludskii, 1992, pp. 18–26.
270 Mills and Hofer, 1998, p. 25.
271 Heptner and Sludskii, 1992, p. 19.
272 Mills and Hofer, 1998, p. 25.
273 Heptner and Sludskii, 1992, p. 19.
274 Igor Khorozyan, Alexander Malkhasyan and Marine Murtskhvaladze (2011) The striped hyaena Hyaena hyaena (Hyaenidae, Carnivora) rediscovered in Armenia, *Folia Zoologica*, 60, 3, 253–261, p. 253.
275 Osborn and Helmy, 1980, pp. 428–9.
276 Ibid.
277 Heptner and Sludskii, 1992, pp. 19 and 21–22.
278 Hervé Monchot and Marjan Mashkour (2010). Hyenas around the cities. The case of Kaftarkhoun (Kashan-Iran), Journal of Taphonomy, 8, 1, 17–32, p. 18.
279 M. B. Qumsiyeh, Z. S. Amr and D. M. Shafei (1993) Status and conservation of carnivores in Jordan, *Mammalia*, 57, 1, 55–62, pp. 58–9.
280 Monchot and Mashkour, 2010, p. 18.
281 Mills and Hofer, 1998, p. 20.
282 Osborn and Helmy, 1980, pp. 428–9.
283 Oldfield Thomas (1900) On the Mammals obtained in South-western Arabia by Messrs. Percival and Dodson, *Proceedings*, 1900, p. 100.
284 Desmond Foster-Vesey-Fitzgerald (1952) Wildlife in Arabia, *Oryx*, 1952, 5, pp. 233–4.
285 Lee Talbot (1960) A Look at Threatened Species. Fauna and Flora Preservation Society, *Oryx*, 5, 4-5, no page numbers.
286 Jim Corbett (1954) *The Temple Tiger And More Mean-Eaters of Kumaon*, Bombay: Oxford University Press, pp. 118–9.
287 Lord Irwin (1929) Wildlife Preservation in India. Memorandum from HE the Viceroy, *Journal*, 9, p. 98.
288 Wabis Ameer (1930) The Fauna of India, *Journal*, 1930, 11, pp. 25–56.
289 Ibid., p. 27.
290 F. W. Champion (1933) Wild Fauna Preservation in the United Provinces, *Journal*, 1933, 18, p. 21.
291 Julie E. Hughes (2015) Royal Tigers and Ruling Princes: Wilderness and wildlife management in the Indian princely states, *Modern Asian Studies*, 49, 4, pp. 1210–1260, p. 1218.
292 Sudipta Mitra (2005) *Gir Forest and the Saga of the Asiatic Lion*, New Delhi: Indus Publishing Company, p. 22.
293 R. I. Meena, Sandeep Kumar and Shmshad Alam (2014) *Action plan for the Conservation of the Asiatic Lion (Pantheraleo persica Meyer, 1826)*, Junagadh: Gujarat Forest Department, p. 10.
294 Mitra, 2005, p. 24.
295 Jim Corbett (1944) *Man-Eaters of Kumaon*. Bombay: Oxford University Press, pp. viii–ix

7

CONTEMPORARY AFRICA, WEST, CENTRAL AND SOUTH ASIA

There has been no recent verifiable estimate of the overall populations of the four hyena species. The estimates given for each species at the start of Chapter 2 based on the IUCN's Red List remain the best starting point for assessing numbers and distribution.[1] The IUCN's Hyaena Specialist Group is carrying out a mapping project to identify the ranges and populations of the four species.[2] Stephanie Dloniak, the chair of the specialist group, told me they hope to have new species maps ready by around mid-2021.[3] It will be interesting to see if the material produced supports Yirga's projection of larger number of spotted hyenas in Ethiopia and therefore in the overall population,[4] and Frank's suspicion that the global striped hyena population is larger than current estimates.[5]

Conflict between people and predators "has a long historical existence, its increasing severity and complex nature has made it a central issue to wildlife management",[6] and "Large carnivores are especially sensitive to human activity; because their requirements often conflict with those of local people, predators have been actively persecuted in most regions of the world".[7] Killing by people remains the greatest threat to the survival of large carnivores.[8] Many large carnivores avoid human-dominated environments, especially urban ones, but spotted and striped hyenas have found ways of living close to humans.[9] Human-wildlife conflict (HWC), the consequences of bushmeat hunting and the contempt or hatred of many peoples for hyenas are the main threats to them. Habitat loss affects all hyenas to some extent, resulting from human population expansion and growing demand for land for crops or livestock.

The preceding chapters show how hyenas have proved flexible, innovative and opportunistic, enabling them to benefit from human presence while limiting danger and limiting, to some extent, the effects of habitat loss. Hyenas "demonstrate greater learning and problem-solving skills in novel conditions" that are found in many wildlife species which adapt to the expansion of urban environments in place of more suitable habitats.[10] Some hyenas vary their foraging behaviour, times and places of activity and denning to make use of feeding opportunities while avoiding periods of human activity; this is now being found to apply in some areas to brown as well as spotted and striped hyenas.[11] But they cannot totally avoid conflict with humans, which remains a major threat to them for reasons within the power of humans rather than the hyenas to eliminate or mitigate.

Hyenas are one of the "most targeted predators in communities affected by livestock depredation".[12] Hyenas are also under threat from illegal human hunting for meat, termed bushmeat.[13] Offtake of wild ungulates by bushmeat hunters reduces prey for carnivores, but offers opportunities for brown, spotted and striped hyenas to feed from snared animals. It also has the danger of hyenas being snared, either as they travel on targeted trails or feed on snared animals. Persecution and killing also occur where fear and hatred of hyenas means they are killed on sight, as has been recorded in Palestine,[14] or captured and used for inhuman sports by having large dogs set on them, as in India and Pakistan.[15] Methods of killing vary from shooting, spearing and snaring to the use of poisons. As Darcy Ogada has already been quoted as establishing, colonialists brought an "extermination culture" to their colonies in Africa and Asia, especially regarding carnivores.

> [By] independence, the use of poisons to eliminate predators and scavengers was widespread...chemicals had spread, and were widely used by pastoral and farming communities...Perhaps more than any other animal in Africa, hyenas have been systematically targeted in poisoning campaigns...where they have also consistently been identified as the most despised predator. Their indiscriminate food choices make them particularly vulnerable to poisoning.[16]

This chapter examines human-hyena conflict and coexistence in the last 50 years on a region-by-region basis. There will be a strong focus on Eastern and Southern Africa, but this will not be to the exclusion of other regions of Africa or Asia, and issues of coexistence and conflict there, though there will be gaps as there is simply no data for some countries on presence, range and numbers – this has not changed substantially since Mills and Hofer had to write Data Deficient against a large number of countries in their survey.[17]

Part 1: Sub-Saharan Africa

East Africa

The largest single population spotted hyenas is found in the Serengeti-Mara ecosystem spanning northern Tanzania and southern Kenya, with about 7,500–8,700–7,200–7,700 in Serengeti and 500–1,000 in the Mara;[18] Arjun Dheer believes the Serengeti figure may be too high, but that the Ngorongoro Conservation Area (NCA) may have 3,000–4,000 hyenas.[19] This accords with Reid's estimate of 7,500 in the Serengeti-Mara system, vastly outnumbering the 2,800 lions there.[20]

Across East Africa, IUCN says "high densities occur in savanna and some open woodlands in Tanzania and Kenya, as well as in montane forests (0.32–2.4 individuals per km^2), such as Selous Game Reserve, Aberdare National Park, Ngorongoro Crater", in addition to Tsavo, Samburu, Amboseli, Tarangire, Ruaha and Katavi NPs, and smaller parks and reserves. Most populations in protected areas in eastern Africa "are considered to be stable".[21] There are significant populations outside protected areas in Kenya and Tanzania and they are found in NPs, reserves and some unprotected areas in Uganda.

Striped hyenas are far less numerous and accurate numbers are not available for populations in specific regions and countries. They are found in grassland, open woodland and hilly habitats, particularly in northern Kenya, northern Tanzania and northern Uganda, and been have been recorded as far south as Ruaha NP.

Kenya

Kenya has a substantial but declining wildlife population, with 35% in protected national parks and reserves and 65% on private or communally owned land[22]. Between 1977 (when hunting was banned) and 2012, wildlife numbers declined by 70%,[23] with an accompanying rise in the numbers of sheep and goats (76.3%), camels (13.1%) and donkeys (6.7%). The major cause of wildlife loss is the change in land use in recent decades, with major increases in livestock on often marginal land, and seasonally-shifting pastoralism being progressively replaced by privatisation of communal lands, fencing-off of rangelands and growth in crop production, which have impeded freedom of movement and led to the exclusion or extermination of wildlife.[24] Wildlife ranges outside protected areas are contracting, leading to declining numbers and increasing HWC. As Frank, Woodroffe and Ogada highlight, large areas are devoid of wildlife because of conversion of land to cultivation, extensive bushmeat hunting and increasing livestock numbers. In the thinly-populated, arid rangelands of Kenya that once had substantial hyena populations, overgrazing, use of poison and the availability of assault weapons have meant that "substantial predators populations persist only in the rangelands north of Mount Kenya (particularly Laikipia District), and in the south close to the border with Tanzania", in the Mara and Tsavo.[25] Overall, Kenya's spotted hyenas are not seriously threatened even though their numbers, according to the IUCN, are declining in some districts.[26] Reid pointed out, though, that spotted hyenas are quite adept at adapting to human presence, including that of pastoralists like the Maasai, by increasing the areas over which they forage and becoming more nocturnal to adapt to the daylight hours presence of herders with their livestock,[27] which may help limit any decline.

Henschel et al. have estimated the Tsavo ecosystem's spotted hyena population at 3,903 (+/-514) and the striped hyaena population at 679 (+/-144)[28] – with the hyenas distributed across the ecosystem. The spotted hyena is Tsavo's most abundant carnivore, but, as Darcy Ogada writes, Tsavo has been "Kenya's poisoning hotspot for decades".[29] There is also substantial illegal grazing of livestock in Tsavo NP, which increases HWC. Henschel et al. identified current threats to large carnivores taking the form of "retaliatory killing associated with illegal livestock incursion and bushmeat poaching, which reduces the prey base and directly kills large numbers of carnivores as by-catch".[30] Human-hyena conflict resulting from predation of livestock has long been an issue on the private and group ranches bordering Tsavo. The Taita and Rukinga ranches are managed as part of a wildlife conservancy, with under 1,000 cattle that acts as a buffer between the park and livestock ranches. When the two ranches still had 4,000–5,000 cattle in addition to goats, sheep, donkeys and some camels, a study was carried out there of predator attacks on stock.[31] It recorded 312 attacks on stock between 1996 and 1999 and the deaths of 433 animals. Lions were blamed for 372 deaths and spotted hyena just 14.[32]

There is concern about Kenya's striped hyenas, which were believed to have declined over much of their global range and had a surprisingly low estimated population in Kenya of around 1,000.[33] The Kenya Wildlife Service (KWS) at one stage reported that Tsavo East NP no longer had striped hyenas. But they were found to be the third most abundant large carnivore there, after the spotted hyena and lion. Henschel et al.'s research in Tsavo East showed that they:

> ...were in fact present throughout the study area. We found them to be common and widespread, at a density of 3.26/100 km2, comparable to the 3/100 km2 documented in the livestock/wildlife rangelands of Laikipia County... These results suggest that this rarely seen carnivore may be more abundant and widespread than previously perceived [both in Kenya and possibly replicated across its global range]...[34]

The Mara ecosystem, covering about 6000km^2 – the Maasai Mara National Reserve (MMNR), Mara Triangle conservancy, smaller conservancies to the north and the Maasai group ranches (75% of the area) – is vital habitat for spotted hyenas, with about 1,000 resident (linked to the larger population in Tanzania's Serengeti), and a smaller striped hyena population. Some group ranches combine pastoralism and tourism or are leased solely for safari tourism.[35] The region has resident ungulates and the huge wildebeest, zebra and Thomson's gazelle migration passes though. There is a large livestock population on group ranches.

There is a major long-term threat to the ecosystem – the sale of communally-owned land and creation of fenced, private farms with inward migration by farming communities, resulting in significant land transformation, particularly on the northern and western borders of the ecosystem. Mechanised wheat production and intensive small-scale agriculture are spreading, which has contributed to the decline in ungulate populations.[36] From the mid-1980s onwards, large areas of grassland/communal grazing land on the Loita Plains north of the Mara, used seasonally by livestock and wildlife, were leased out by Maasai to farmers for wheat cultivation. According to Lamprey and Reid, due to this "habitat loss, poaching and other disturbances, resident wildlife populations in the Mara have declined by over 70% over the last 20 years".[37] Pastoralists are banned from the MMNR, but frequently take their livestock into the reserve to graze. The reserve is unfenced and open to the adjoining conservancies and communal grazing areas. The MMNR senior warden for the Trans Mara, David Ole Seur, warns:

> ...as communities become more sedentary and change their lifestyles and as populations increase, there is an inevitable increase in conflict with wildlife over access to resources...It is unlikely that conflict can ever be totally eradicated, but it needs to be controlled at a level that local people can tolerate, and...people need to see a benefit from wildlife to offset those costs of conflict...[38]

and through that help wildlife conservation. Spotted hyenas are present across the Mara, occupying large territories populated by clans that can reach 100. While they will take maximum advantage of the presence of the migration – feeding on young wildebeest and zebra – they don't rely on just one or two prey species.[39] When the migration moves on, according to the study of the 65-strong Talek clan by Cooper, Holekamp and Smale, they hunt the remaining wildebeest, resident topi and Thomson's gazelles. About 40% of hyena intake was carrion, some from lion kills.[40] Perhaps because of the presence of abundant natural prey, no killing of livestock illegally grazing in the reserve was recorded during the study. Stock were guarded all day and taken back out of the reserve before night.[41] Between 1988 and 1998, the number of pastoral settlements along the reserve boundary increased, as did incursions into the reserve. By 2001, settlements were concentrated along the reserve boundary, with researchers seeing livestock in the reserve on 90% of days they were monitoring the areas.[42]

> Monthly livestock censuses recorded an average of 991 cattle and 1,038 sheep and goats utilizing the area within the Talek West territory, and 515 cattle and 257 sheep and goats within the Talek East territory.[43]

There was also a large and regular human presence through safari tourism.[44] Although tourists did not seem to bother hyenas, as they hunted mainly after dark, Boydston et al. found the Talek hyenas stayed closer to dens when lots of tourist vehicles were present.[45]

The increase in the human population around the MMNR continued through the opening decades of 21st century, with increases in HWC. Ogutu et al. recorded an increase in predation by hyenas, leopards and lions on livestock, especially in the wet season.[46] Predators as a group were blamed by the Maasai for greater economic losses than disease, despite more stock being lost to the latter.[47] Interestingly, there appeared to be advantage for hyenas of persecution of lions. In areas where lions were killed or driven out, such as Koiyaki Group Ranch, hyenas were present in larger numbers than elsewhere as they did not have to compete with lions, suggesting that hyenas could survive in areas with extensive pastoralism.[48] This was emphasised by Kolowski and Holekamp, who "documented little avoidance of livestock use areas in general".[49] Areas used heavily by livestock on a daily basis were avoided, though,[50] and there were killings of hyenas at Koiyaki. In a separate study, they demonstrated that refuse pits were attractions for hyenas, especially for low-ranking females (with poorer access to kills or carcasses).[51] Utilisation of refuse pits made it harder to reduce livestock predation and human-hyena conflict.[52]

Between 1988 and 2006, 20 of 83 deaths among the Talek West clan were attributed to humans (spearing, snaring and poison).[53] The Mara River clan, located far from pastoral communities had no human-related deaths. On the Koiyaki and Siana group ranches, Kolowski and Holekamp found that between 2001 and 2005, at least nine hyenas were killed at bomas during livestock attacks and 130 incidences of carnivore predation of livestock were recorded from March 2003-April 2004. Hyenas were involved in 69 of the incidents (53%), with leopards and lions involved in 32% and 15%.[54] Hyenas attacked animals grazing during the day and at night in bomas – with sheep and goats the favoured prey. During the study period, predation of stock cost US$6049, with hyenas responsible for 45%.[55] Although hyenas were wary of humans during the day, the authors' observations suggest:

> ...at least in darkness, hyenas do not appear concerned about humans. Groups of hyenas sometimes slept for extended periods within 150 m of large bomas and hyenas were seen to walk calmly within 50 m of humans, only fleeing from those carrying flashlights.[56]

They concluded:

> ...hyenas, as opportunistic feeders, are making regular visits to bomas not for livestock primarily, but rather for discarded food and other edible items. Large bomas, with more human activity, would thus be attractive to hyenas interested in exploiting refuse and opportunistic attacks on livestock should therefore be more likely to occur at these bomas...secure refuse disposal at bomas may reduce hyena attack frequency.[57]

Recent studies reinforce these findings. Green's 2018 paper on anthropogenic disturbance of the Mara's spotted hyenas and lions noted continued killing of hyenas by pastoralists using spears or poison in retaliation for livestock losses.[58] Humans constituted the greatest single source of mortality for hyenas in areas of the MMNR used by pastoralists. Green also found that despite human threats, the Talek West clan was able to cope with human presence better than lions and increased in size, reaching 113 members by 2013 – the largest ever documented. Lion numbers in the reserve declined by 40%.[59] Both lions and hyenas were recorded as leaving the reserve at night, entering areas with a greater human and livestock presence, leading to conflict and mortalities.[60] The Kenya-based conservation scientist Holly Dublin said that in decades of conservation research in Kenya she was aware of very extensive poisoning of spotted hyenas by the Maasai, using chemicals from cattle dip. She said young Maasai men and boys would try to spear hyenas if they came across them and if they found a den would destroy it.[61]

Green and Holekamp recorded that the growth in human and livestock numbers in a private conservancy bordering the Talek section of the MMNR occupied by several thousand Maasai pastoralists constituted the greatest source of hyena mortality there, and in the eastern part of the reserve. Pastoralists armed with spears and clubs accompanied their livestock into the reserve daily.[62] Comparing the speed of movement of hyenas in the Talek area (inside and outside the reserve) and two clans on the western side of the reserve at Serena North and South, they found that those from Talek both within and outside the reserve travelled faster, with the reasonable inference that this resulted from human/livestock presence and the potential threat to hyenas.[63] The varying of behaviour, such as speed of movement in areas with greater human presence, may help explain why spotted hyena can survive and even increase numbers in areas of high human disturbance, flexibility being a key to more successful coexistence for hyenas compared with lions and other predators.[64]

To the north, in the Laikipia and Samburu districts, savannahs support a diversity of wildlife and significant livestock numbers. A mixture of conservancies, commercial and communal ranches engage in pastoralism, some combining it with wildlife tourism, the latter accounting for up to 50% of income on some ranches.[65] Large predators present are spotted and striped hyenas, lions, leopards, cheetah and wild dogs. Tolerance of them on land with livestock varies greatly. A study carried out in community grazing lands and community sanctuaries in Samburu district found that predation of livestock was damaging to community incomes.[66] Analysis of scat samples showed that wild prey made up 81.1% of spotted hyena food and domestic animals 18.8%. Hyenas were the most problematic of the carnivores for livestock owners; striped hyenas fed on carcasses of livestock but only occasionally killed stock.[67] The killing of hyenas and other predators, especially lions and cheetahs, has reduced the potential for tourist income on ranches where retaliation is highest,[68] which is part of a vicious circle as tourism income can encourage coexistence.[69] Retaliatory killing involves shooting, spearing, trapping and poison. Carnivore predation does cost farmers but as mentioned in relation to other regions, livestock predation is lower than loss from disease.[70] Frank et al. found that hyenas took sheep and goats more than cattle but would kill multiple sheep and goats if they got into bomas at night. Lions took more cattle. The cost to commercial ranches, based on losses and number of predators present, are $35 per hyena against $360 per lion.[71] But most farmers wanted to see a reduction in hyena numbers rather than lions or other big cats, presumably because of the higher tourism potential of cats. There was slightly more tolerance of spotted hyenas where there was tourism, and a high level of tolerance of striped hyenas overall.[72]

A survey of 416 community members and commercial ranchers in the Ewaso district of Laikipa found 90% of commercial ranchers in favour of having predators on their land, but only 20% of community members wanted spotted hyenas on their land, though 40% wanted other predator species.[73] Romanăch et al. found that one out of 48 Laikipia commercial farmers surveyed had a shoot on sight policy for hyenas, while 79% of pastoral communities said they'd kill predators on sight; 15% of pastoralists admitted to killing a predator in the previous year.[74] Frank et al. found that the number of hyenas, lions, leopards and cheetahs killed on commercial ranches correlated closely with the numbers of stock killed by those predators. Many Laikipia commercial ranches are also wildlife conservancies and none have used poison for decades. However, on neighbouring communal land, poisoning carcasses killed by predators is common and indiscriminate, killing any carnivore feeding from a carcass.[75] Threats to livestock lessened the better the husbandry; dogs were useful, because they alerted herders to the presence of predators.[76]

Another cause of antagonism towards spotted hyenas comes from the attempts to conserve the rare Grevy's zebra. On small reserves it is believed that high predator numbers reduce the zebra's chances of survival.[77] On the Lewa Wildlife Conservancy and Borana Conservancy, Grevy's were a favoured prey of lions and hyenas, which was "a complicated and potentially uncontrollable threat to the survival of small populations" of the zebras.[78] There were more hyenas than lions on the conservancies, with an estimated 38–78 hyenas and 17 lions at Lewa and 31–57 hyenas and 17 lions at Boranna.[79] Davidson et al. said there wasn't a clear management strategy for lions and hyenas.[80] In July 2020, the KWS announced it was implanting hormonal contraceptives into lionesses at Lewa to limit the lion population and cut predation of Grevy's zebras; no mention was made of controlling hyena numbers.[81] Frank found, though, that the hyena impact on Grevy's was lower than would be expected given their abundance, and concluded that the situation on Lewa is probably an artefact of fencing and unnaturally high Grevy's density, especially as its open grasslands are not the natural Grevy's habitat of semiarid bush savannah.[82]

Wagner et al. studied striped hyenas on private ranches and communal lands in Laikipia between 2000 and 2003[83] and found about 25 adults and sub-adults in the study area, but did not report evidence of stock predation or significant striped hyena-human conflict.[84] But striped hyenas are frequently, usually inaccurately, blamed for stock-killing and killed as a result. Ndeereh detailed the rehabilitation of striped hyena cubs rescued in northern Kenya. The mother was killed in April 2016 by local people in an area where wildlife habitat was converted into small farms.[85] After being reared by humans until they were large enough to be released, the cubs were set free in Meru NP. The female cub left the park and had to be recaptured and taken back, but then moved out of the park into grazing land and was killed by herders.[86]

Leakey et al.'s 1999 study of striped hyenas in northern Kenya revealed that they had a significant level of dietary dependence on the waste produced by the local Turkana communities. Examination of bone accumulations by the hyenas showed they consumed dog and human remains as well as those of livestock, and hunted small livestock as well as scavenging their remains.[87] The accumulations contained fifteen mammalian prey species composed of 63.4% ungulates, 25.7% carnivores and 9.9% humans.[88] The territory west of Lake Turkana is inhabited by Turkana pastoralists. The habitat is very degraded because of overgrazing but supports gazelles, hares and porcupines as well as both species of hyena, caracals and wildcats. Turkana herders told the researchers that striped hyenas killed goats and sheep, usually ones not enclosed at night – examination of bone accumulations supported the view that some livestock had been killed rather than scavenged. Local people also claimed that hyenas killed babies and dug up human remains from graves.[89] Monbiot, in his account of life among the Turkana, said that he visited a cattle camp that had been destroyed by Toposa raiders. He wrote:

> The Turkana do not bury people who are violently killed, in case, by acknowledging their misfortune, they encourage it to reappear. So the remains of all the people of the cattle camp still lay where the hyaenas had dragged them.[90]

The Turkana freely admitted shooting hyenas to protect their stock.

The outskirts of Nairobi had a substantial spotted hyena population in the early 20th century. While it is likely that hyenas are still foraging at the waste dumps at night and in the remaining areas of bush and forest, they are not often reported as being seen or coming into conflict with people. There are a few, again rarely seen, in Nairobi NP – on a dozen visits to the park over two decades I have seen many lions but no hyenas; but there are believed to be about 20 resident. One regular visitor reported in 2014 of his surprise on seeing a spotted hyena in the park, adding that he was aware that they did sometimes leave the park and forage for animal carcasses and other waste and were sometimes the victims of poisoning.[91]

There has been more of a problem in Kajiado district, south of Nairobi NP and north of Amboseli NP. In October 2019, it was reported that 103 sheep had been killed in one night by a pack of spotted hyenas thought to have moved out of one of the parks A resident told the *Daily Nation* newspaper that "If they [hyenas] attack, let's say one sheep, we poison the carcass and put it in their path. Once they eat this, they'll die".[92] Residents of this district and others close to Nairobi NP want the park to be completely fenced – which would stop the seasonal movements of ungulates and commuting of hyenas in search of food.[93]

On 6 August 2020, the Tourism and Wildlife Cabinet Secretary Najib Balala said that the government and KWS was launching a lion and spotted hyena recovery plan, noting that both carnivores were facing "complex threats" and were two of Kenya's most threatened species – though no supporting evidence was given.[94] The complex threats were not identified nor has the KWS website any reference to the plan. No further details were given, beyond saying that he admitted that in the past no proper planning had been carried out to reduce human-wildlife conflict ministry, despite a similar exercise 10 years ago. The main point of his launch seemed to be an appeal to NGOs and donors to support Kenya's conservation efforts during the COVID-19 pandemic. Searching for the plan, I found a report by KWS that gave no confidence in their grip on issues relating to hyenas. It concerned the relocation of two striped hyenas. The KWS headlined it, "Efforts to conserve threatened stripped hyena species", and throughout the report referred to stripped hyenas. Naked ignorance and incompetence don't bode well for conservation.[95] And the new plan itself contained nothing of substance. just well-intentioned waffle with, as Frank put it to me:

> ...lots of plans for lots of meetings, workshops, committees, education, awareness creation, new positions, etc., with little practical on the ground action. It is encouraging, however, to read that there are now serious penalties on the books for poisoning. However, very few poisoning cases have been prosecuted, and fewer found guilty.[96]

The report admitted that Evaluation scores of the existing strategy revealed that "no activity or objective was fully and effectively achieved" and nothing in the new plan suggested the results would be any different.[97]

Tanzania

Tanzania is one of the most important countries for spotted hyenas, with a quarter of the current estimated world population. There are no reliable estimates of striped hyena numbers there. Spotted hyenas are widely distributed, while striped hyenas are far less common, occurring only as far south as Ruaha NP. Tanzania is believed to have 10–12,000 spotted hyenas, with 7,000+ in the Serengeti-Ngorongoro ecosystem.[98] Large spotted hyena populations can be found in Ruaha NP and surrounding game management and farming areas, Katavi NP, Selous GR, Tarangire and in smaller parks, reserves, hunting concessions and unprotected areas.

Given the ability of spotted hyenas to adapt to different environments, persecution is a greater threat than habitat loss, "with people killing hyaenas to protect stock, for fun, for target practice, or out of fear of the animal".[99] Hyenas are an important part of Tanzanian folklore, belief in the supernatural and traditional medicine – with some healers having captive hyenas for the latter purpose.[100] Kruuk recorded that many Tanzanians believe:

> ...that witches keep hyenas in chains...that witches ride hyenas at night and that this explains why hyenas have sloping backs... that witches milk their hyenas, hyena milk having magical powers. Witches make butter out of hyena milk, and they fuel their torches with this butter.[101]

Kruuk also wrote that he once found a group of people around a dead hyena on a road; they were removing the tail, ears, genitals one of the legs and some of the skin. They said the skin would be dried and fed to cattle, or rubbed on people and cattle to ward off evil[102].

Hyenas thrive in these regions because of the massive population of ungulates – especially wildebeest, zebra and gazelles – supporting one of the highest concentrations of predators in the world.[103] The annual migration of wildebeest, zebra and Thomson's gazelles provides a substantial food source for hyenas and other predators, but resident populations of Grant's gazelles, topi, impala, warthog and buffalo, help support large hyena clans as well as lions, leopards and cheetah.[104] Hyenas outnumber lions by about three to one.[105] Ngorongoro Crater, with a very high density of hyenas, has resident wildebeest and zebra populations. In many areas on the boundaries of the parks, there are significant livestock populations. These often graze illegally in the protected areas. In some conservation areas, pastoralists have the right to graze cattle at certain times of the year, leading to increased human-hyena conflict. The Maasai pastoralists on the boundaries of the Serengeti and in the Ngorongoro Conservation Area still conduct ritual lion hunts as part of manhood transition rites. They also kill hyenas in retaliation for stock deaths, though these do not bring the same prestige or have the same cultural significance. Poison is frequently used to kill predators.[106] A Maasai safari guide from a pastoralist family near Loliondo told me that in the NCA lions are killed in retaliation for stock predation, but hyenas may be killed on sight if young men can catch up with them.[107] This supports the much earlier work on conflict with predators in the NCA by Homewood and Rodgers, who found that Maasai pastoralists were the chief culprits in poaching in the conservation area.[108] Snaring is a threat to hyenas (spotted and striped). Hofer and East found that adult mortality rose from 6.2% for females to 19.8% and from 9.7% to 22.7% for males in the dry season, with "incidental snaring of hyenas in snares set for herbivores" a major cause.[109]

On the Maasai steppe in northern Tanzania, the large populations of hyenas, lions and other carnivores "face substantial pressures from the increasing human population and ongoing land use changes including conversion of rangelands into agricultural farms, as well as increased settlements".[110] Hyenas come into conflict with pastoralist communities in the wet season when they hunt over greater distances: "[T]hese seasonal movements may contribute to high levels of human–carnivore conflicts and retaliatory killing of carnivores".[111] Kissui et al. found that between 2009–2013, 89% of livestock attacks were by spotted hyenas, with the rest (each under 5%) by leopards, jackals and lions. During that period 106 carnivores were killed. The annual average predator killing was about 20 lions and 0.2 hyenas, despite hyenas being responsible for most raids on bomas and livestock deaths.[112] Spotted hyenas here frequently commuted up to 50 miles to prey on the migrating herds, returning to the clan territory every few days.[113] Female hyenas with young might take foraging trips that last 3–4 days, while for females without young and males it might be 6–10 days.[114]

In the eastern Serengeti, human-hyena conflict centred on spotted hyenas killing cattle and other livestock in grazing land close to the NP boundary. Maasai pastoralists say hyenas are the most common stock killers.[115] Livestock were guarded during the day and generally enclosed at night, but the daytime guards were young boys and girls, poorly equipped to deter large carnivores.[116] Mbise's research showed that spotted hyenas killed 28.2% (51) of Maasai stock taken during his study period and 50% of predated stock owned by the Sonjo community situated further from the park boundary – which suggests that hyenas are willing to travel further from protected areas to seek food close to human habitations.[117] Holmern et al. found that livestock losses in areas bordering the north-western boundary of the NP remained a serious problem, with households reporting losses of 4.5% of their livestock to predators and spotted hyenas blamed for 97.7% of predation;[118] 73.4% of respondents approved the retaliatory killing of predators.[119]

The large hyena population in the Ngorongoro Crater has a large resident prey population and is far less affected by conflict with humans. The Crater floor is divided between eight clans, whose size has varied over time according to the availability of prey. Territories have shifted in size and shape according to prey numbers, though the number of clans did not change even when hyena numbers fell – going down from 385 in the 1960s to 117 in 1996 before climbing back up to 333 in 2002, by which time an eighth clan was being formed.[120] Some hyenas may leave the Crater (notably dispersing males and low-ranking females), while males may disperse from clans in the NCA, where there are good numbers of hyenas, into the Crater.[121]

In the Ruaha ecosystem, encompassing the national park and adjacent wildlife management areas (WMAs), where hunting is permitted alongside pastoralism, there is significant human-predator conflict, which includes the killing of hyenas, lions and other carnivores by Maasai and Barabaig pastoralists.[122] It is an area where livestock represent the main source of income for 71.3% of households.[123] On village lands around the NP, livestock losses to predators were estimated at 0.32% of livestock (compared with 4.4% for disease), but were seen as a major cost, generating hostility to carnivores[124] and retaliatory killing – 27 lions were killed between January 2011 and May 2012; no numbers were kept for hyenas killed. Livestock were let out to graze during the day. Local people said they were always accompanied by herders (these were often children). This did not deter attacks by predators, with 97.1% of daytime attacks taking place when herders were said to be present. The presence of dogs did reduce losses.[125] At night, stock were kept in thorn bomas, but hyenas and lions would still break through and kill animals – 35% of night attacks were by spotted hyenas.[126] Hyena depredation on livestock can be eliminated by strong bomas, even if the walls are only four feet high. Hyenas cannot jump, and wire netting surrounding thorn bomas would help exclude hyenas.[127]

Most people who took part in a survey in one of the areas, the Pawaga-Idodi WMA adjoining the south-eastern border of the NP, wanted spotted hyenas (95% of respondents), wild dogs (96%), lions (94%), cheetah (95%) and leopards (94%) to be reduced in numbers or exterminated,[128] because of livestock predation. Respondents said that there had been a few attacks on people – 60% were by lions and 30% by hyenas; 50% were on children.[129] People from all the communities questioned, apart from Sukuma villagers who were agro-pastoralists, admitted to retaliatory killings of predators. Despite this, according to Dickman, most people had a surprisingly positive view of wildlife, excepting hyenas, which were disliked because of attacks and strongly held beliefs that they were "ugly" scavengers.[130] Analysis of reported attacks during Dickman's study period for her 2008 thesis showed that of 469 reported attacks, 165 (35.2%) were by hyenas, compared with 135 (28.8%) by lions.[131]

Uganda

The period of rule of Idi Amin (1971–79), the war that overthrew him and then seven years of instability and civil war that followed had a disastrous effect on Uganda's wildlife. Elephants were poached for ivory and ungulates for meat. Wildlife numbers fell dramatically, including those of large carnivores.[132] The ability of the wildlife department to stop poaching, much of which was by the army or well-armed insurgents, was practically non-existent.[133] Spotted and striped hyenas probably survived better than lions and other species because of their adaptability, nocturnal habits and willingness to scavenge animals remains from poaching and human remains from armed conflict, even in areas of close to human habitations.

Under the authoritarian rule of Yoweri Museveni from 1986 to the present, security improved in national parks and wildlife numbers began to recover. A survey carried out in Queen Elizabeth Protected Area, Murchison Falls Conservation Area and Kidepo Valley National Park estimated that between them they had 408 lions and 324 hyenas – a tiny number of the latter compared with populations in Kenya and Tanzania.[134] The Queen Elizabeth Protected Area had the largest populations of hyenas with 211. Omoya believes that spotted hyenas had once been widespread in Uganda but by the second decade of the 21st century "rarely occur outside protected areas", and were just found in Queen Elizabeth, Murchison Falls, Lake Mburo, Kidepo Valley, Mgahinga Gorilla and Mt Elgon NPs and in Kigezi, Kyambura, Bugungu, and Karuma wildlife reserves.[135] Areas of the Kidepo Vally NP could not be surveyed because of poor security, but it was thought few carnivores survived there because of hunting of their prey species by Karamojong hunters, with perhaps as few as 75 hyenas present.[136] Omoya also reported that poisoning of hyenas with Carbofuran in retaliation for stock killing was common.[137]

Horn of Africa

Despite growing human populations, serious depletion of wildlife through habitat loss, bushmeat hunting and decades of armed conflict making conservation impossible in many areas, striped and spotted hyenas still have a wide distribution. The spotted hyena is the most common large carnivore in Ethiopia, Somalia and Eritrea.[138] It is found in Ethiopian national parks such as Alatash, Awash, Gambella and Omo, but is also abundant in unprotected areas, where it lives in surprisingly high numbers in areas heavily populated by people. In much of the region scavenging waste from human habitations, combined with some stock theft, is more important than hunting wild prey as a source of food, and spotted hyenas live close to villages and towns, including Addis Ababa. In many urban areas they are tolerated "as efficient sanitation units", removing waste and carrion from towns, despite their periodic attacks on livestock and people and the digging up of corpses from graves.[139] Famine, drought and warfare have provided opportunities for feeding on human remains and dead livestock.

Yirga et al. believe that Ethiopia has a very large population and that the generally accepted estimate of about 2,000 spotted hyenas is an underestimate. They project that the population in Tigray province alone may be 28,620 – a number equal to the bottom of the range of the entire African population. Bauer, one of the co-authors, says this is an educated guess based on the research in Tigray around heavily populated areas, but that it raises questions about the size of the Ethiopian and overall spotted hyena population.[140] They believe Tigray is not exceptional within Ethiopia and the country may have a huge hyena population, which could have a major effect on estimates of the entire wild population;[141] this has not yet been verified by other studies. Ethiopia has a livestock population exceeding 80 million – providing hyenas with a prey base and regular supply of carcasses from disease, drought and slaughtering by people for meat.[142] Hyenas have reacted flexibly to human waste disposal, scavenging in streets where waste is not cleared, and habituating waste dumps outside towns.[143] In heavily populated areas, hyenas stay out of sight in bush/dens during the day, commuting into villages and towns at night.

Yirga and Bauer's study of spotted hyena diet in Ethiopia's Tigray province, carried out using scat analysis (which gives a rough idea of consumption), found they ate only domestic stock (killed and scavenged), sheep being the most common but also goats, donkeys and cattle.[144] Scats had sheep remains in 259 cases (21.58%); horse 200 (16.67%); donkey 173 (14.42%); cattle 137 (11.42%); goat 115 (9.58%); dog 100 (8.33%); human 66 (5.5%); and camel 20 (1.67%).[145] Of 362 spotted hyena attacks that were reported during the study period, 31.87% were on donkeys, 14.56% on goats and 10.7% for sheep – the discrepancy with the scat analysis put down to the greater likelihood of people reporting lost donkeys than sheep.[146] The human remains do not mean that they hunted humans, but rather consumed waste that contained human hair. No wildlife hairs were found in the scats, which is significant.

While in many areas hyenas are tolerated to a surprising extent, the heavy predation on stock means that in parts of Tigray, "there is an intense persecution of these animals on the part of livestock owning people" and sustained attempts to eradicate them.[147] Even in areas where livestock depredation and retaliatory killings are high, there were areas in which hyenas could find sanctuary. These are the remnants of forests around Ethiopian Orthodox Churches. Religious belief and centuries-old traditions protect the forests, usually in a circle of approximately 50m radius, and the protection extends to hyenas using them to rest out of sight during the day.[148] This limited form of protection even seems to extend to areas where attacks on humans have been reported. In a study of hyena conflict near Mekelle in Tigray, ten attacks on humans (all adults) were reported and respondents to survey questions did not want hyenas in the area, but they did not try to kill them in the church forests.[149]

In Yirga's study of hyena-human coexistence in the Wukro district of Tigray, he noted the extent of land degradation, with a consequent decline in wild prey for carnivores.[150] The district had 24,583 households with 21,908 oxen, 30,588 cows,

15,431 goats, 82,950 sheep, 9416 donkeys, 1333 mules, 79 horses, 54 camels, and 47,265 poultry in 2009, and rain-fed cultivation of grain crops.[151] Using a call-up survey, in which hyenas respond to broadcasts of animal calls, Yirga estimated that there were 535 spotted hyenas there, a density of $52/100km^2$. [152] Over a period of five years prior to 2013, 203 livestock had been killed by hyenas. The losses made up about 0.7% of annual household income – loss from disease was 1.6 times higher. In most of the district, hyenas depended solely on domestic stock – killed or scavenged. Only in two areas did wild porcupine and hare figure in the diet. The availability of the carcasses of domestic stock or the opportunities to hunt livestock, Yirga concluded, had a "substantial impact on hyena density and supports a viable population of spotted hyena".[153] Few livestock were killed during the study period, suggesting that the majority of livestock remains found in hyena scats came from the remains discarded on waste dumps, or of livestock that died of disease, starvation or from old age and were left where they fell. Yirga reported seeing hyenas routinely foraging on waste around urban areas.[154] Unlike in the area of Tigray in Yirga and Bauer's study above, in this area there were no reports of retaliatory attacks on hyenas, demonstrating:

> …a rare case of coexistence, where spotted hyenas benefit from waste disposal and human communities benefit from the waste clearing service by spotted hyenas. It also demonstrates the high adaptability of hyenas which in our case specialize entirely in waste consumption.[155]

It is the ability of hyenas to adapt to different foraging strategies and diets that appear to enable a very substantial population to thrive in Ethiopia, even in areas where natural prey is absent or depleted.

There is also evidence from research by Yirga and his colleagues that spotted hyenas adapt feeding according to seasonal variations in human disposal of animal carcasses. Given the dependence of hyenas in Tigray on scavenging carcasses from villages, towns and waste dumps, one would expect major problems to arise during the annual 55-day fasting period of the Ethiopian Orthodox Church. During that time, adherents of the Church, which account for the "vast majority of people in northern Ethiopia", do not eat animal products.[156] Yirga et al. studied the diet of hyenas during the fast, when there is a sharp reduction in the animal waste available. They found an increase in hyena predation on donkeys to compensate for the lost scavenging opportunities.[157] Donkeys are common in Tigray and are not usually enclosed at night. Old and sick donkeys are frequently abandoned, rendering them vulnerable to attack. Reports of attacks on people, even during the fast, are very rare in this area.[158] This reinforces the view that spotted hyenas "are able to change between scavenging and hunting as the opportunity arises".[159]

In parts of Ethiopia, attacks on people are not uncommon. Gade recorded that between October 1998 and January 1999, hyenas attacked 50 people near El Kere and Bare in south-eastern Ethiopia; 35 were children. At Fedis, they killed three people and injured three others during the same period.[160] Gade also reported, warning that the details had not been verified, that during a campaign to clear street children from Addis Ababa, many of them were dumped in forests outside the city that had hyena dens, leading to the belief that the hyenas killed them.[161] There were also reports of hyenas feeding on the bodies of young people killed in Addis Ababa and Jijiga during the Red Terror period after the Ethiopian revolution of 1974. Those killed by landmines during the Ethiopia-Eritrea war in 1998–2000 were left where they died because of the danger from mines and so were available as food for the hyenas abundant in northern Ethiopia and southern Eritrea.[162] There are numerous records of hyenas attacking people, especially the homeless, sick or those sleeping outside their houses on hot nights, either killing them or leaving them with horrific facial injuries, as hyenas often tear at the faces of sleeping or unconscious people.[163]

There have been attempts by the authorities, though, to cull hyenas where they have become a danger to people and have taken up residence very close to substantial urban populations. Ludwig Siege worked with the Ethiopian wildlife authorities from 2008–16 and told me he was involved in an attempt to get rid of hyenas that set up dens in an area of overgrown bush known as Bulbula on the south side of Addis Ababa.[164] These weren't hyenas that entered the city at night to forage among the garbage and for livestock carcasses, but were a resident urban population. The police told Siege that at least 40 hyenas lived in the area. They had been tolerated as they cleaned up the carcasses of donkeys, sheep, cattle and dogs. But when they killed a baby and injured its mother, Siege was charged with helping exterminate them. To his surprise, he found them far from being shy of people and he saw several around Bulbula in daylight and was able to find their dens. In 3.5 hours they shot 13 hyenas, and two more two days later.[165] The local people then cleared the bush from Bulbula and the remaining hyenas left the area. Siege also noted that there were large populations living in the city in culverts under roads, especially near hospitals, and on the large site of the Ethiopian Television Academy, where Siege shot five after attacks by them on pedestrians. Siege estimated, on the basis of his work around Addis Ababa with the wildlife authorities, that the capital had a population of between 300 and 2,000 spotted hyenas, with a density of up to $5/1km^2$. [166] Siege agrees with the views of Yirga et al. detailed above that Ethiopia's spotted hyena population is far larger than the 1998 IUCN estimate of 1,000–2,000.[167] Siege was also deployed to reduce the number of spotted hyenas in the Senkelle Swayne's Hartebeest Sanctuary, where a large spotted hyena population was killing too many of the rare hartebeest for them to increase their numbers. He shot 20 hyenas in four days. Siege noted that striped hyenas were "pretty widespread" in Ethiopia but caused no obvious problems.[168]

Over the last 60 years, the curious relationship between spotted hyenas and humans on Harar – mixing coexistence, mutual advantage and periodic bouts of conflict – has continued. They remain the dominant predators in the region, occupying dens outside the walls of the town, foraging outside the town, visiting the town's refuse dumps and taking part in "shows" through which some Hararis make money by encouraging people to pay to watch them feeding hyenas by hand. Kruuk witnessed this 50 years ago[169] and it continues. Kruuk noted that the ability to scavenge from waste dumps and the feeding displays seemed to discourage hyenas from attacking livestock.[170] The continuation of the historic Harari-hyena relationship and the feeding "shows" in Harar are given by Baynes-Rock in his fascinating book, *Among the Bone Eaters*. [171] I will give an outline of the curious interactions, but those interested in a detailed account backed up by extensive research of the context should read the book.

Baynes-Rock was guided to the "hyena holes" in the walls by local people, who believed the accounts that they were deliberately made hundreds of years before to allow hyenas to enter the city to clear up waste, as part of a truce between people and hyenas. Camilla Gibb questions this account of the holes in the wall, suggesting they are culverts to drain water from the town during heavy rain.[172] Whatever their provenance, they are used by hyenas to enter parts of the town. Baynes-Rock said they were big enough "to admit a hyena and were in sheltered areas where hyenas could hide to ensure that it was safe to proceed without people around".[173] He met one of the men who regularly fed groups of spotted hyenas, which had become habituated to being fed by hand in the presence of Hararis and tourists, who paid to see the spectacle. Baynes-Rock said that during the time he spent at the hyena feeding in Harar, he spoke with tourists from overseas. While many said that they knew about the hyena feeding before travelling there, "no-one with whom I spoke had come to Harar specifically to see the hyenas".[174]

As Baynes-Rock walked through the town to one of several hyena feeding areas, he passed an area where the bodies of the homeless and destitute town were "unceremoniously deposited in shallow graves, easily-accessible to grave-robbing hyenas".[175] At one feeding spot, he watched an Oromo man feeding hyenas meat from a bucket. He put the meat on a stick which he held in his teeth.[176] For a payment, tourists were allowed to use the stick to feed the hyenas. The feeding ceremonies, according to local residents. had been going on since the 1950s. Hyenas also gathered near a butcher's shop, where they fed on remains of butchered livestock, the shop owner driving away domestic dogs so the hyenas could feed.[177] Baynes-Rock recounted the many folktales in Harar about hyenas being "transmitting stations" for Muslim saints and beliefs about the ceremonies mentioned in previous chapters about local people feeding porridge to the hyenas being linked to the prediction of famines – if the hyenas ate all the porridge there would be a famine.

In contrast to the human-hyena coexistence in Harar, Baynes-Rock detailed the conflict around the town of Kombolcha in eastern Ethiopia in 2010, involving the killing of hyenas.[178] In April 2010, 10 hyenas died after eating the carcass of a goat laced with poison.[179] At the time there had been instances of hyenas killing goats and people driving hyenas away. After the poisoning, it was reported that hyenas entered the village and killed two children, leading to the killing of another hyena.[180] Local people were said to have called for a villager who could supposedly communicate with hyenas to talk to them, but the local administration got the people to cut down weeds and other vegetation that was providing cover for hyenas during the day. Baynes-Rock said that this worked at first, but while he was there another hyena appeared and was killed. Local people then cut off the ears and patches of skin to use in traditional medicine and for religious purposes.[181] He added that the events in Kombolcha weren't unusual, with periods of coexistence punctuated by hyena predation on stock or attacks on people followed by retaliatory attacks on hyenas.[182] One interesting aspect of the relationship with hyenas that Baynes-Rock noted was the higher tolerance of farmers for hyenas in areas where the narcotic khat (also written chat or qat) was grown for sale – there was a huge market locally but particularly in Somalia and among Somali communities outside the Horn of Africa. The crop is a profitable one and the presence of hyenas may deter thieves from going into the fields at night to steal plants and also get rid of small antelopes, like dik-dik, which eat the leaves.[183]

As late as 1967, Somalia (still then unified) was viewed as having well-preserved wildlife populations in semi-arid and arid areas, but with the looming threat of depletion of ungulates through hunting and the degradation of vegetation by overgrazing and desertification. The growing availability of firearms (a result of Somali government encouragement of Somali irredentist movements in northern Kenya and Southern Ethiopia), loss of forests and hunting for ivory, horn and skins threatened biodiversity. One account warned that if prompt action was not taken, "Somalia will be as bare of wildlife as any of the poorest countries in Africa".[184] There are no functioning protected areas in Somalia or the self-declared independent state of Somaliland. Since the Ogaden war of the late 1970s, the revolt against the Siad Barre dictatorship in 1990 and the continuing Somali civil war, the area has been flooded with military weapons which can be used for livestock protection and illegal hunting.

Sudan, the Sahel and Central Africa

Spotted hyenas are present in good numbers in much of Sudan, from south of Khartoum and across South Sudan (which became independent in 2011). Striped hyenas are present in most of eastern Sudan, from just west of Khartoum over to the Red Sea coast and in eastern South Sudan. In a survey of protected areas in Sudan in 1973, Cloudsley-Thompson recorded spotted and striped hyenas as being present in Dinder NP, on the border with Ethiopia, and in the Southern NP in Bahr el-Ghazal province.[185] Outside the protected areas, he said wildlife was scarce and rarely seen, but both species of hyena were found outside the parks, given their ability to subsist on waste from human settlements where wild prey and their carcasses were unavailable. Albert Schenk, the Wildlife Conservation Society country director for South Sudan, didn't put a number on the spotted hyena population there now, but told me it "is very common, often heard and caught on camera traps in Boma, Badingilo and Southern NPs; the striped hyena is less common or maybe just less often observed" and there are only a few records of it in South Sudan.[186]

In much of Sudan and South Sudan – such as Darfur and Kordofan in Sudan, and across most of savannah and open woodland areas of the South – the raising of cattle, camels, sheep and goats is the main form of subsistence. Many South

Sudanese communities have cattle at the centre of their economies and cultures. The Nuer, Dinka and some other communities engage in transhumant pastoralism, moving their cattle to take advantage of seasonal rains and floodplains.[187] The decades of conflict in South Sudan, Kordofan and Darfur have made extensive research into hyenas there and their relations with pastoral peoples difficult if not impossible. Bauer et al. in 2017 noted that the main threat to carnivores in Sudan had arisen from the war, which forced people to hunt more to supplement food from their livestock and from trading cattle and dairy products for food grains. Bushmeat hunting, encroachment of pastoral communities on protected areas and retaliatory killing of carnivores for stock losses had become serious problems for carnivores.[188]

While accurate accounts of human-hyena conflict in the Sudans are scarce, Clarke in his work on dangerous wildlife cites an FAO report that in poor areas of Sudan, children forced to sleep outside because of lack of shelter or destitution are killed by hyenas. In the Nyamlell region of South Sudan, which housed refugee camps for those displaced by the civil war which led to the secession of the South, 280 orphaned children died, many through being preyed on by hyenas as the children had nowhere to sleep that was safe from predators.[189]

Both spotted and striped hyenas are found across Chadian territory. The sparsely populated country has suffered years of civil war. Much of the country is desert. Hyenas are found on the fringes of the desert and in the remaining areas of woodland and savannah in the south and south-east. Zakouma National Park, in the south-east, is the most important wildlife refuge in a country whose wildlife and its habitats have been severely damaged by deforestation, desertification and the effects of ivory and bushmeat poaching, exacerbated by the difficulties of anti-poaching operations in a war zone that has been flooded with small arms suitable for poaching. The park is a mix of thornbush, grassland and acacia woodlands. Both spotted and striped hyenas are found, but little mention is made of them in accounts of the park's fauna.[190] They are likely to have survived the ravages of poaching better than many other mammals and even to have benefitted from the thousands of elephant carcasses left by ivory poachers in the worst period of killing from 2002 to 2010, when about 4,000 were killed for their ivory. When management of the park was handed over to the conservation organization African Parks, there was a transformation into a better-managed park with substantially reduced ivory and bushmeat poaching.[191]

The main threat to hyenas and other carnivores in the Zakouma ecosystem is conflict with pastoral communities that live in villages or nomad camps around the NP boundary. Three of 11 villages there were found to have predation problems, resulting from hunting of livestock by lions and hyenas.[192] Daytime herding of cattle in regions around the park is usually by children, who have limited ability to ward off large carnivores and may themselves be in danger.[193] Most predation of stock by spotted hyenas occurs at night, when stock is enclosed in poorly constructed bomas, or when animals are brought in at dusk.[194] Hyenas are the most prolific stock-killers, taking livestock throughout the year. They are particularly active in the rainy season, when wild prey disperses from permanent water sources. They are capable of breaking into thatch and thorn-bush bomas.[195]

In the neighbouring Central African Republic (CAR), years of civil war and uncontrolled ivory and bushmeat poaching have not prevented the country from retaining a substantial wildlife population. It is listed in a 2017 study of global conservation performance as being a major performer in terms of conserving wildlife,[196] ranked seventh in the world. Despite major conservation problems resulting from civil war, availability of guns, widespread bushmeat and ivory poaching, poor law enforcement and corruption, it has great biodiversity and significant areas of protected grassland and forest. CAR's vast wildlife refuge, Chinko Nature Reserve, covers 5.9 million hectares in the south-east of the country near the borders with South Sudan and the DRC. It is part of approximately 100,000 km^2 of continuous intact habitat between the Congo rainforest and Sahel savannah. It has important primate, elephant, ungulate and carnivore populations – with 24 carnivore species (including spotted and striped hyenas). Until recently it was plagued by bushmeat poaching and anti-poaching efforts were hindered by inadequate management, corruption and the influx into the reserve of refugees fleeing the civil war. African Parks took over management of the reserve in 2014 under a partnership arrangement. The management and anti-poaching regime put in place combatted the rampant bushmeat and elephant poaching and worked to remove from the park herders and their livestock who were illegally using it for grazing and coming into constant conflict with the resident carnivores.[197] But well-armed nomadic pastoralists, some of whom have moved into the area from conflict zones while others regularly move through the region in search of seasonal grazing, remain a threat to predators like hyenas, through retaliatory and pre-emptive killing to protect livestock.[198] Aubischer has detailed declines in hyena, lion and wild dog numbers in the region as a result of conflict with pastoralists using guns and poison to kill them – with spotted hyena numbers reduced to 50–70% of their original level (no precise numbers are given).[199]

To the west in Cameroon, substantial areas were set aside as national parks, reserves or hunting zones under French colonial rule and maintained by the post-independence governments. Spotted and striped hyenas were and still are found in suitable habitat, but there are no verified population estimates. The main threats to hyena numbers across their range in Cameroon are habitat loss and conflict with pastoralist communities, especially nomadic pastoralists who often graze their animals in NPs and hunting concessions. Some spotted hyenas are shot as trophies in hunting zones, of which there are 25 in northern Cameroon, especially around Waza NP, with 42 recorded as shot over 20 years.[200] In 2001, Bauer and Kari reported that the Waza NP staff estimated that there were between 80 and100 hyenas in the park (a mix of spotted and striped) but that reliable accounts were not available; nor were there reliable figures for predation by carnivores on livestock.[201] Hyenas were said to attack domestic stock at night but were easily chased off by local people if their presence was detected.[202] Hyenas, along with lions and jackals were recorded by Tumenta et al. as being a serious problem for local pastoralist communities. While Bauer and Kari had noted problems along the southern border of the park, Tumenta et al. showed that localities west and east of the Park also faced intense conflict with predators and that the increased presence of large herds of livestock, many owned by nomadic Fulani herders, especially on the western boundary of the park, had increased conflict.[203] Spotted hyenas were reported to be responsible for most attacks on livestock (50% – 91 cattle, 799 sheep and 227 goats, totalling 1,11), followed by jackals (28% – all sheep and goats), and lions (22% – 425 cattle and the rest sheep or goats).[204] Tumenta et al. reported a decreasing tolerance of predators and an increased likelihood of retaliatory killings. There was also evidence of increasing intrusions into Waza by pastoralists as a consequence of weak protection, "resulting in an almost permanent presence of livestock within the Park during the dry season", again serving to increase human-predator conflict.[205]

The Bénoué Complex, an area of protected habitat and hunting blocks, is located in the wooded savannah region of northern Cameroon and comprises three national Parks (Bénoué, Bouba Ndjida and Faro). It is connected with similar habitat in eastern Nigeria. It has 28 hunting zones, where the operators maintain wildlife habitat and try to deter poaching and pastoralist encroachment. Some of the income from hunting operations is given to local communities to encourage tolerance of wildlife, especially carnivores.[206] The estimated population of spotted hyenas in the Bénoué Complex is 131–237, with a higher density (2.43 per 100 km^2) than lions or leopards – no estimate is given for striped hyenas.[207] Spotted hyenas regularly take livestock from villages within the region, prompting retaliatory killings. The presence of nomadic Fulani pastoralists, often moving illegally into protected habitat, increases the level of hyena-human conflict.[208] Croes put the number of goats killed by spotted hyenas in Bénoué at 28 (plus 8 chickens) in 2004–6. Local people saw spotted hyenas as the most problematic carnivores in the area and admitted poisoning predators using a pesticide called Landrin.[209] Spotted hyenas (more so than the shyer striped) are in danger of decline through excessive retaliatory killing and the possibility that trophy hunting does not account for illegal killing in setting quotas, which could lead to excessive off-take.[210]

In Bouba Ndjida NP, Bauer estimated the spotted hyena population at about 120 in 2004 – double the number of lions present.[211] A later survey by Kirsten et al., between 2005 and 2014, noted the general decline of carnivores, pointing out that in northern Cameroon, cheetahs and wild dogs were believed to be extinct and there was concern over the numbers of spotted hyenas, lions and leopards.[212] In Bouba Ndjida, as elsewhere in the region, human-wildlife conflict and the related encroachment on wildlife areas by nomadic pastoralists were the chief causes of decline. Using call-ups surveys, Kirsten et al. estimated a very broad range of 87–618 hyenas in the southern part of the NP, but a spoor survey in 2015 gave the figure of 498 for the whole of the NP and the surrounding hunting zones.[213]

To the south, in the Republic of Congo, much of the habitat is dense rainforest and less suitable for hyenas, although spotted hyenas are found in areas of savannah and open woodland on the margins of denser forest.[214] They are present in the Odzala-Kokua NP, which has a mixture of dense forest, woodland and grassland.

Henschel et al. carried out camera trap surveys and estimated from their identification of 46 different animals that the park could have between 72 and 104 hyenas with a density of about 15.89 per 100 km^2. [215] Odzala is now managed by African Parks, and camera trap surveys they carried out in 2018 revealed three hyena clans in areas of dense forest, numbering about 20 animals, but didn't give an estimate of the population for the whole park.[216] The population of hyenas was said to number around 70. In neighbouring Gabon, spotted hyenas are rare but are still believed to inhabit small grassland areas and surrounding forest on the Batéké Plateau, which stretches from Congo into Gabon. In 1997, it was believed that spotted hyenas and lions were still to be found in the Batéké area.[217] Phillipp Henschel of Panthera told the author that in 2015 a spotted hyena had been identified in the area between the Mwagne and Ivindo NPs in Gabon, across the border from Odzala-Kokua NP, with a camera trap taking a shot of a female hyena in 2019.[218] Henschel said perhaps 2–3 hyenas were present in the area, seeming to be resident there, and were unlikely to be a viable population.

Southern Africa

Southern Africa has the only brown hyenas and a large, widely distributed spotted hyena population. Spotted hyenas do not appear to be under threat, but the brown hyena is listed by the IUCN as near threatened, with a population between 4,365 and 10,111[219] – at one stage it was considered in danger of extinction.[220] Better conservation measures improved surveys and reintroductions in small, fenced reserves have led to a reassessment and the possibility that the overall population is higher than thought. Brown hyenas were once seen occasionally in Malawi and southern Zambia but by the late 1970s had disappeared.[221] Botswana has the largest brown hyena population, at about 3,900 animals,[222] followed by Namibia with between 566–2,440 and South Africa with 800–2,200. There are suggestions that there could be a free-ranging South African population that may number over 2,500.[223] No reliable population estimates are available for Zimbabwe, Angola, Mozambique, Lesotho or Swaziland.[224] Outside protected areas in Southern Africa, brown hyena population trends are unknown but they face anthropogenic threats – shooting, poisoning, trapping and hunting with dogs. The main driver of killings is the perception of farmers that brown hyenas kill livestock, "a concept that has largely been found to be untrue with very few cases of reported livestock predation".[225]

Angola

All three species are found in Angola, though decades of liberation war followed by civil war have meant little research has been carried out on wildlife there until recently, and conservation efforts were prevented by war and the extensive use of landmines in many areas of the country (particularly the once wildlife rich south-east around the battle site of Cuito Cuanavale and the Mavinga and Luengue-

Luiana NPs). There was significant poaching of wildlife, especially elephant, rhino and buffalo, in the south-east when apartheid South African military units occupied large areas in support of the rebel UNITA movement, which was fighting Angola's MPLA government.[226]

The brown hyena is found mainly in arid areas of southwestern Angola, in Kaokoveld desert and dry savannah/woodland. This includes the Skeleton Coast Transfrontier Park between Angola and Namibia. Beja et al. said that recent surveys indicate that it is widespread in the Iona NP, but it was not detected in Luengue-Luiana NP, despite its presence in the nearby Bwabwata NP of Namibia.[227] It is not clear whether the population is stable or in decline. The spotted hyena is found mainly in the south of the country in savannah and woodland in protected and unprotected areas and "is one of the few large carnivores species that shows no evidence of recent population declines",[228] being present in Bicuar, Mupa, Quiçama, Luengue-Luiana and Mavinga NPs.[229]

The main threats to hyena populations come from human activity – including direct persecution, bushmeat snaring, death on the roads and being killed by uncleared landmines. Huntley, who worked for the Angolan wildlife department, said that in the 1960s hyena and other predators appeared to be "very few in number" given that game rangers and cattle herders waged "an effective anti-predator campaign through the 1960s, using strychnine-baited carcasses to kill predators they failed to shoot". He added that conflict with predators was increased by the illegal grazing of cattle in Quiçama NP, and illegal hunting by foreign dignitaries during the last decades of Portuguese rule. South African soldiers also poached around the UNITA HQ at Jamba in Cuando Cubango province and in Iona NP on the coast.[230] During and since the war, bushmeat hunting has been vital to the survival of people displaced by war or whose grazing areas or fields have been sown with landmines or littered with unexploded bombs and shells. Snaring or shooting animals for meat is common, with bushmeat being sold at roadsides in many rural and urban areas.[231] A recent survey by Gonçalves reinforced the view that Angola's fauna was greatly depleted during 41 years of war. He noted the prevalence of bushmeat hunting and its sale across most provinces, concluding that, "Despite being considered a subsistence activity for inhabitants in rural areas, it is concerning due to the increasing number of people becoming dependent on bushmeat trade for income generation and demand for bushmeat in the main cities", which is a major threat to biodiversity.[232]

Demining has been going on in Angola since a ceasefire in the civil war in 1995, but as I witnessed first-hand in that year in Kuito and Malanje, it is a slow and dangerous process. It has been continuing, with work by groups such as HALO Trust and the Mines Advisory group to help the Angolan government and UN organizations to clear an estimated 250 minefields, with millions of mines.[233] In June 2019, the Angolan government and HALO announced a $60m plan to clear landmines to aid the recovery of conservation and promote tourism. In addition to mines, the availability of combat weapons is a problem for wildlife, enabling poaching for ivory, hides and meat. Roland Goetz, an experienced game ranger who worked for two years as a technical advisor in the Mavinga and Luengue Luiana National Parks, told the author that there are still "many illegal weapons even though the Angolan Government are now doing their best to remove them, but in remote areas (hunting) bushmeat remains an important way of [people] feeding themselves".[234]

In his 2015 survey of lions in Luengue-Luiana and Mavinga NP, Funston found spoor of lions, leopards, wild dogs, leopards and spotted hyenas. Hyenas were the most abundant large carnivore in the two parks, with an estimated population of 776 and a density of $0.92/100km^2$. [235] Funston found substantial evidence of human encroachment in the park, particularly between the between Longa and Cuito Cuanavale rivers, with people engaged in illegal logging and bushmeat poaching.[236] It is perhaps the ability of hyenas to adapt to human presence that made, as Funston observed, the spotted hyenas the most successful carnivores in the two parks, though even the hyenas avoided areas of heavy human activity.[237] In Iona NP, spotted hyenas are still present, and brown hyenas are found there and in surrounding arid coastal and inland areas.[238] Elizalde failed to find spotted hyenas in Quiçama NP or the Luando Reserve in the most recent survey, but did detect them in Moxico province in the east and Bicuar NP, with evidence of conflict with livestock herders in both.[239] Leopards and spotted hyenas are the most numerous carnivores in Bicuar, reasonably common and spread throughout the park.[240] In the core areas of the park, spotted hyena density was $10.8/100km^2$ with an estimated population of 145.[241]

Botswana

Botswana has one of the best records in Africa for conservation of large predators, with substantial populations of spotted and brown hyenas, lions, leopards, wild dogs and cheetah. Major wildlife concentrations are found in the Okavango Delta; Chobe NP; the adjacent Linyanti, Selinda and Savuti districts; and, though in lower densities, in the Tuli bloc to the north-east, the Central Kalahari Game Reserve (CKGR) and the Kgalagadi Transfrontier Park (KTP – split between Botswana and South Africa), in the centre and south of the country. There is considerable wildlife in unprotected areas in a country with low densities of human populations outside the towns and main livestock areas.

In the immediate pre-independence period, white settlers involved in farming or hunting and Tswana leaders in the Okavango region came together to get game reserve status for the Moremi area of the delta, which had large and diverse populations of ungulates and predators, including substantial numbers of spotted hyenas. Brown hyenas are found on the fringes of the delta around the Boteti river, which flows from Maun down to the Makgadikgadi Pans.[242] The Okavango region has the greatest density and diversity of wildlife and abundant prey for spotted hyenas. In a survey published in 2017, spotted hyenas were shown by camera traps to be the most numerous and widely distributed carnivores in Moremi.[243] Both species are found in the more arid regions or ones with purely seasonal water, notably the CKGR and KTP, and in the Tuli Block, bordering

South Africa and Zimbabwe, though brown hyenas appear scarce in the latter. In Tuli, it is believed that spotted and brown hyena numbers have been declining for several years.[244] Kuhn wrote in 2012 that spotted hyenas had a reputation for killing domestic stock on farms around the Tuli game reserve, where 7,954 people maintained large livestock herds.[245] The reserve has a diverse predator population with spotted hyenas, lions, leopards, wild dogs, jackals and caracals. There was one confirmed sighting of a brown hyena between 2002 and 2012.[246] A survey of faunal remains of animals eaten by spotted hyenas showed that domestic livestock made up less than 3%, and many of these remains were thought to have been scavenged from human settlements rather than being killed by hyenas.[247]

Clearer signs of hyena predation of stock can be found in the Botswana Department of Wildlife and National Parks (DWNP) Problem Animal Control Registers, which records farmers' reports of livestock losses, to enable them to claim compensation. Examples from Khutse GR, south of the CKGR, show that in areas adjacent to the reserve with significant livestock populations there is regular killing of livestock by predators, with brown hyenas blamed for some.[248] Most attacks were blamed on leopards (857) and lions (588), making up 64% of stock killed. Brown hyenas were reported as killing 269 animals. Spotted hyenas didn't appear in the list. Compensation was only given for predation by lions and leopards.[249]

The Ghanzi region, an area of semi-arid rangeland west of the CKGR, became a major livestock area after the arrival of Afrikaner trekkers at the end of the 19th century. Cattle now outnumber the previously plentiful wild ungulates. The erection of fences prevents seasonal movements of wildlife, leading to die-offs during dry seasons and drought, something worsened by the construction of veterinary fences in the 1970s to prevent the spread of cattle diseases. This resulted in rangeland degradation, the disappearance of most free-ranging wild ungulates, a growth in poaching and increasing conflict between farmers and predators.[250] Ghanzi farmers perceive leopards, cheetah, brown hyenas, wild dogs, caracals and black-backed jackals to be the most problematic carnivores[251]. While brown hyenas may take some stock (small numbers of sheep, goats and poultry), they are not reliant on hunting livestock as a major part of their diet, and livestock remains made up no more than 5% of estimated food intake according to scat surveys.[252] They may be blamed when they or their spoor are found where other predators have killed stock and hyenas scavenged the carcass. Their labelling as stock thieves has led to widespread shooting, poisoning and trapping of brown hyenas in the Ghanzi region.[253]

Spotted hyenas are not a major threat in Ghanzi, as the majority, along with lions, have been exterminated by farmers.[254] Camera trap surveys in the region showed black-backed jackals and brown hyenas to be the most abundant predators, with no signs of spotted hyenas. Tempest Kent estimated there were 428–432 brown hyenas in the Ghanzi farming block, based on spoor surveys, but 547 to 736 based on camera images – concluding that 430 was a prudent figure to work from. The evidence suggested that brown hyenas existed at higher densities outside protected areas.[255] Boast and Houser's survey of Ghanzi predators found that brown hyenas were the most common large predator with an average density of 2.18/ 100km^2 (similar to densities in KTP and Makgadikgadi Pans NP); no spotted hyenas were detected.[256] The brown hyenas were distributed across all the survey sites they used in Ghanzi.[257]

The disappearance of spotted hyenas from most of Ghanzi may have assisted brown hyenas to multiply and range widely. Whether this will continue is unclear, as there was some evidence that in north-east Ghanzi spotted hyenas were recovering and were identified by farmers there as a growing problem, partly as a result of the recovery of wild ungulate populations.[258] Game farmers in the region were less concerned than cattle farmers about most predators but many did not want to see an increase in spotted hyenas or lions, though believed brown hyenas were ecologically important as they cleaned up carcasses.[259] This comparative tolerance of brown hyenas may account for the large population believed to be present outside protected areas and for the growing realisation that commercial farmland "may be advantageous to the species", reinforcing the view that hyenas are able to adapt to human environments better than other large predators (perhaps excepting leopards).[260] The Ghanzi surveys suggested that they prospered more on livestock than game farms, again perhaps due to the extirpation of apex predators, and to the availability of livestock carcasses resulting from death from disease, old age or drought.[261]

Up to 2014, when hunting (trophy and commercial) was suspended by the Khama government, there was evidence that the elephant carcasses produced by trophy hunting in northern Botswana and around the boundaries of the protected areas of the delta were a useful source of food for hyenas, especially spotted ones. Carcasses from hunting influenced the hunting/foraging behaviour of hyenas, with hyenas often spending up to 12 days at a time in the region of large carcasses, reducing the need to hunt.[262] The data was derived from tracking hyenas fitted with radio collars in the south-eastern section of Moremi GR, an unfenced protected area of 4,871 km^2, and the adjacent WMAs where hunting took place – this area had seven spotted hyena clans averaging 30 adults per clan. During the study period, a maximum of 14–17 elephants could be shot in the designated hunting areas.[263] The carcasses were a substantial addition to the food available to spotted hyenas, contributing 8–10% of hyena intake annually.[264] A related study observed 244 spotted hyenas in the area, with a density of between 12.3 and 16.9 per 100/km^2.[265]

Human-wildlife conflict was a problem in the WMAs and adjacent farms. This was partly a result of the increasing presence of livestock in the WMAs, usually seen as buffers between protected areas and grazing land.[266] Gussett et al. reported a total of 938 predator attacks in September–October 2005, with lions and spotted hyenas the major predators, mainly in the Shorobe region where predation on livestock has for years been a major cause of concern for farmers. The region has over 80 cattle posts operated on communal land by several hundred community members – separated from the neighbouring WMA, where hunting was allowed until the 2014–19 suspension and will be allowed again once lifting COVID-19 restrictions allow the return of foreign hunters.[267] Most farmers say they enclose cattle at night, but only 20% had herders with stock during the day; 11% used dogs to guard their stock at cattle posts and two-thirds of them said they suffered losses to predators – but most predation took place out in the bush and at night, indicating that substantial numbers of cattle and other stock were unprotected at night.[268] More cattle die from disease than predation but the latter was viewed as a more serious threat because of the belief that the government could do something about it.[269] Spotted hyenas were the most abundant large carnivore in the area and were said to be a threat to all types of stock, from cattle through donkeys down to poultry.[270] The response to predation was reporting losses to the wildlife department in the hope of compensation, but 24.3% of farmers admitted shooting or snaring hyenas and other predators.[271] Livestock predation in WMAs and livestock areas increased after the suspension of hunting in January 2014. Chief Timex Moalosi of Sankuyo village, just north of Shorobe, told the author that the suspension of hunting meant a buffer was removed between the delta and farming areas and predator incursions (particularly spotted hyenas and lions) had increased; he felt it was partly due to the absence of carcasses of elephants, buffalo and large antelope from trophy hunting.[272]

There are both brown and spotted hyenas in the Makgadikgadi Pan NP. The prey species there fluctuate, with zebra and wildebeest present in substantial numbers during the wet season (November to April); young zebra and wildebeest are an important food source for spotted hyenas, and zebra carcasses (particularly from lion predation) are an important food source for brown hyenas.[273] Food availability falls in the dry season, when most zebra and wildebeest migrate eastwards in search of water and grazing. For hyenas of both species living outside the park, livestock are an important source of prey or carcasses. During the dry season both species of hyena in the NP ranged further afield than during the wet season in search of prey or scavenging opportunities, and clan territories seemed to be flexible. Clans present in livestock areas had more stable food sources all year.[274] Maude and Mills believe that the brown hyena population in and around Makgadikgadi is stable and viable, despite retaliatory killings by livestock farmers.[275] These result from suspicions of stock killing, though the evidence is that brown hyenas scavenge livestock remains and kill few livestock. A survey of potential food for brown hyenas in WMAs found that 70% of the available food consisted of cattle carcasses that could be scavenged, while inside the park 61% of food consisted of zebra remains.[276] In the WMAs, hunted animals accounted for just 3.7% of food and this was mainly springhares. Observations of brown hyena hunting attempts showed them to be poor hunters but able to steal from predators like cheetah.[277]

There was evidence of frequent spotted hyena, lion and jackal predation on stock in WMAs, but no evidence of brown hyena predation.[278] This indicates that retaliatory killing of brown hyenas by farmers is not justified. Overall, Maude and Mills found that the brown hyenas in cattle areas:

> ...benefited greatly from the presence of farmers by eating livestock carcasses... that there was a viable population of brown hyaenas within the cattle area...and stable populations in other cattle areas throughout southern Africa,

suggest that the hyaenas generally benefit from the presence of farmers in spite of their persecution...[279]

This led to a high density of brown hyenas in these areas. The ability of brown hyenas to adapt to human-settled areas also means that in some places they have proved able to survive close to towns and even on the edges of them, as well as in farming areas. Like striped hyenas and spotted hyenas in some areas, they have been able to cope with persecution because of available food and their ability to remain inconspicuous.[280]

In Botswana's central Kalahari, the studies by the Owens found a healthy population of brown as well as spotted hyenas. The former relied on finding carcasses; stealing prey from leopards, cheetah, jackals and caracals; a small amount of hunting of small mammals, birds and reptiles; and consumption of water-bearing fruits. The Owens observed the patience of brown hyenas, who would sit for hours until lions had vacated a kill.[281] Spotted hyenas are present, but in scattered and relatively small clans in the central Kalahari, though in greater numbers and often in larger groups in the southern Kalahari's KTP. In and around the KTP, the brown hyena is the most common large carnivore (density of 1.8/100km^2), with spotted hyenas at half the density, mostly likely as a result of the brown hyena's more eclectic diet and lesser dependence on water.[282]

Malawi

Given its higher human population density than any of the other southern African states, it is not surprising that most of the larger wildlife species are limited to protected areas (such as Kasungu, Liwonde and Nyika Plateau NPs and game reserves like Majete and Nkhotakota). Many of these have suffered poaching for decades; ungulate numbers and diversity have been reduced, along with predator populations. Taking Majete as an example, between its establishment in 1955 and the early 2000s, most large mammals disappeared, mainly as a result of poaching and human encroachment. Only a few ungulates and a small population of hyenas survived[283] – again demonstrating the ability of hyenas to adapt to anthropogenic threats. Kasungu experienced heavy poaching, too, but its larger area and movement of animals to and from Zambia meant that while the elephant population was substantially depleted, some large ungulates survived, along with a small population of carnivores, including spotted hyenas, lions and leopards. Some wildlife is to be found in farming and even urban areas, such as small ungulates, monkeys, hyenas and even servals. My first sighting of a spotted hyena was in the central of the capital city, Lilongwe. There are still hyenas living in urban areas, around villages and on farmland.

As noted in the previous chapter, there have been episodes of attacks on people by spotted hyenas, especially in the south of the country. This continued into the modern era with periodic attacks, though not at the level of the 27 killed in the late 1950s and early 1960s around the southern Mt Mulanje region.[284] In 1972, there was a report of three hyenas chasing a teacher as he cycled to work early in the morning – he was pulled from his bike and badly mauled. Game wardens tracked down the hyenas and killed two of them.[285] When I lived in Lilongwe from 1981–2, there were periodic reports of hyena attacks on people around the capital, often put down by the government to rabies, but many ordinary Malawians believed witchcraft was involved. In 2002, a five year-old boy was among six people reported as killed by hyenas on the outskirts of the commercial centre of Blantyre in southern Malawi. The government quickly denied rumours that the deaths involved witchcraft and said shortages of wild prey in the hilly areas around the city had forced hyenas to forage around the city.[286]

It is hard to estimate the country's hyena population. According to Brennan PetersonWood of Conservation Research Africa/Carnivore Research Malawi (CRM), "there are so many living outside protected areas and in the forestry plantations, an estimate would be wildly inaccurate".[287] On farmland and even within towns, where there are dens where they can stay out of sight during the day, hyenas are often found. PetersonWood told me that in order to try to survey the populations outside protected areas, the researchers were trying to get funding for drones "to do surveys of the isleburgs and other rocky areas where they tend to den and, from that and some ground surveys, create a predictive map of where hyenas should be and in what numbers". He added that 5–7 hyenas live in or around the Lilongwe Nature Sanctuary and move out from there at night and are frequently seen in Kumbali and in the area of state house (aka Kamuzu Palace), where there is another den. There are also hyenas in the area of Bunda Agricultural College just south of the city. PetersonWood believes the wider Lilongwe district had nearly 30 hyenas overall – with 21 individually identified and possibly another eight, with most of them denning outside the urban areas and entering the city at night to forage.[288] They are believed to feed on waste dumps, around markets and butcheries, but also on dogs that roam the city streets, and occasionally vervet monkeys, common duiker, mice and birds. They are vulnerable to attack if spotted by people and one was killed by a mob in Lilongwe city in July, and they often fall victim on the roads at night.[289]

CRM reported in January 2020 that a radio-collared female had led them to a new den site near farmland and a small village just outside Lilongwe. With the permission of the local chief, they set up a camera trap and captured images of a young hyena pup at the site.[290] A month earlier, a sub-adult male was killed by a car in Lilongwe – he had been part of the urban clan that was being monitored by the researchers. His tail had been cut off for use in witchcraft or traditional healing. CRM noted that the expansion of Lilongwe and the building of new roads was a problem for the urban hyenas and those entering at night, as it blocked safe movement corridors and foraging routes, increasing the likelihood of death in road accidents. Hyenas are also killed or injured in snares set on game trails around towns and villages to catch animals for meat.[291]

CRM not only researches the numbers, locations and diet of Malawi's hyenas, but also tries to educate people about them and give advice on how to avoid conflict, how to build better stock enclosures and dispose of waste safely.[292] When a woman was killed by hyenas at Ntcheu, south of the southern end of Lake Malawi, in July 2020, the CRM put out information about how to avoid being attacked and how to act if you come across a hyena, telling people not to run but to look as big as possible and yell or clap loudly to scare it off.[293] The woman killed had been drinking and had decided to sleep out in the bush, where she was attacked by several hyenas. A young man chased away the first hyena that attacked, but when he went to get help the other hyenas attacked and killed her. There had been attacks in Ntcheu two years earlier in which a child and several adults were attacked and injured.[294]

Mozambique

Spotted hyenas are widely distributed in protected and unprotected regions of Mozambique. Spotted hyenas are well represented in the Greater Limpopo Transfrontier Park (GLTP), Niassa Reserve, Gorongosa NP and other protected areas and hunting concessions and are also found outside these areas in farmland and around towns. In the Niassa reserve, hyenas are numerous but shy of people and are rarely seen, except around safari camps and villages, where they forage for waste. It is believed they might be declining in the reserve as a result of poisoning (after attacks on livestock) and bushmeat snaring.[295]

One positive development has been the project to reintroduce spotted hyenas to the Zinave NP part of the Greater Limpopo Transfrontier Conservation Area, to the east of the northern part of the GLTP and Zimbabwe's Gonarezhou NP.[296] In August 2020, four spotted hyenas were captured in the Sabie Game Park, on the southern edge of the GLTP, and introduced into Zinave, to start the process of restoring large carnivores to the park by the Peace Parks Foundation. Over 2,000 plains game from 13 different species have already been reintroduced to Zinave. According to Bernard van Lente, Peace Parks' Project Manager in Zinave:

> Zinave's herbivore population is growing extremely well…so much so, that we have decided that it is now time to reintroduce predators such as hyenas into the system. Animals die of natural causes, which is completely normal and part of nature's cycle, but because there are no large predators and very few scavengers in Zinave's sanctuary where most of the wildlife was introduced, the carcasses are not being properly recycled.[297]

Prior to the reintroduction, jackals were the only predators regularly seen – lions will be reintroduced at some stage.

Namibia

Namibia is home to the three southern African hyena species and is second only to Botswana in numbers of brown hyenas, with a population estimated at 566–2,440 animals,[298] with the top end of the scale seeming to be more realistic given their distribution across most of the country. The spotted hyaena population was estimated in 2012 at 7,198–13,092 and no more recent estimate is available.[299] Both hyena species were under threat for many decades because of the bounties paid by the South African-controlled South West Africa administration for killing predators viewed as livestock killers. Some predators were killed for doing "nothing worse than [digging] holes in fields or under the 'Vermin' jackal-proof fences enclosing the sheep".[300] In 1965–66 the records showed that 22,242 jackals, 160 hyenas (no details on whether they were brown or spotted), 156 wild dogs, 824 caracals and 39 leopards were killed.[301] The interests of commercial, white-owned livestock farming were put above conservation. These removals disrupted food chains and often had negative effects on livestock by not having predators to remove other animals that ate the grasses.[302]

Livestock depredation is a cost to commercial (usually white) and communal (black) farmers in Namibia, with the average cost in 2013 calculated at US$3,641 per farm, with some smallstock farmers struggling to make a profit because of losses to hyenas, cheetahs, leopards, lions and jackals.[303] Agriculture is the largest source of employment in Namibia, so such costs are felt nationally. In rural areas, many small farmers are among the poorest in the population, with 85% of the poorest households in rural areas.[304] The conservancy movement – based on the concept of community-based natural resource management (CBNRM) – was launched after independence with the first four communal conservancies established in 1998. There are now 86, with 227,941 inhabitants, out of a total national population of 2.448m (9.3%). They have control over the use of their resources on the conservancies and mix photographic tourism, hunting, game cropping and pastoralism. The aim is to provide rural communities with income while increasing wildlife populations outside protected areas.[305] This has helped a recovery in ungulate numbers across suitable habitat and the survival of predators, though not without conflict with livestock owners in the conservancies and outside, which remains a problem.[306] In 2010, the Ministry of the Environment (MET) established a payment system (but not compensation for every animal lost) to offset losses suffered on state land, which includes conservancies. Farmers do not get the full market value of lost stock, but receive payments if claims are verified by MET. Commercial farmers do not get this off-set payment, so they will usually kill carnivores that take stock.[307]

Conservancies and commercial farms are found across the country and provide habitat for wildlife and livestock, as well as safari and hunting areas. HWC is an issue for all farmers because of the presence of wild ungulates and predators, whose numbers have increased in the 30 years since independence. Some farmers tolerate wildlife and even welcome it in order to engage in income generation from tourism or hunting, but persecution of predators outside protected areas has been one of the reasons for declines in numbers (cheetah being the worst affected but also with regular killing of spotted and brown hyenas, jackals, lions, leopards, caracals and even aardwolves). This is legal in unprotected areas in Namibia, including conservancies, if the predators pose a threat to livestock.[308] Rust and Marker's survey of 147 respondents on conservancies and farmland on which communities had been resettled south and south-east of Etosha NP since 1990 found that smallstock were the most frequently predated, with jackals the chief culprits (with spotted hyenas involved, too). Cattle predation was chiefly by spotted hyenas.[309] Livestock losses led to the killing of jackals, cheetah, lions and leopards either by farmers or trophy hunters – no brown or spotted hyenas were reported as killed.[310]

A study in north central Namibia, on 30,000km^2 of rangeland in the districts of Gobabis, Windhoek, Okahandja, Otjiwarongo, Omaruru, Outjo, Tsumeb and Grootfontein found a wide distribution of brown hyenas in addition to spotted hyena, cheetah, leopard, lions, jackals and a few wild dogs.[311] Over 80% of all farmers thought predator conservation on farmland important. Conservancies in the area placed greater importance on the survival of large carnivores (especially cheetahs and brown hyenas) than did non-conservancy farmers. All farmers were less enthusiastic about having spotted hyenas and lions.[312] Over 65% of the respondents to a survey of conservancy and non-conservancy farmers reported livestock losses to large predators and about 50% of respondents, even in areas tolerant of predators, said that they used shooting or box traps to remove predators from their farms; a few admitted to using leg-hold traps or poison.[313]

The use of some conservancies/farms for tourism or trophy hunting increased tolerance of predators, given the income generated, the perceived value of predators, even lions and spotted hyenas, and the perception that trophy hunting controlled numbers.[314] In some farming areas, the existence of waste dumps, offal pits (for remains from slaughtering livestock for meat) and livestock carcasses in the bush all proved attractive food sources for spotted and brown hyenas on conservancies and commercial farms.[315] The ability of all three species of hyenas to survive on farm and conservancy land was demonstrated, at least for the study area in southern Namibia, by Edwards's survey of 24 farmers. Spotted hyena were found on 50% of cattle farms, none on smallstock farms, 37.5% of cattle and smallstock farms and 100% of game farms; brown were found on 83.33% of cattle farms, 100% of smallstock farms, 31.25% of cattle+smallstock farms and 100% of game farms.[316] Spotted hyenas were more prevalent on conservancies and farms in central-northern Namibia, and brown hyenas less so, most likely because spotted hyenas will dominate carcasses and attack brown hyenas. Brown hyenas were common and widespread in central to southern Namibia, where spotted hyenas were less abundant.[317] In areas of northern Namibia near the boundaries of Etosha NP, spotted hyenas were denning in the NP but foraging on farmland at times.[318] An earlier study by Trinkel et al.[319] of a spotted hyena clan in Etosha NP showed that all clan members followed migratory prey during the wet season, when ungulates dispersed from dry season concentrations near water sources. Etosha was estimated in 2009 to have 340 spotted hyenas, the largest concentration in Namibia, with clans ranging in size from 11–30 adults.[320] The IUCN Hyaena Specialist Group held a meeting at Ongava Reserve (bordering southern Etosha) to work on mapping new ranges for all four hyena species in

2019. They found that "a single clan composed of about 30 adults was living on Ongava at a density of 8.1 hyaenas per 100 km^2, which is much higher than in the neighbouring Etosha (2.7 hyaenas per 100 km^2)". This was thought to be due to higher prey densities found on the private reserve than in Etosha.[321]

Brown hyenas are more numerous than spotted in the southern half of the country, on many areas of farmland and in the Namib Desert and Namib-Naukluft NP. They are also found in higher numbers along coastal areas with seal colonies. There is conflict with livestock farmers and with those wanting to protect feral horses that breed in the wild. South-western Namibia has the country's only population of feral horses in the Namib-Naukluft NP. A long-running battle has been fought over whether the feral horses should be treated as rare wildlife and be protected, if necessary by culling or even exterminating the region's spotted hyenas. Hyenas prey on the horses and both are attracted to the area's single water point in the NP. By early 2019, it reached crisis point as horse numbers plummeted from 286 to only 77.[322] At the urging of an NGO called the Wild Horse Foundation, MET took the bizarre decision to shoot some of the hyenas after the only foal in the herd was injured by hyenas – thereby choosing conservation of a feral species over an indigenous one. Three spotted hyenas were shot and plans were made to trap and relocate three more out of a possible 11 in the area. While welcomed by horse lovers, the decision was rightly condemned by the Namibian Environment and Wildlife Society, which said "MET appears to be contradicting its own tourism and wildlife policies by advocating removal of the hyenas from the national park… Spotted hyenas are classified as vulnerable and are therefore a conservation priority in Namibia". A carnivore specialist who wished to remain anonymous, was clear that "To have a natural species killed in favour of a feral species in a national park is a very, very sad day for carnivore conservation".[323] After the killing of most of the hyenas, it was reported in December 2019 that the two remaining hyenas there had produced a cub, with the prospect of the clan expanding again.[324]

On commercial farms near the town of Aus on the edge of the Namib-Naukluft NP, Edwards found that gemsbok, springbok, ostrich and small mammals were the main prey species existing alongside livestock, with boreholes providing water for the stock. Spotted hyenas were present, in small numbers, but brown hyenas were more numerous. Some farmers tolerated predators, but others didn't and used lethal control as the first response to losses.[325] Brown hyenas were often blamed for stock losses, partly because they may be observed or their tracks found by carcasses of stock killed by other predators. On many farmland areas, brown hyenas scavenged from cheetah and leopard kills. Hyenas can easily take prey from cheetahs and will often take carcasses from leopards.[326] Brown hyenas are subordinate to spotted hyenas, wild dogs and lions but in areas where they don't occur, brown hyenas are at the top of the carnivore guild (though not always 100% dominant over leopards).[327] Although brown hyenas may be blamed and persecuted as suspected stock killers, "farmland areas may be strongholds for brown hyaenas because, as scavengers, they do not pose a critical threat to livestock operations but are able to access carcasses as food".[328] The ability to adapt to living on farmland may explain why "Namibia's brown hyaena population appears stable, if not increasing".[329]

Brown hyenas appear to be thriving along the coast, especially where there are large Cape fur seal colonies. The hyenas hunt the young seals born at the colonies and where they stay until big enough to take to the ocean. The colonies also produce fish and other detritus and the carcasses of a variety of marine animals are washed up on to the beaches. Wiesel et al. recorded that the large breeding colonies "form the basis of a year-round reliable food source" for brown hyenas.[330] The hyenas hunt and kill the young seals, frequently killing in excess of their need because of the availability of helpless prey animals – sometimes eating only the brains. The ability to predate the seals and scavenge carcasses of other marine animals along the coast is an important part of the hyenas' diet and enables females to raise cubs more successfully.[331]

Brown hyenas have also been successful in increasing numbers in some enclosed reserves. In 2018, a brown hyena survey was carried out at Okonjima reserve in central Namibia, which is fully enclosed, has no lions and only one spotted hyena, but numerous leopards.[332] Camera traps at latrine sites and areas of known hyena activity identified 48 different hyenas, with the highest brown hyena density ever recorded, at 24.01 brown hyaena/100km^2.[333] Edwards believes this results from:

> ...a combination of protection from human persecution, the perimeter fence stopping hyaenas moving out of Okonjima, a high density of herbivores providing food resources in the form of non-violent mortalities and hyaenas stealing or scavenging kills from Okonjima's high density leopard population.[334]

There were six clans there, with average core ranges of 40km^2, and two nomadic males – given that the whole of the reserve was used by the clans, it is not surprising that there were frequent fights, as the nomadic males could not avoid clan territories.[335] There is the potential for overcrowding and releasing brown hyenas into neighbouring farmland is not a good solution, according to Edwards, as the hyenas are quite habituated to people and the danger of conflict is high. What is more likely is finding other fenced reserves into which dispersing animals, particularly males, could be relocated.[336]

The probable need for management of numbers was demonstrated by the first ever evidence of infanticide by a brown hyena in October 2019 at Okonjima. A den with two cubs was discovered by researchers and a camera installed. On 7 October, an unidentified female brown hyena entered the den and brought out one cub which was killed. No sign was ever seen of the second cub. After a few days, the mother abandoned the den. Edwards et al., who were monitoring the den, postulate that:

> the main driver for the infanticide stems from the high population density and enclosed nature of the study site...The resource competition hypothesis might also provide an explanation: the individual may have killed the cubs in order

to secure resources as food or, if the individual was a female, the den site for her own current or future offspring.[337]

In northern Namibia, spotted hyenas are the most numerous of the species, notably in Etosha and north and north-east to the Caprivi Strip. Prior to 2010 no extensive research had been carried out of the hyena population of Caprivi, but they were thought to be present in their thousands and to be a major cause of HWC, through livestock predation.[338] One way in which this conflict was being mitigated to an extent, by appeasing local communities, was for the conservancies in the region to each have a quota of one hyena that could be hunted annually.

To try to assess the population and possible effects of HWC and hunting, Lisa Hanssen of the Kwando Carnivore Project (KCP) carried out surveys in Bwabwata NP (BNP), which is divided into a fully protected area and multiple-use conservancies, where there is a substantial human population with substantial numbers of livestock. In these areas there are also two hunting concessions with quotas of three spotted hyenas which can be hunted. In the park, hyenas were found to be largely restricted to the core conservation area and there were none resident in the multiple use areas. The survey suggested a density of 0.6–1.5/100 km^2, with a population of 15 to 25.[339] The survey also found that hyenas generally avoided villages, but livestock straying away from human habitations and with no herders suffered predation on an opportunistic basis by hyenas. The simple answer to reduce this loss would be better livestock husbandry techniques with herders accompanying animals and stock being securely enclosed at night. Hanssen concluded that trophy hunting of hyenas was not sustainable in the BNP because of the small population and low density; she added that income annually from hunting hyenas would be $1,500 and hyenas were not a main attraction for foreign hunters.[340] In June 2014, representations from some of the BNP communities led to the hunting quota being removed and the national quota for hunting hyenas cut by half.[341] The spotted hyena was also taken off the hunting quota for the conservancies at the Mudumu and Mamili NPs in the Caprivi, where the spotted hyena population was at a slightly higher density than BNP and there was also hyena predation of poorly guarded livestock.[342]

South Africa

South Africa has substantial populations of spotted and brown hyenas. Spotted hyenas are the most abundant with 1,300–3,900 in Kruger NP and significant numbers in the KTP, Hluhluwe-iMfolozi in KwaZulu-Natal, Madikwe, other game reserves, private game areas and in some unprotected areas.[343] Brown hyenas have been estimated at 800–2,200, though there are suggestions, not yet supported by research, that there could be a free-ranging population that may number over 2,500.[344] The depletion of wildlife outside protected areas and culling in parks like Kruger right up to the late 1970s had reduced the range and numbers of spotted hyenas, but they did survive and gradually increased in numbers. The major threat to brown and spotted hyenas in South Africa is anthropogenic – either killings related to HWC and livestock predation or snaring, with hyenas caught in snares set to harvest bushmeat from wild ungulates. Brown hyenas, under South African law, are protected species and should not be killed without special permission, even if a problem animal is killing stock.[345]

Another potential threat is the use of hyenas in *muti* (traditional medicine and "magical" charms), which has a long and continuing importance to many South Africans.[346] A study of animals parts in *muti* in KwaZulu-Natal put hyenas 31st in the list of animal parts most used.[347] Brown hyena parts, usually dried and powdered, are used to ward off the danger of "stepping over dangerous tracks laid by sorcerers" and spotted hyena parts used to treat painful limbs.[348] Other uses recorded include fat from hyenas being used to combat arthritis, parts of hyena noses as an aphrodisiac and dung being used for magical purposes. Many of the bodies used for *muti* come from roadkill or animals caught in bushmeat snares.

Sam Ferreira, a senior scientific officer for SanParks at Kruger, believes the 50 years of annual culling means Kruger has a smaller spotted hyena population than would naturally be the case and that it has still not fully recovered, despite an increase in numbers in the last 30 years.[349] Despite the years of culling and the slow recovery, in the 1970s it was again suggested that hyenas and lions be culled to reduce predation on young wildebeest and zebra, whose numbers were falling. Smuts, who took part in the cull, said that target animals would be darted and euthanised with lethal injections.[350] Between 1974 and 1978, 297 hyenas were killed. After the cull, populations took at least 17 months to start recovering because of the slow reproduction rate of hyenas.[351] The culling stopped when surveys showed no long-term change in wildebeest calf survival and indications that hyenas preyed on all age groups, and that they killed more impala than wildebeest or zebra.[352]

A 1989 hyena survey estimated 3,667 spotted hyenas in the park,[353] at the top end of the range given by the IUCN.[354] Ferreira and Funston noted major issues affecting the park's spotted hyenas. One is the major overlap in areas occupied and prey hunted with lions (hyenas outnumber the Kruger NP lion population)[355], with the consequent conflict and lions periodically killing hyenas. The main threat in the park, though, is disease. Canine distemper can be spread from domestic dogs kept by the increasing human population that rings the South African side of the park and the adjacent private reserves.[356] But the human presence in and around Kruger can also be of advantage to spotted hyenas, with access to refuse dumps and water sources. Hyenas regularly frequent areas around the rest camps. At the park HQ and largest rest camp at Skukuza I've seen and heard them on each night that I've stayed there – patrolling outside the fence looking for ways of getting food. Hyenas also hang around the fenced-off tourist areas in the hopes that people will break the rules and throw food to them, which many do.

There is a clan that has its territory in the area surrounding Skukuza, and another at Afsaal picnic site and restaurant.[357] The Skukuza clan is kept outside the rest camp but has access to the staff village, the golf course, a shop and some communal areas. About 2,300 staff live in the village and individual houses and their refuse bins are fenced off, but larger waste sites are not, and damage to fences enabled hyenas to gain access to many areas. The Afsaal site is unfenced. Because visitors have to leave the park or be back in their camps by sunset, hyenas have access to many areas around the camps and at picnic sites without human disturbance.[358] Belton's two year study of hyena use of anthropogenic resources showed 29 individually identifiable hyenas using Skukuza and 25 at Afsaal.[359] At Afsaal, the local clan frequently established its den site 500m from the shop, restaurant and picnic site.[360] A former chief warden of Kruger, Bruce Bryden, wrote that hyenas would come into the unfenced areas of camps and take anything remotely edible, even old tyres, hosepipes and pots.[361] Bryden also noted that many Mozambicans who infiltrated Kruger in an attempt to find a new life in South Africa were killed by lions and hyenas in the park every year. A total of 13,380 has been suggested for the number of Mozambicans who died crossing Kruger between 1960 and 2005, the majority killed by lions, but many by hyenas.

The Kgalagadi Transfrontier Park (KTP) houses both spotted and brown hyenas and was the site of one the earliest major research projects on brown hyenas and comparison studies of spotted and brown hyena behaviour by Mills. In 1990, he estimated that there were 77–99 (0.8–1.0/100km^2) spotted hyenas in the park and about 175 (1.8/100km^2) brown hyenas[362] – numbers will have undoubtedly fluctuated since then but there is no reason to suspect a drastic change. The spotted and hyenas, Mills said, rapidly became habituated to research vehicles, but the brown were more elusive and harder to count.[363] Within the park there was no conflict with people, but on livestock farms around the park both species were blamed for losses and persecuted as a result. In the park, the spotted hyenas killed over 70% of their food intake and scavenged 27%, with large and smaller ungulates like gemsbok, wildebeest, springbok. hartebeest, eland and kudu as favoured prey.[364] Brown hyenas scavenged most of their food (large and medium-sized ungulates) and foraged for vegetable matter such as tsamma melons. They hunted small mammals (bat-eared fox, springbok young and springhares mainly) but with low success – only 6 from 128 hunts being successful.[365] Around the fringes of the park, where it is fenced off from farmland, mortality of hyenas through HWC is most evident. In 1981, Mills recorded that brown hyenas that were killed by people met their fate along the fence between the farmland and both the South African and Botswana sections of the park. He said that 30 hyenas died during his study period, 12 of them as a result of human actions – many being poisoned by sheep and goat farmers or caught in gin-traps.[366]

In the Hluhluwe-iMfolozi Park in KwaZulu-Natal, there was thought to be a population of around 200 spotted hyenas. But a survey published in 2009 revealed a much higher and healthier population of 321. The hyenas competed with resident lions and had a density of 0.357 individuals/km^2,which was relatively high compared to other southern African conservation areas.[367] Leopards and jackals were also present and attempts had been made to reintroduce wild dogs, cheetahs and brown hyenas, the latter failing to establish themselves and dying out.[368]

As noted with Okonjima in Namibia, brown hyenas appear to thrive in fenced reserves. Pilanesburg NP was opened in 1980 after the relocation thousands of animals to add to the game which had survived on farms in the region. The park was enclosed by 110km of electrified fences and a total of 6,000 ungulates, predators and other species introduced.[369] Thorn estimated that in a small study area in the park totalling 36km^2 (out of the park's total of 552km^2), there were 12 brown hyenas, with the possibility that there could be a maximum of 22.[370] The density was 2.8/100km^2, which exceeded that of KTP and Mkgadikgadi NP. But density of hyenas outside the NP in farming areas was low, presumably because of persecution.[371] In contrast, in South Africa's North West province, Kent and Hill found that brown hyenas were doing well on areas given over to livestock production.[372] There appeared to be relatively peaceful coexistence of people and hyenas, with the presence of hyenas 57% higher on farms that had resident wildlife in addition to stock. Farms given over to game showed the greatest tolerance of carnivores.[373] The tolerance of brown hyenas on unprotected and farmland varied, and the density of brown hyenas was largely governed by the attitudes of landowners.[374] That sufficient landowners either tolerated or were unaware of the presence of brown hyenas (given their nocturnal foraging and use of dens during the day) may explain why they appear to have spread from the northwest of the country (Limpopo, North West, Mpumulanga and parts of Gauteng province) and from the Northern Cape (around KTP) to northeast KwaZulu-Natal and the Eastern Cape (some being introduced on private reserves in the latter, such as Shamwari, Kwandwe and also Mountain Zebra and Addo Elephent NPs).[375] Reintroduction at Mokala NP, in the Northern Cape, was less successful, with one of the two hyenas released being shot by a farmer after leaving the park and killing livestock, and the other then relocated.[376]

Across areas of South Africa where brown hyenas are believed to be resident on farmland or other unprotected land, a survey of farmers showed 28% thought they were increasing in numbers, 28% remaining stable, 39% did not know. Only 5% thought they were in decline. Many thought they were a danger to livestock (especially calves) but some did not, with 60 farmers answering that they were and 106 that they weren't. But 148 to 19 thought they were part of the healthy ecosystem.[377] Where brown hyenas were blamed for stock killing, notably in North West province, they were shot at night using spotlights or trapped. But spotted hyenas were viewed by most farmers in this province as a greater threat, though they were not found in as many areas as brown hyenas.[378]

In Limpopo province a study of brown hyena diet in Zingela GR, showed that though the reserve is unfenced and brown hyenas could move out into livestock farming areas, no evidence was found in scats of livestock remains.[379] Faure et al. concluded from this research that:

> Given our results suggesting that brown hyena rely mostly on natural prey species, we advocate the necessity for increased community engagements

efforts to create awareness of the valuable roles scavengers and other carnivores play within ecosystems in order to increase tolerance of brown hyaenas.[380]

Limpopo is a mainly agricultural province with a large game farming and hunting industry. This means that the possibilities for HWC involving hyenas are substantial outside protected areas. Game and livestock farmers interviewed by Thorn et al. in 2011 said that carnivores killed 1.4 % of total game and domestic livestock. But this level was "generally not sufficient to threaten farming livelihoods or the provincial economy. Farmers reportedly killed 303 carnivores in the year prior to the interview", 44 of which were threatened species (mainly wild dogs).[381]

The study sites in Limpopo were in areas where caracals, black-backed jackals, leopards, cheetahs, brown hyaenas, African wild dogs and occasional nomadic spotted hyaenas and lions were found. The 95 farm managers interviewed said that stock killers were killed, 92% of the predators killed were shot, and 79% of those who took part said they used enclosures designed to exclude predators.[382] Brown hyenas were believed to have killed 58 out of 934 game or livestock killed on farms in the previous year and 27 brown hyenas were killed, out of the total of 303 predators, despite 92% of farmers saying they would tolerate brown hyenas. The rate of killing exceeded the hyena's toll of stock; no figures were given for killing by or of spotted hyenas.[383] The implication was that "carnivores were persecuted according to their availability and not in relation to their perceived culpability for predation" which persecuted brown hyenas disproportionately.[384]

In a number of fenced reserves, brown hyenas have been introduced after having been driven to near extinction in the early 20th century,[385] and have adapted well, increasing their numbers, notably in the Eastern and Western Cape. The growth of private reserves (like Shamwari and Kwandwe) with considerable income from safari tourism has prompted the reintroduction of predators and ungulates. Most of these reserves are enclosed and so offer protection, while bringing up issues of managing metapopulations of predators like brown hyenas across a range of fenced reserves. For them to play a positive conservation role for brown hyenas, "careful monitoring and subsequent management" has to be used to prevent in-breeding, overstocking and the ecological consequences of that.[386] The Kwandwe Private Game Reserve in Eastern Cape, which is surrounded by an electrified fence to separate its wildlife from the adjacent farming areas, reintroduced six brown hyenas in 2013 in a reserve with a high density of lions, leopards and cheetah, as well as prey species like kudu. By 2013, the hyenas had increased to at least 28 adults, with a high density and substantial food available from the carcasses produced by the predators, and the absence of competing spotted hyenas and anthropogenic threats.[387] High reproduction rates could produce hyenas that could be moved by reserve managers between the various fenced reserves to prevent in-breeding and overcrowding in any one area, and restock areas where they are not currently found.[388]

At the private reserve at Shamwari, 4 adult females and 2 adult males were introduced in 2002, while Mountain Zebra NP had 3 hyenas relocated there.[389] At Shamwari, hyena territories are small and the density high (14/100km^2).[390] John

O'Brien, the chief ecologist at Shamwari, studied in detail the problems of managing the carnivores on a small fenced reserve with high densities of brown hyenas, lions and leopards, noting the need for careful monitoring and management of predators to prevent overpopulation, inbreeding and too great a reduction in prey.[391] Wild dogs were introduced along with the other predators, but removed because of the high toll they were taking of ungulates. But the brown hyena population has prospered and appears to be largely self-regulating, while the lions have bred "too well" and need population management.[392] The 6 hyenas increased to 23, but in the long-term are not genetically viable without means of artificially dispersing some animals through relocation and refreshing bloodlines through introduction of animals.[393] The Shamwari hyenas have found ways of breaking through or under the fences and mixing with free-ranging hyenas in the surrounding unprotected areas and other nearby protected areas (Phumba Private reserve and Addo Elephant NP). The Shamwari brown hyenas have always returned to the reserve. There are no spotted hyenas at Shamwari but they are found on the other two nearby reserves. One managed to enter Shamwari for a few days but then went out into farming land, killed some calves and was shot.[394]

Spotted hyenas are present in some fenced reserves. At Selati Reserve in Limpopo province, which is surrounded by an electrified fence to stop predators getting into surrounding farmland, spotted hyenas compete for prey with lions and have a more varied food intake than them – 16 mammal species (not counting small rodents and also birds) were found in hyena scats and 11 in lions scats.[395] Spotted hyenas at Selati have a density of 12.52/100 km^2, compared with 1.53/100km^2 for lions.[396]

Zambia and Zimbabwe

Spotted hyenas are found in both countries. There are brown hyenas in south-western Zimbabwe but no recent confirmed sightings in Zambia, though it is just possible that they are present in the south-west in Sioma Ngwezi NP and the surrounding game management areas (GMAs). The Hyaena Specialist Group's mapping project, still in the draft stage, may be able to confirm the status soon. The main threats to the hyena species are, as in the rest of their ranges, anthropogenic, related to retaliatory killing for actual or presumed livestock predation and bushmeat snaring.

Spotted hyenas in Zambia are found in the major national parks and reserves (such as Kafue, both North and South Luangwa, Liuwa Plain, Mosi-oa-Tunya, Nsumbu and Sioma Ngwezi, in hunting concession, and GMAs and in some unprotected areas). In farming areas and some of the management areas where game and livestock species are farmed, the former for cropping or trophy hunting, hyenas are frequently blamed for predation. Between 2002 and 2010, there were no reports of human fatalities from hyena attacks but numerous reports of stock losses. There were 305 instances of stock predation during this period with lions, crocodiles and spotted hyenas blamed the most – hyenas blamed for 44 of the attacks, killing 16 cows, 22 goats and 6 pigs.[397] Spotted hyenas thrive in Liuwa

Plain NP, where lions had disappeared and were reintroduced but are only present in very small numbers. This makes hyenas the most numerous and apex predator, with at least 233 identified in 2015.[398] The presence of abundant prey, low carnivore competition and a low level of HWC have meant that hyena population is not vulnerable and is expanding.[399] African Parks, which manages Liuwa, put the population at 350–400 in mid-2020.[400] Wildebeest make up 92% of hyena kills there and there is little predation of livestock even though 10,000 people reside legally in the park and many more in the GMA and farming areas surrounding it. M'soka's study of Liuwa's hyenas did not:

> ...detect any significant effects of proximity to people on hyaena survival...and human-wildlife conflict surveys conducted throughout the GLE indicate that despite human presence, carnivore conflict is low, owing in part to good livestock husbandry practices that include use of effective bomas and herders...anthropogenic impacts are relatively low and likely ameliorated by the high density of prey and low levels of competition from lions in the intensive study area.[401]

The planned and organised killing of game and predators on a huge scale in Southern Rhodesia/Zimbabwe as part of both tsetse fly and rabies eradication campaigns and the clearing of land of woodland and bush for crop cultivation and the livestock industry had, by 1960 when the eradication campaigns was limited to game-free corridors between wildlife areas and commercial farmland, limited most large mammals and carnivores to NPs, game reserves and the few hunting areas.[402] This began to change in the 1970s, when some white farmers moved from cattle or crop production into wildlife farming, game cropping, sport hunting and other game-related activities. Some scattered game and predator populations, like brown hyenas and spotted hyenas, had survived in small numbers in inhabited/farming areas. In the areas of lower quality land set aside for Zimbabwe's black population, wildlife survived but under pressure from subsistence hunting, practised to complement the meagre living the inhabitants could eke out from their poor-quality farmland. Large carnivores like lions had been practically wiped out in unprotected areas.[403] Numbers and diversity only began to pick up in the 1970s, when some white farmers with substantial areas of land that was relatively unproductive under cattle or crops began to turn to game cropping, commercial and sports hunting and, after independence in 1980, tourism.[404] Often farmers joined together and created large conservancies, taking down dividing fences, restocking with wildlife and developing greater income from wildlife-related activities, while enabling the recovery of wildlife numbers, including large predators.

The 1975 Parks and Wildlife Act (enacted by the white minority Smith government) had given landowners rights over wildlife, which enabled utilisation for a range of activities including ranching, trophy hunting and tourism. The utilisation of wildlife in this way only spread to the areas of land allotted to black communities, known under the Smith regime as Tribal Trust Lands and then labelled Communal Areas, after the end of white minority rule in 1980. These were communally owned rather than under individual ownership. They suffered from poor quality land, inadequate water resources and poor inputs (such as fertilisers, good quality seeds or supplementary feed for livestock). They had some wildlife present, but this was often seen by struggling peasant farmers as a threat rather than a resource and they did not have the utilisation rights accorded to white landowners in 1975. Carnivores were frequently killed as they killed livestock and ungulates were hunted or snared for food. In many communal areas few or no large carnivores survived, as Williams et al. found in the communal areas around the Save Valley Conservancy, despite a substantial and diverse carnivore population in the conservancy itself (including 13 brown hyenas and 114 spotted hyenas).[405] In 1988, with strong urging from conservationists like Rowan Martin, the Mugabe government instituted the Communal Areas Management Programme for Indigenous Resources (CAMPFIRE), which was intended to enable people in the communal areas to utilise wildlife through game cropping, sale of quotas for trophy hunting and development of hunting or photographic tourism.[406]

Hwange NP (HNP) and other NPs – such Matusadona, Mana Pools, Gonareazhou, Victoria Falls and Matopos – have viable populations of spotted hyenas and those in the south and west have small numbers of brown hyenas, with no estimate of the latter's population beyond a guestimate by Mills and Hofer of around 100, with the largest population believed to be in Hwange.[407] Spotted hyenas are far more numerous with a total probably around 5,600 for the country as a whole – with Mills and Hofer in 1998 estimating 3,350 in national parks, reserves, hunting safari areas and conservation areas; 1,150 in communal areas; 800 on commercial farms; and 300 on state land.[408] There has been no national census since. The density of spotted hyenas in HNP is put at 0.9/100km^2.[409] The chief threats are from retaliatory killing for predation of livestock and casualties from bushmeat snaring. Loveridge et al. recorded in 2020 that there was a greater likelihood of hyenas being snared by bushmeat hunters in hunting concessions than national parks. In a study covering the 16 protected areas – five National Parks, five Safari Areas and six Forest Areas – in 2017–8, between 52 and 255 hyenas were estimated as having died in snares set for bushmeat. Spotted hyenas made up 92% of snared animals.[410]

Outside protected areas and hunting concessions, Mhlanga et al. found considerable antipathy towards spotted hyenas, with the general view that they were problem animals, perhaps because hyenas, more than other carnivores (with possible exception of leopards), had extended their ranges into areas of substantial human settlement.[411] People living closer to protected areas had greater dislike of hyenas, but most respondents in communal areas also had negative views of hyenas because of the danger to stock. Even in areas with low stock losses, farmers were against the conservation of spotted hyenas and "wanted their numbers reduced or the species exterminated from the area".[412] Attitudes towards hyenas did not seem to be improved if people in communal areas were part of the CAMPFIRE community-based natural resources programme.[413] On communal land about 80km south of the Save Valley Conservancy in south-eastern, Mbiba et al. found considerable antipathy towards spotted hyenas because of stock losses. Between 2012–2015, 211 livestock were killed in 74 attacks, 69 of which were said to be by spotted hyenas.[414]

The Save Valley Conservancy was estimated to have a population of about 209 spotted hyenas, with unknown numbers in the communal lands around it.[415] Nearly a quarter of Save was resettled at the end of the 20th and beginning of the 21st centuries in the fast-track land resettlement programme, with landless black Zimbabweans being given land on areas that had been part of the conservancy. Surveys of carnivores in the area since resettlement showed that in the remaining conservancy areas 12 brown hyena and 106 spotted hyena spoor were found, with an estimated population of 13 brown and 114 spotted hyenas – only six spotted hyenas were believed to be present in the resettled areas, where extensive bushmeat snaring and killing of carnivores appears to have taken place.[416]

West Africa

Spotted and striped hyenas are present across the belt of woodland and savannah between the denser West African forests and the southern Sahara. These populations have been studied less than those in East and Southern Africa and so this narrative and the level of details are scantier. It appears that hyenas are nowhere abundant, despite being widely distributed. The draft maps being prepared by the IUCN's Hyaena Specialist group suggest only scattered populations of both species in the areas shown on the maps at the start of Chapter 2. Spotted hyena ranges theoretically cover most West African countries, but the draft being worked on shows small and quite isolated populations in southern Mauritania, Senegal (one pocket near the coast and the other in the east bordering southern Mali), southern Mali, Burkina Faso, southern Niger, northern Ghana, northern Benin and possibly northern Togo. They are also present in a couple of widely separated populations in Nigeria. Striped hyenas are shown on the drafts as more scattered, with populations in Senegal Mauritania, Mali, southern Niger, northern Benin and perhaps Togo and possibly north-western Nigeria near the border with Benin and Niger, Burkina Faso and northern Ghana.[417] The Hyena Specialist Group cautions that the maps on which I've based the above detail are "first drafts of hyaena distribution maps reviewed during the 2019 workshop. These remain a work in progress as additional observations are being added and updated distributions are refined accordingly".[418]

The scattered nature of the populations has several causes – chiefly the greater density of human populations and the relatively narrow band of suitable habitat running from the Atlantic coast through to central Africa. Another influence, in the former Francophone countries, is that in Senegal and other territories the French colonial administration systematically poisoned predators with strychnine over a period from 1950–60, and in 1970–72 poisoning was carried out in Burkina Faso by its government. The aim was to reduce the threat to livestock from predators, including both species of hyena.[419] Conflict deriving from actual or perceived predation on livestock by hyenas currently constitutes the greatest threat to both species, allied to bushmeat snaring and encroachment on protected areas and hunting zones by nomadic Fulani pastoralists.[420] Where hyenas survive, they appear to have adapted to living in close proximity to human settlements, while facing persecution. In and around the Pendjari reserve in northern Benin (linked in what is known as the WAP complex with the W (Niger) and Arly (Burkina Faso) NPs), human-carnivore conflict is a serious problem, as is human encroachment. A study of livestock predation around Pendjari from 2000–7 found that the main culprits were spotted hyenas (53.6%), baboons (24.8%), and lions (18.0%), preying primarily on sheep and pigs – hyenas being the major problem as lion numbers were low due to previous retaliatory killings, poaching and regulated hunting.[421] Sogbhossou et al. analysed the predation and concluded that the reasons for high levels of predation on farmland bordering the park was inadequate husbandry and low awareness of predator behaviour.[422] Pendjari and WAP are situated in areas of extensive livestock production. There are hunting concessions around parts of WAP – the Pendjari and Konkombri hunting zones in Benin and hunting concessions in Burkina Faso – though hyenas do not figure on the lists of animals that can be hunted, according to the websites of major hunting operators there.[423] But hunting does bring in a substantial part of the income for conservation across the WAP, given low rates of photographic tourism.[424]

Spotted hyenas occur at reasonable densities in Pendjari and the linked reserves, with 1.5/100km^2 being recorded.[425] They compete with lions for wild prey and coexist alongside small and fragile populations of cheetah and wild dogs and an unknown number of leopards. The predators coexist uneasily with the villagers who live around the edges of the WAP complex. The level of HWC is exacerbated by the presence of Fulani pastoralists who encroach on the park and hunting zones during the dry season. The park is underfunded and is unable to maintain effective anti-poaching and anti-encroachment activities. Around the boundary of the W NP in Niger there is regular predation of livestock and consequent killing of predators. A survey of 154 people in 32 villages on the edge of the park indicated that between 2000–6 3,271 livestock were killed, averaging 468 a year. In one 7 month period, 593 attacks took place, of which 267 were by caracals or jackals; 193 (166 on smallstock, 17 on cattle, 8 on donkeys and 2 on pigs) by spotted hyenas; 125 by lions; and the rest by leopards, cheetah or wild dogs.[426] Of the 154 people interviewed, 81.53% had a negative attitude towards predators, 4.28% confirmed they would kill predators, 30.51% indicate they have no means to stop attacks and only 6% actively defended their livestock.[427]

Both striped and spotted hyenas are found in small numbers in northern and parts of western Nigeria, away from dense forest and large urban centres. During the colonial and early independence period, there is evidence of regular persecution of hyenas. A former colonial official described how hyenas were causing a nuisance around a mining camp in Gashaka, near the Gashaka-Gumti NP in west-central Nigeria on the border with Cameroon. The miners set up a spring trap, but caught a local villager rather than a hyena. In another incident, a hyena was shot and had its jaw badly injured – villagers then claimed to have found a local "wizard" dead with identical wounds to the jaw.[428] Perhaps because of regular persecution, hyenas of both species are wary of people. Happold, in his survey of Nigerian wildlife, refers to hyenas being rare because of the payment of bounties for dead hyenas by local administrators over many years.[429] There are reasonable numbers in some national parks – Andrew Dunn of WCS says that a recent estimate for Yankari NP was of 100–200 spotted hyenas based on a 2019 spoor survey, but that there was no evidence of striped hyenas there.[430] Dunn said that a few years ago a man was arrested in Yankari for trying to trap live hyenas for use by travelling entertainers and traditional healers. These have often utilised hyenas, baboons, pythons and other animals to attract audiences both to get paid for performances and to sell traditional cures; there are also unverified accounts of debt collectors keeping hyenas to intimidate debtors.[431] Pictures of muzzled spotted hyenas kept on chains and taken around towns to give performances have been pictured by Pieter Hugo.[432] Some of these performers/healers were the subject of a film by Chinese television, which is available on YouTube. Hyenas are captured when young, trained and kept captive to draw crowds. The film shows them performing in Abuja, Zaria and Kano.[433]

Senegal, Mali and Mauritania all have scattered populations of both species of hyena. The Niokolo-Koba NP in Senegal has probably the largest population of spotted hyenas, though estimates are not presently available of the numbers or of the presence of striped hyenas.[434] In southern Mauritania, striped hyenas are present in the Pare National du Bane d'Arguin and scavenge the beaches for remains of sea turtles and other marine animals.[435]

Part 2: North Africa, West, Central and South Asia

From the Atlantic coast of Morocco, across the Mediterranean littoral, through the Middle East and into West, Central and South Asia, the striped hyena is the sole hyena species present. Its distribution is extensive, but it is nowhere recorded as present in large numbers. Its nocturnal habits and denning in large holes, caves and hidden rocky clefts have made it hard to estimate ranges, numbers and densities.[436] Across the range it suffers varying levels of persecution because of its alleged livestock predation and the danger many peoples believe it poses to man. It is blamed for the killing of dogs, sheep, goats and poultry, and this is certainly likely but without being a main source of food. While it is accused of taking live cattle, donkeys and horses, it is most likely that these are cases of scavenging mistaken for predation. It is also persecuted because of its damage to crops, particularly to melon and date palm crops.

Across its North African and Asia range, it is vilified as a grave robber, a contemptible, cowardly scavenger with links to magic and the occult. To add to the striped hyena's problems, many peoples believe its body parts are efficacious as cures for diseases or mental conditions, sexual impotence, as an aphrodisiac or as charms to prevent harm to the wearer. In some areas – India and Pakistan, in particular – it is caught alive and has large dogs set on it as a form of cruel entertainment. Even where tolerated as nature's waste disposal team, hyenas are despised and counted as unworthy of the respect that people in the region accord leopards, tigers, bears or even wolves. One of the threats to its survival in many areas is the decline in large hunting carnivores like tigers, cheetahs, leopards and wolves, which produce the carcasses that hyenas scavenge, as well as declines in the prey of those carnivores due to human actions, mainly habitat loss and killing for food or livestock protection. Their low densities and scattered populations mean there is the danger of small, separated populations becoming unviable and vulnerable to extinction.[437]

The western boundary of the striped hyena range is Morocco. There are few recent accounts of their numbers or locations. In 1998, Mills and Hofer put the population at 400–500, but with the proviso that it could be under 100, a figure also applied to the Algerian population, which was threatened by deforestation and hunting.[438] No reliable data is available for Tunisia. Osborn and Helmy did not record it as present,[439] while the IUCN's Red List of Threatened Species said hyenas were declining there.[440] Libya's wildlife population has been studied from an historical angle by Masseti, who details the steady decline in biodiversity and wildlife numbers in Libya, including hyenas and noting that hyenas are rarely seen but when sighted are often dead, having been killed by cars on roads at night, adding that they are blamed for killing sheep.[441] Osborn and Helmy record them as present in Libya, notably on the edge of Egypt's Western Desert, while Mills and Hofer say there is a lack of data on hyenas in Libya.[442]

Mills and Hofer put the Egyptian population at one of the largest, at more than 1,000.[443] They are widely distributed from the Western Desert, to the Nile Valley, Nile Delta, the Red Sea coast and the Sinai;[444] in deserts; where there is sufficient wildlife and vegetation; around oases; and on the fringes of farming areas, where they raid crops and water sources at night – the poisoning of hyenas has been recorded around some oases to protect stock and melon/date crops.[445] In areas where camel caravans are still part of the trading system, hyenas may be found in the vicinity, taking advantage of any camel mortality on the routes.[446] The southern Sinai massif is a stronghold for Egypt's hyenas, as well as for ibex and wolves, particularly around the St Katherine protected area. The human population is small, consisting mainly of nomadic Bedouin pastoralists, and there are no details of conflict with the hyenas and other predators.[447] A survey of carnivores in this region published in 2017 reported that while wolves were found in the three surveyed areas of the protected zone, only one area had striped hyenas. That area, Sheikh Awad, produced 472 hyena images from camera traps, but no estimate was produced of hyena numbers.[448]

Palestine and Israel

The area of the Palestinian territories (the West Bank and Gaza) and the territory of the state of Israel has a small but seemingly stable striped hyena population, which has been quite extensively researched. In 1976, Kruuk reported their presence around the Dead Sea and in the Negev desert, and along the border of the

West Bank with Jordan, but gave no indication of likely numbers.[449] They were generally believed to be present in small, scattered populations, whose numbers had been progressively reduced in the 30 years of the British mandate over Palestine (1918–48), during which they carried out a concerted poisoning campaign against predators and other wildlife capable of spreading rabies. Numbers are thought to have recovered when the British departed and Palestine was split between the new state of Israel, the Palestinian West Bank (then under Jordanian control) and the Gaza Strip (under Egyptian control).[450] In 1977–8, Skinner and Ilani estimated that there were around 130 hyenas in Israel, mainly in arid regions south of Jerusalem.[451] Macdonald studied the behaviour and ecology of the Israeli hyenas in the late 1970s and found they differed from the East African striped hyenas in that they gathered in larger groups where there were abundant food resources, such as waste dumps, and scat analysis showed a high intake of offal scavenged from such sites, and from feeding sites on nature reserves established particularly to feed vultures.[452]

Examination of a bone assemblage of a striped hyena in the Negev desert indicated a high level of remains of donkeys, camels, sheep and goats, which appeared to have been scavenged rather than hunted.[453] Skinner et al. examined hyena den sites in the Judean and Negev deserts and monitored feeding at feeding sites at En Gedi and Sde Boqer. On one occasion, five donkey carcasses at one of these feeding sites attracted four individual hyaenas, a pack of six wolves and 20 foxes. The Sde Boqer site was visited by three adult and three subadult hyaenas, three wolves, six foxes, a caracal and at least 20 griffon vultures.[454] One study looked at the issue of consumption of human remains by striped hyenas. It estimated a population in Israel of 170 striped hyenas, mostly south of Jerusalem. It supported Macdonald's view of a greater level of social activity among the Israeli population and larger home ranges (about $58km^2$), resulting from human presence, terrain and food availability.[455] Excavation of two hyena dens in the Negev revealed human bones, along with ones from dogs, sheep, goat, ibex, camels, equids, pig, cattle, porcupine, hare, birds and fox.[456] The hyenas scavenged animal carcasses left by Bedouin pastoralists, refuse from waste sites, animal remains placed at nature reserve feeding sites and hunted small animals, but there was no indication of predation of large mammals or livestock. The human remains included skull fragments from eight different adults, one juvenile and one from a child, along with some long-bone shafts.[457] There was no suggestion of humans being killed so the inference was that these were remains removed from graves or of people who had died out in the open and not been buried.

The main threats to hyenas in Israel and Palestine include retaliatory attacks for attacks on wildlife, and road accidents, which Mills and Hofer said accounted for at least 20 deaths in 1982–5 and 24 in 1988–91 on one road to the Dead Sea. They believed 20–30 were killed annually on the roads.[458] Shamoon and Shapira believe that anthropogenic factors are key in limiting hyena numbers and their range in Israel.[459] Hyenas did occur close to human population centres, but always in lower densities than in areas with few people. One thing that attracted hyenas to villages and towns were waste dumps and landfill sites.[460]

Hyenas coexist with and scavenge from wolf packs in Israel and Palestine. They have been filmed together by camera traps at feeding sites and waste dumps. In 2016 it was reported that spoor found by researchers in 1994 showed a striped hyena that appeared to be running with the wolves. At first it was thought that it was following the wolves as they hunted in order to scavenge, but it emerged that the hyena was travelling in the middle of a pack of wolves, and this was supported by evidence from Israeli zoologist Beniamin Eligulashvili, who saw seven wolves and one hyena, with the hyena moving in the middle of the pack.[461] Dinets and Eligulashvili wrote that it could have been a symbiotic partnership:

> The hyaenas could benefit from the wolves' superior ability to hunt large, agile prey, while the wolves could benefit from the hyaenas' superior sense of smell and their ability to break large bones…and dig out fossorial animals such as tortoises.[462]

In the Palestinian West Bank, the striped hyena is nowhere that common but is distributed in all habitats, including the desert, the Jordan valley and the central high lands zones.[463] The growth of illegal Israeli settlements, the building of the wall to divide Israelis from Palestinians, water abstraction by the illegal settlers and increased grazing of livestock by them have all affected biodiversity on the West Bank.[464] A study carried out in the central highlands areas of the West Bank in 2014–5 found a general decline in wildlife, and only one image was captured of a striped hyena. Albaba believed this was a result of logging, quarrying for stone and forest fires, which had reduced suitable habitat and den sites, and both hunting and poisoning to kill predators.[465] Killings of hyenas were often reported in the local media – ten reports being recorded since 2010, including the killing of four hyenas near Hebron. As in many other parts of the ranges of the spotted and striped hyenas, there were still beliefs among many Palestinians that witches rode on hyenas and this led to attacks on them if they were found by people. Overall, anthropogenic actions were the threat to survival of the small population – for while hyenas are adaptable to human presence, they still need "natural habitats free of anthropogenic disturbances to serve as refugee for source populations".[466]

Lebanon and Syria

The striped hyena is the largest carnivore remaining in Lebanon, is the country's national animal and for a long period has proved capable of surviving in human-dominated landscapes. The years of conflict in Lebanon (1975–1990) and in the territories on its borders, according to Ab-Said, were helpful to hyena survival because of the displacement of people from their villages and landmines along the borders between the combatants serving to give hyenas space in which to survive.[467] Since the end of the war, people have returned to their villages, urbanisation has increased, squeezing hyenas into smaller territories and creating great HWC, not helped by "the intense feelings of people towards carnivores arising from the myths that surround hyaenas".[468] Extinction is now a greater threat, Lebanon already having lost the Syrian brown bear, the lynx and fallow deer. Few large mammals are found there because of hunting, overgrazing by domestic animals and intense cultivation of land.[469] Wolves remain in some areas, but the striped hyena survives, albeit in small numbers, because of the adaptability of its lifestyle and foraging habits. They forage around waste dumps and any sites with animal remains, may take some livestock and hunt for small animals such as rabbits, squirrels, mice and small wild ungulates, though they are not the most adept of hunters.

There are only a few protected areas in the country, and they are surrounded by human populations, and even within reserves hyenas have been known to have been killed by people, according to Abi-Said. Between 1999–2001, the Lebanese press reported the killing of 23 hyenas, all during the winter months when hyenas forage for food closer to villages.[470] There is little organised collection or disposal of waste in much of the country and this attracts hyenas to dumps near towns and villages. They also forage around poultry and pig farms, where animals are slaughtered and offal disposed of nearby. One scat study showed that domestic mammal remains make up 77.8% of hyena intake, with wild mammals 14.3% and birds 7.9%.[471]

Between 1952 and 2004, when more protection was decreed, the country's hunting laws permitted the hunting of hyenas without any close season.[472] The end of officially sanctioned hunting may have lessened mortality rates, though persecution and other human actions remain the main threat to Lebanon's hyena population, despite little evidence that they kill livestock.[473] Little information is available, even today, on its numbers, ranges within Lebanon or the stability of the population. Because they are both persecuted and need to live close to humans, hyenas are elusive and almost entirely nocturnal, choosing den sites in rocky areas with caves or ravines and in forests. There are some areas where striped hyenas appear to be surviving well, one being the Jabal Moussa Biosphere Reserve on the coast of central Lebanon, north of Beirut. There, camera traps captured 220 images, but this does not mean 220 hyenas, as animals may be pictured several times without clear identification. Local people believe hyenas to be common there.[474] The best estimate for the population, according to Dr Stephanie Dloniak, is that the country has a few hundred hyenas.[475] Dr Abi-Said, who has written the most detailed accounts of Lebanese hyenas, is working to improve the image of the hyena and prevent persecution on the basis of incorrect views about livestock predation and superstitions about hyenas being evil, mesmerising people to attack and eat them and killing children.[476]

In Syria, striped hyenas have survived while the brown bear, leopard, cheetah, Asiatic lion and Caspian Tiger have become extinct.[477] Numbers and locations are not known, and research is prevented by the continuing civil war. The picture for neighbouring Jordan is clearer, with striped hyenas surviving in reserves and some unprotected areas. By the mid-1960s, hunting had exterminated the populations of oryx, wild asses, roe deer, fallow deer, bears and cheetah.[478] But some Arabian and Dorcas gazelles survived along with striped hyenas and wolves, despite great losses in vegetation and increasing desertification. Hyenas were found in habitats from extremely arid desert to Mediterranean-type forest. New hunting laws in the late 1960s and the founding of the Royal Jordanian Hunting Club, which had a conservation and enforcement role, brought greater regulation of hunting and a reduction in the loss of game species.[479] Despite this, Mills and Hofer said in 1998 that they were in decline because of persecution,[480] with small, scattered and unstable populations subject to persecution by pastoralists.

A captive breeding unit was established to increase gazelle populations and reintroduce oryx, with reserves set up to provide full protection in many areas. Hyenas weren't top of the conservation agenda, but any increase in wild ungulates would improve the lot of hyenas, providing carcasses from natural mortality and the remains of kills by wolves. In 2004, Qarqaz et al. reported that while they only saw one hyena during a survey of their status in nature reserves, they found 30 active dens across the regions of Jawa, Dana Reserve, Tall Al-Ashqaf, Tall Humaylan, Jebel el Aritein, Wadi al Uwayni and Faydat al Dahikiyah – while local people in these areas reported 17 recent sightings.[481] Bones found at dens included cattle, camels, donkeys, dogs, foxes, domestic goats, sheep, gazelles and ibex.[482] Qarqaz et al. noted that the hyena was associated in Jordan with a range of superstitions, including that they killed and ate people (humans supposedly being their favourite prey), that their skins and body parts had curative properties and they were animals associated with the occult.[483] The authors found 12 hyenas that had been killed, some of them poisoned, during their study, despite the hyena being a protected animal, and evidence of dens being destroyed. A recent study by Edwards et al. in Dana Reserve recorded the presence of striped hyenas, wolves, caracals, red foxes, porcupines, hares and ibex.[484] A study of waterhole use at Dana suggested a population of 10 there.[485]

In neighbouring Iraq, research and estimates of the striped hyena population have been rendered impossible by decades of warfare, occupation and civil conflict. Mills and Hofer suggested in 1998 that the population was in decline and the best estimate was between 100 and 1,000.[486] The abundant wildlife of Iraq's southern marshes was more or less exterminated when Saddam Hussein drained the marshes as part of a campaign against rebel groups. As the marsh dried out, the resident striped hyenas were forced out, died out as food sources disappeared or were hunted out.[487]

The Arabian Peninsula

This region has a widely distributed but relatively small population, with the largest concentration in the Dhofar Mountains in Oman. Mills and Hofer estimated no more than 100 in Saudi Arabia (a result of decades of persecution – including shooting from vehicles and deliberate running-over of hyenas by car drivers), an unknown number in Yemen, the United Arab Emirates and Qatar, but a population between 100 and 1,000 in Oman, where there is no protection but the government discourages killing because of the hyena's useful role as a scavenger of dead ungulates.[488]

Numbers have declined across Arabia and the Gulf since the 1960s, when they were described as "relatively abundant throughout the peninsula except in the interior of the desert".[489] Hunting animals from vehicles to shoot them on the move has been responsible for a loss of biodiversity, the disappearance of the cheetah, onager, ostrich, Arabian oryx and Saudi gazelle from the peninsula and declines in the population of striped hyenas, most of which are found in hilly areas of Yemen, the south-west Saudi Arabia and Dhofar.[490] Threats to striped hyenas come from persecution as suspected stock thieves but also because of the superstitions about the occult nature of the hyena. Cunningham et al. in 2009 gave an optimistic estimate of the population of 300–400 animals but warned of steady decline because of shooting, poisoning and death on roads. They reported finding dead wolves and hyenas hanging from trees, presumably as some sort of warning; hyena body parts had been removed for use in traditional medicine.[491] The animals were also hunted from vehicles at night using spotlights. Poisoning is common in areas where there have been livestock losses – often blamed on the Arabian leopard. When carcasses are poisoned, it is often striped hyenas that are killed.[492] Deforestation in the valleys of mountainous areas is another threat, removing safe habitat for denning. This is occurring in south-west Arabia and Yemen, where the Wadi Rijaf is home to striped hyenas, wolves and prey species.[493]

In the Dhofar region of Oman there have been more concerted efforts to protect forest and valley habitats, which has benefited hyenas along with other carnivores and ungulate species. The mountains are a narrow 23km belt extending for 400km along Oman's Indian Ocean coast. They receive the highest rainfall in the peninsula, with annual monsoon rains. They are a refuge for hyenas, leopards, caracals, wolves, ibex, gazelles, wild camels and hyrax.[494] A survey of the area's biodiversity in 2013 recorded that in the Wadi Sayq valley in Dhofar there was still a diverse carnivore population with striped hyenas present in reasonable numbers, though with good reason to believe that the population was declining as disturbance from new roads and human encroachment increased and the prey base for predators declined.[495] Striped hyenas caught on film by two researchers and conservationists, Khalid and Hadi al-Hikmani, were shown in the BBC series *Wild Arabia*. Part of the film showed a striped hyena successfully taking a goat's carcass (killed by a leopard) from 12 wolves. The two biologists were described as campaigning to save the Dhofar habitats and carnivore species and this was a rare TV depiction of hyenas as anything but evil, skulking cowards or threats to more charismatic carnivores.[496]

In the United Arab Emirates, attempts are being made to breed the striped hyena at a conservation project on Abu Dhabi's Sir Bani Yas Island in Abu Dhabi Sea. In 2009, a female striped hyena gave birth to two cubs, the first born in the Emirates for decades, where the hyena was last seen in the wild in 2002. The island has seen the reintroduction of hyenas and cheetah, in attempt to control the wild antelope and gazelle population.[497] One presumes that hyenas would steal or scavenge cheetah kills, forcing the cheetahs to hunt more.

Iran, Turkey, the Caucasus and Central Asia

Iran has a widely scattered population of striped hyenas, often in hilly, forested or mountainous areas. It is believed to have declined substantially since the 19th century, when it was common across the country. This has been caused by habitat loss and persecution.[498] There have been no national surveys and Mills and Hofer in 1998 estimated the population at anywhere between 100 and 1,000.[499] Hyenas are present in a number of protected areas and their immediate vicinities, but have proved hard to monitor,[500] with political circumstances and security concerns obstructing research – eight cheetah researchers were jailed on charges of spying because they were using camera traps in what was deemed to be a sensitive area.[501] A study of four reserves in 2012 in the Jajroud Complex Region close to Tehran revealed ten hyenas spread across three protected areas there.[502] Hyena researchers in Iran lament the paucity of ecological and biological data on Iran's population, its numbers and distribution.[503] But there is evidence of threats to the population from anthropogenic causes – notably proactive culling, snaring/trapping, death on roads at night, and fewer carcasses of wild animals due to a decrease in predators within its range. Foraging in areas of human population make them vulnerable to persecution or road accidents. Hyenas are also frequently caught in traps that are set to catch wolves suspected of attacking livestock, and poison is used frequently to kill predators.[504] Hyena body parts are widely used in traditional medicine in Iran, and hyena carcasses discovered by researchers have often been cut up to remove body parts. Some of those dead hyenas were killed in road accidents, and Tourani et al. record that at least 12 died on roads between 2006–9.[505]

In their survey of hyenas living near cities, Monchot and Mashkour found evidence of multiple hyena dens at Kaftar Khoun (which means the hyena house) on Iran's central plateau area, but no estimate is given of the number of hyenas that resided there or how recently, though evidence suggested the bones had been collected over a ten year period.[506] Old bone assemblages at the site indicated a diet including the remains of cattle, sheep, goats, donkeys and mules, as well as those of domestic dogs, jackals and wolves. Some camel bones and also a fragment of a human skull were found, confirming beliefs in Iran that striped hyenas will eat human remains. A survey of people's attitudes to hyenas carried out at Khojir NP east of Tehran in the Alborz mountains found that people had a positive attitude towards hyenas in and around the park. It appeared to be tolerated in Khojir except if it attacked livestock. Elsewhere in Iran, they were "viewed with contempt and fear, and are frequently associated with witchcraft as their body parts were used in traditional medicinal treatments. They are thought to influence people's spirits, rob graves, and evoke many superstitious fears".[507]

While Mills and Hofer reported that there was too little data to estimate striped hyena numbers in Turkey, the westernmost outpost for them in Eurasia, research in the last 20 years has shed more light on their presence. Hyenas survive in a number of regions alongside brown bears, wolves, lynx and caracals, which may provide carcasses for them to scavenge, and prey species of red deer, roe deer, goitered gazelle, chamois, wild goat, mouflon and wild boar.[508] In many areas, hyenas are treated by farmers as a pest. Over the last 65 years they seem to have disappeared from some areas where they were commonly found, such as the Termessos NP and the Beydağlari Mountains, both in south-west Turkey, though some local people reported them as still being present in the 1980s and some conservationists believe it is possible they are still in Termessos in small numbers.[509] WWF Turkey reported in 2002 that reports that the hyena had died out in Turkey were incorrect and that they were still surviving in south-eastern Anatolia and the Amanos Mountains and possibly in isolated populations on the south-western Mediterranean coast.[510] Kasparek et al. believe there is evidence for them being present in the Kazdağı, Edremit, Ezine, Ayvacık, and Balıkesir regions of Marmara region, near Istanbul, where villagers are said to reinforce graves to prevent hyenas digging them up.[511] But verified sightings seem to be few and far between, even there, the most convincing being from two hunters who found a hyena den about 40 years ago, and hunters who said they shot hyenas in 1986 and one who shot a hyena in 1999 while hunting wild boar.[512] In eastern Turkey, three young hyenas were trapped near the border with Syria in 2002, taken to a zoo and in 2004 released into the wild.[513] The likely range in Turkey seems to stretch from the Sea of Marmara coast in the west along the south coast as far as Hatay province on the border with Syria, though it is unlikely that this is an unbroken range and is most likely split into smaller ones that aren't inter-connected.[514] There are no current estimates of the size or viability of the population.

The striped hyena is still present in the Caucasus mountains, Turkmenistan, Tajikistan and Uzbekistan, in remaining areas of savannah, semi-arid scrub and mountainous areas with ravines, deep valleys and caves to provide dens, but not at high altitudes or in dense forest.[515] During the first half of the 20th century they were found in southern Russia bordering the Caucasus but disappeared as intensive agricultural development in the Soviet Union and the virgin lands project to bring previously uncultivated areas under grain crops increased the areas of southern USSR under the plough in the 1950s, exterminating wildlife.[516] Heptner and Sludskii give considerable detail of the retreat of the hyena within the USSR in their survey of mammals there.[517] They record population ranges and a rough idea of abundance through records of the Soviet fur trade. Hyenas were trapped or shot for their skins in the Caucasian and Central Asia republics of the USSR. They detailed the ranges as being in the coastal hills and valleys of the Caucasus along the Caspian Sea in southern Dagestan and the Azerbaijan republic, but also present in reasonable numbers in Armenia and some in Georgia. In Central Asia they were found in the Turkmen, Tajik and Uzbek republics.[518]

The decline in numbers in the Caucasus and Central Asia is put down to excessive hunting for skins but also persecution because of fear that the animal was responsible for the disappearance of children; and the disappearance of prey species such as goitered gazelles, wild goats and deer that were hunted or forced out by agricultural expansion. Populations in each of the Soviet and then independent republics of Armenia, Azerbaijan, Georgia, Tajikistan, Turkmenistan and Uzbekistan were hard to estimate, with most put around 20 animals per republic but with Armenia and Georgia put at 50–100, and Turkmenistan having the broad range of 100–1,000.[519] Heptner and Sludskii noted that by the 1970s, less than ten hyena hides were being received from each republic by the fur trade.[520] A more recent study noted that in Armenia hyenas were found in small and scattered populations in areas of semi-desert, arid foothills of the mountains and some river valleys.[521]

Afghanistan, Pakistan, India and Nepal

Striped hyenas are still found in Afghanistan in foothills of the mountains, open woodland, dry scrub and arid regions with sufficient cover for dens. They have long been persecuted as stock killers but also captured alive for use in brutal "entertainments" in which large packs of dogs are set on captive hyenas in front of large groups of people.[522] Frembgen recorded incidences of hyena baiting happening frequently in the city of Kandahar.[523] The four decades of civil war and invasions has rendered both research and conservation measures almost impossible, and there are no current estimates of numbers. It is very likely that they are found along the borders with Pakistan, where they are found around Quetta and western Baluchistan.

In Pakistan, the hyena is recorded as living in arid scrubland and the foothills of mountainous areas with woodland to provide shelter, especially in Baluchistan, being most common around Quetta in the north of the province.[524] There is no estimate available of the overall national population. A survey in 2001 in Kirthar NP found just eight hyenas, which were under threat from poaching.[525] As in Afghanistan, hyenas are captured and then baited and killed using packs of dogs.[526] Men go out into the hills and search for hyena dens, pulling the hyenas out of their dens, keeping captive them until a dog fight is organised. The hyenas may be tied up so they have limited movement and can't run away and then dogs – singly or in groups – are set on them.[527] The Wild Pakistan conservation group frequently posts online and on Twitter details of these brutal fights as part of its attempts to draw attention to the need to conserve the hyena.[528]

Striped hyenas are the most widespread carnivore in India, even though numbers are believed to be falling; there may be fewer than 1,000 remaining in scattered populations.[529] They are threatened by habitat loss, expanding urban areas and cultivation of land, deliberate persecution by humans and deaths on roads. The emphasis of successive post-independence governments on increasing food output and then India's massive industrialisation have been the background against which these threats have developed.[530] Located in at least 54 different population groups spread across India, Kartick Satyanarayan, the CEO of Wildlife SOS, says that they "occur in arid and semi-arid ecosystems, as well as in the extremely wet regions of the south-western coast of India. They can be found in Rajasthan, Uttar Pradesh, Madhya Pradesh, Maharashtra, Gujarat, and Karnataka", but are found now in only about 40% of the terrain that could support them.[531]

Satyanarayan says hyenas do have a serious image problem in India, noting that:

> Due to their scavenging habits, people tend to see them as harbingers of evil. They are considered to be mounts of witches in some parts of the country. Seen as malicious, devilish, and often associated with witchcraft, hyenas have been widely misrepresented in culture, folklore, and even in movies.

Persecution is frequent with poisoning, hunting, and poaching for body parts to be used in traditional medicine the biggest threats to the hyena populations in India.[532] He also records that villagers will often carry out mob attacks on hyenas if they are seen around villages and in farmland. Wildlife exists in unprotected areas, but is under constant threat there from anthropogenic actions, and relies on the 89 national parks and 489 wildlife sanctuaries for protection from poaching and human encroachment.[533] Indirect anthropogenic threats to wildlife, particularly carnivores like hyenas, comes from the widespread and largely uncontrolled use of pesticides and livestock drugs, like diclofenac – which is used to treat cattle and has been blamed for the deaths of millions of vultures.[534] An insecticide called endrin and another called folidol have been used by farmers to poison carcasses of livestock killed by tigers,[535] but consumption of the remains will kill hyenas. No research has been published estimating hyena mortality. On the plus side of the relationship between tigers and hyenas, the launching of Project Tiger in 1973 has benefitted hyenas by protecting habitat and tigers, thereby ensuring a supply of carcasses for hyenas to scavenge.[536] Human-wildlife conflict is a serious problem for farmers and wildlife in India, but hyenas do not figure as serious problems, the main problem animals being elephants, tigers, leopards, macaques and sloth bears.[537]

According to the best field guide to Indian mammals, hyenas are found, usually in small scattered populations, across most areas of India, excepting the densest forest and high mountains, with tiger reserves, national parks and the Gir Forest.[538] Gir, in Saurashtra district of Kathiawar peninsula in Gujarat, is a major area for the survival of predators in western India – particularly the Asian lion. The conservation of the lion's forest refuge has had a positive effect on the survival of striped hyenas, given their reliance on scavenging carcasses. The national park and surrounding areas have 39 mammal species recorded as present, with chital deer, sambar deer, four horned antelope, nilgai (blue bulls) and wild boar now providing ample prey for lions and scavenging opportunities for the hyenas.[539] The Gir lions also prey on the cattle owned by the Maldhari community in the forest and adjacent unprotected forest areas, providing further carcasses, with deaths of the cattle from accidents, disease and old age adding to the available food. There is HWC around the Gir Forest, relating mainly to human encroachment, damage to habitat and loss of wild prey.

Census figures given by Mitra show an estimated 71 hyenas inside the Gir Protected Area in 1874, 84 in 1979, 92 hyenas in 1984, 97 in 1990, and 137 in 1995.[540] Alam et al. used camera traps to survey hyenas in Gir and identified 24 individuals, with a desnity of 0.02–0.12/km^2 over the national park – compared with 0.15/per km^2 in the Sariska tiger reserve, 0.03 to 0.05/km^2 in Rajaji National Park and 0.03 and

0.06km^2 in Kumbhalgarh and Esrana sites in Rajasthan.[541] Like the Gir forest protected area, tiger reserves across India have proved safe areas for hyenas and they are found in most, including those mentioned above and Ranthambore, Corbett, Achankmar, Kanha and Bandhavgarh reserves. They are also found in wildlife reserves, forest reserves and some unprotected areas of the Western Ghats in south-central India and Rajaji National Park, Uttarakhand, in northern India.

Outside the protected areas, like Gir, there is a greater danger of human-predator conflict affecting hyenas. The greatest danger to scavengers like hyenas is the use of poison by farmers when they lose stock to a carnivore. Carcasses are laced with poison to kill the animal that killed it if it returns to feed on the carcass. This is totally indiscriminate and scavengers like hyenas will very often be the innocent victims when they feed from a livestock carcass resulting from lion, leopard, dhole or tiger kills.[542] Hyenas are also directly persecuted as stock killers, sometimes mistakenly and sometimes with evidence of hyenas hunting livestock. A hyena diet survey in 2006–7 in north Gujarat, the districts of Mehsana, Patan, Banaskantha and Sabarkantha showed that hyenas relied heavily on livestock remains (with no clear indication of how many stock were killed rather than scavenged). Scat analysis indicated 58% of the remains were of domestic livestock, followed by rodents (11%), hares (7%) and dogs (2%) – it is thought that hyenas would occasionally kill young cattle and smallstock when the opportunity arose.[543]

In Maharashtra state, hyenas have adapted to living in areas with substantial human populations. In Ahmednagar district of western Maharashtra, Athreya used camera traps to assess wildlife numbers and identified 12 individual hyenas and suggested a density there of 5.03/km^2. [544] This was an intensively farmed areas with sugar cane and maize fields – crops that could give shelter to hyenas and leopards, which are common there. Despite persecution and the danger of mortality on roads, hyenas are still surviving near to large human populations, utilising carcasses of dead livestock and food from waste dumps. In Rajasthan, Singh found hyena populations present in arid zones of the state but also close to large towns and villages, but believes them to be under threat from degradation of habitat due to a variety of anthropogenic causes, given that Rajasthan is one of the most densely populated arid regions in the world.[545] Hyenas there appear to survive mainly by scavenging the carcasses of livestock that have died of natural causes – as inadequate veterinary care leads to a high death rate from disease – and old age or accidents, with very little consumption of meat in the region thereby producing more entire carcasses for hyenas.[546]

Singh et al. surveyed the hyenas in the area in 2006 and 2007, focusing on protected areas and adjacent land at Kumbhalgarh and Esrana in northern Rajasthan. They observed considerable human encroachment through livestock grazing in protected areas and cutting of trees for firewood.[547] Using baited camera traps they identified 15 hyenas at Kumbhalgarh and 8 at Esrana.[548] In the study area the density of livestock was massive, with 22,304 stock in Kumbhalgarh, with a density of 73/km^2 and 67,842 animals at 311/km^2 in Esrana. This provided a substantial food source in the form of carcasses for hyenas, which also had safe denning sites in the hilly terrain.[549] Hyenas were also found in the Thar desert of Rajasthan, along with jackals, wolves caracal and jungle cats. The hyenas survive mainly on scavenging livestock carcasses and carcasses of wild prey killed by wolves, but also on fruit and vegetables. There is persecution by local pastoralists and because of reported attacks by hyenas on children, which resulted in the retaliatory killing of a hyena.[550] In Odisha state on the coast of the Bay of Bengal, in a parallel to brown hyenas in Namibia, the striped hyena has been found to scavenge the shoreline for the carcasses of marine animals and to prey heavily on Olive Ridley Turtles lay their eggs in the sand – about 80,000 use beaches in areas with hyena populations.[551] Hyenas dig up the eggs and eat young turtles when they hatch.

In Nepal, the hyena is found in hilly, but not high mountain, areas of southern Nepal, but is considered endangered in the country and has a high protection status.[552] Poaching is a problem in Nepal and includes the killing of hyenas, sometimes in snares set for other wildlife. Habitat loss and human encroachment on protected areas to poach, cut wood and carry out other illegal activities are the major threats to Nepal's habitats and wildlife.[553] There are no recent surveys of the population, the only estimate is a broad one of 10–100, with 50 thought to be the likely maximum.[554] The population is distributed in small groups across the Terai region of Nepal within the protected areas of Bardia National Park, Chitwan National Park, Parsa Wildlife Reserve and Shukla Phanta Wildlife Reserve. Outside of protected areas they have been recorded in the districts of Bara, Kailali, Kapilbastu (Niglihawa VDC), west of the Bagmati to Kanchanpur and one dead striped hyena was found in Udayapur District in August 2003.[555] The main threats are loss of food sources due to lower wild and domestic ungulate populations, habitat loss due to clearing of forests for cultivation and persecution of predators through poisoning carcasses of livestock that have been killed.

Notes

1 https://www.iucnredlist.org/search?taxonomies=100451&searchType=speciesaccessed accessed 23 July 2020.
2 https://www.inaturalist.org/projects/hyaena-distribution-mapping-project-2018 accessed 15 July 2020.
3 Personal communication with Dr Steph Dloniak.
4 Gidey Yirga et al. (2016) Spotted hyena (Crocuta crocuta) concentrate around urban waste dumps across Tigray, northern Ethiopia, *Wildlife Research*, 42, 7, 563–569, pp. 566–7.
5 Personal communication with Laurence Frank.
6 Shaurabh Anand and Sindhu Radhakrishna (2017) Investigating trends in human-wildlife conflict: is conflict escalation real or imagined? *Journal of Asia-Pacific Biodiversity*, 10, 2, 154–161, p. 154.
7 Rosie Woodroffe (2000) Predators and people: using human densities to interpret declines of large carnivores, *Animal Conservation*, 3, 165–173, p. 165.
8 Woodroffe, 2000, p. 165.
9 P. W. Bateman and P. A. Fleming (2012) Big city life: carnivores in urban environments, *Journal of Zoology*, 287, 1–23, pp. 1 and 5.

10 Helene Lowry, Alan Lill and Bob B. M. Wong (2013) Behavioural responses of wildlife to urban environments, *Biological Reviews*, 88, 537–549, p. 538.
11 Ibid., p. 540; G. Maude and M. G. L. Mills (2005) The comparative feeding ecology of the brown hyaena in a cattle area and a national park in Botswana, *South African Journal of Wildlife Research*, 35, 2, 201–214, pp. 212–3.
12 Mlamuleli Mhlanga et al. (2019) Influence of Settlement Type and Land Use on Public Attitudes towards Spotted Hyaenas (Crocuta crocuta) in Zimbabwe, *African Journal of Wildlife Research*, 49,1, 142–154, p. 142.
13 P. A. Lindsey et al. (2011) Ecological and financial impacts of illegal bushmeat trade in Zimbabwe, *Oryx*, 45, 1, 96–111, p. 96.
14 Penny Johnson (2019) *Companions in Conflict. Animals in Occupied Palestine*, Brooklyn, New York: Melville House, p. 42.
15 Jurgen W. Frembgen (1998) The Magicality of the Hyena: Beliefs and Practices in West and South Asia, *Asian Folklore Studies*, 57, 1998, 331–344, pp. 336–7.
16 Darcy Ogada (2014) The power of poison: pesticide poisoning of Africa's wildlife, *Annals of the New York Academy of Sciences*, 1–20, pp. 3 and 10.
17 See, for example, Mills and Hofer, 1998, p. 69.
18 T. Bohm and O.R. Höner (2015) *Crocuta crocuta. The IUCN Red List of Threatened Species*, https://www.iucnredlist.org/species/5674/45194782 accessed 10 June 2020.
19 Personal communication from Arjun Dheer.
20 Robin S. Reid (2012) *Savannas of Our Birth. People, Wildlife, and Change in East Africa*, Berkeley, Calif.: University of California Press, p. 168.
21 Bohm and Höner, 2015.
22 S. Romañach, Peter A. Lindsey, and Rosie Woodroffe (2011) Attitudes Toward Predators and Options for Their Conservation in the Ewaso Ecosystem, in Nicholas J. Georgiadis (ed) *Conserving Wildlife in African Landscapes Kenya's Ewaso Ecosystem*, Washington D.C.: Smithsonian Institute Scholarly press, 85–95, p. 86.
23 Glen Martin (2012) *Game Changer. Animal Rights and the Fate of Africa's Wildlife*, Berkeley: University of California Press, p. 7.
24 Ibid.
25 Laurence G. Frank, Rosie Woodroffe and Mordecia Ogada (2005) People and predators in Laikipia District, Kenya, in Rosie Woodroffe, Simon Thirgood and Alan Rabinowitz (eds) *People and Wildlife Conflict or Coexistence*, Cambridge: Cambridge University Press, 286–304, p. 286.
26 Bohm and Höner, 2015.
27 Reid, 2012, p. 142.
28 P. Henschel et al. (2020) Census and distribution of large carnivores in the Tsavo national parks, a critical East African wildlife corridor, *African Journal of Ecology*, 00, 1–16, DOI: 10.1111/aje.12730 accessed 28 July 2020, p. 1.
29 Ibid., p. 2.
30 Ibid., p. 10.
31 Bruce D. Patterson et al. (2004) Livestock predation by lions (Panthera leo) and other carnivores on ranches neighboring Tsavo National Parks, Kenya, *Biological Conservation*, 119, 507–17, p. 509.
32 Ibid., pp. 510 and 513.
33 David Ndeereh (2019) Towards saving an endangered species: Rehabilitation of hand-raised striped hyaena cubs in Meru National Park, Kenya, *African Journal of Ecology*, 57, 144–147, p. 144.
34 Henschel et al., 2020, p. 12.
35 Matt Walpole et al. (2003) *Wildlife and People: Conflict and Conservation in Masai Mara, Kenya, Wildlife and Development Series no 14*, London: International Institute for Environment and Development, pp. ix–xi.
36 Walpole et al., 2003, pp. x–xi.

37 Richard H. Lamprey and Robin S. Reid (2004) Expansion of Human Settlement in Kenya's Maasai Mara: What Future for Pastoralism and Wildlife? *Journal of Biogeography*, 31, 6, 997–1032, p. 999.
38 David Ole Seur (2003) Opeing address, in Walpole et al., *Wildlife and People: Conflict and Conservation in Masai Mara, Kenya, Wildlife and Development Series no 14*, p. 2
39 S. M. Cooper, K. E. Holekamp and L. Smale (1999) A seasonal feast: long-term analysis of feeding behaviour in the spotted hyaena (Crocuta crocuta), *African Journal of Ecology*, 37, 149–160, p. 149.
40 Ibid., p. 152.
41 Ibid.
42 Erin E. Boydston et al. (2003) Altered behaviour in spotted hyenas associated with increased human activity, *Animal Conservation*, 6, 207–219, pp. 214–5.
43 Joseph M. Kolowski et al. (2007) Daily Patterns of Activity in the Spotted Hyena, *Journal of Mammalogy*, 88, 4, 1017–1028, p. 1023.
44 Boydston et al., 2003, pp. 214–5.
45 Ibid., p. 216.
46 Joseph O. Ogutu, Nina Bhola and Robin Reid (2005) The effects of pastoralism and protection on the density and distribution of carnivores and their prey in the Mara ecosystem of Kenya, *Journal of Zoology*, 265, 281–293, p. 282.
47 Ibid.; Keith Somerville (2020) *Humans and Lions Conflict, Conservation and Coexistence*, London: Routledge/Earthscan, p. 151.
48 Ogutu et al., 2005, p. 284.
49 J. M. Kolowski and K. E. Holekamp (2009) Ecological and anthropogenic influences on space use by spotted hyaenas, *Journal of Zoology*, 277, 23–36, p. 33.
50 Ibid.
51 J. M. Kolowski and K. E. Holekamp (2008) Effects of an open refuse pit on space use patterns of spotted hyenas, *African Journal of Ecology*, 46, 3, pp. 341–49, p. 346.
52 Ibid.
53 Wiline M. Pangle and Kay E. Holekamp (2010) Lethal and nonlethal anthropogenic effects on spotted hyenas in the Masai Mara National Reserve, *Journal of Mammalogy*, 91, 1, 154–164, p. 156.
54 J. M. Kolowski and K. E. Holekamp (2006) Spatial, temporal, and physical characteristics of livestock depredations by large carnivores along a Kenyan reserve border, *Biological Conservation*, 128, 529–541, p. 533.
55 Ibid.
56 Ibid, p. 534.
57 Ibid., p. 539.
58 D. S. Green (2018) Anthropogenic disturbance induces opposing population trends in spotted hyenas and African lions, *Biodiversity Conservation*, 27, 871–889, pp. 872–3.
59 Ibid., p. 883.
60 Ibid., pp. 884–5.
61 Personal communication with Dr Holly Dublin.
62 David. S Green and Kay E. Holekamp (2019) Pastoralist activities affect the movement patterns of a large African carnivore, the spotted hyena (Crocuta crocuta), *Journal of Mammalogy*, XX, X, 1–13, p. 2.
63 Ibid., pp. 8–9.
64 Ibid., p. 11.
65 William O. Ogara et al. (2010) Determination of carnivores' prey base by scat analysis in Samburu community group ranches in Kenya, *African Journal of Environmental Science and Technology*, 4, 8, 540–546, p. 541.
66 Ibid., p. 542.
67 Ibid., p. 544.
68 Ibid.

69 Laurence G. Frank, Rosie Woodroffe and Mordecai Ogada (2005) People and predators in Laikipia District, Kenya, in Woodroffe et al., *People and Wildlife. Conflict or Coexistence?*, Cambridge, Cambridge University Press, 286–304, p.288.
70 Mordecai O. Ogada et al. (2003) Limiting Depredation by African Carnivores: The Role of Livestock Husbandry, *Conservation Biology*, 17, 6, 1521–30, p. 1522.
71 Frank et al., 2005, p.289.
72 Ibid., p. 295.
73 Romañach, Lindsey and Woodroffe, 2011, pp. 86–8.
74 Stephanie S. Romanăch, Peter A. Lindsey and Rosie Woodroffe (2007) Determinants of attitudes towards predators in central Kenya and suggestions for increasing tolerance in livestock dominated landscapes, *Oryx*, 41, 2, April, 185–195, p. 188.
75 Ibid., pp. 293–4.
76 Ibid., p. 299.
77 Zeke Davidson et al. (2019) Borrowing from Peter to pay Paul: managing threatened predators of endangered and declining prey species, *PeerJ*, 15 October, https://peerj.com/articles/7916/?fbclid=IwAR3IFyKPacgaA1CrA62G3rC5zaSnAAIV2IGJniplvJikoEPqOTHNW13gzCE accessed 15 October 2019, no page numbers.
78 Ibid.
79 Arjun Dheer (2016) *Resource partitioning between spotted hyenas (Crocuta crocuta) and lions (Panthera leo)*, thesis MRes University of Southampton, August, pp. 11–2.
80 Davidson, 2019.
81 Nairobi News (2020) KWS gives six lionesses contraceptives, Kenyans are not amused https://nairobinews.nation.co.ke/news/kws-gives-six-lionesses-contraceptives-kenyans-are-not-amused 17 July 2020 accessed 21 July 2020.
82 Personal communication with Laurence Frank.
83 Aaron P. Wagner, Laurence G. Frank and Scott Creel (2008) Spatial grouping in behaviourally solitary striped hyaenas, Hyaena hyaena, *Animal Behaviour*, 75, pp. 1131–1142.
84 Ibid., p. 1136.
85 David Ndeereh (2019) Towards saving an endangered species: Rehabilitation of hand-raised striped hyaena cubs in Meru National Park, Kenya, *African Journal of Ecology*, 57, 144–147, p. 144.
86 Ibid.
87 L. N. Leakey et al (1999) Diet of Striped Hyaena in Northern Kenya, *African Journal of Ecology*, 37, 3, 314–26, pp. 314–5.
88 Ibid., p. 318.
89 Ibid., p. 323.
90 George Monbiot (2003) *No Man's Land. An investigative journey through Kenya and Tanzania*, Totnes, Devon: Green Books, p. 42.
91 Gareth Jones (2014) Hyena the cleaner, *Star* (Kenya), 12 April, https://www.the-star.co.ke/sasa/2014-04-11-hyena-the-cleaner/ accessed 22 July 2020.
92 *Daily Nation* (2019) Death of Nairobi National Park: Why wildlife risks being wiped out, 22 October, https://www.nation.co.ke/health/Why-Nairobi-National-Park-is-in-danger/3476990-5319682-13dhihdz/index.html accessed 23 October 2019.
93 Ibid.
94 Lion and spotted hyena recovery and action plan launch (2020) http://www.kws.go.ke/content/lion-and-spotted-hyena-recovery-and-action-plan-launch accessed 10 August 2020.
95 Kenya Wildlife Service (2016) Efforts to conserve threatened stripped hyena species, http://www.kws.go.ke/content/efforts-conserve-threatened-stripped-hyena-species 3 November 2016, accessed 10 August 2020.
96 Personal communication with Laurence Frank; Kenya Wildlife Service (2020) *National Recovery and Action Plan for Lion and Spotted Hyena in Kenya (2020–2030)*, Nairobi, KWS.
97 Ibid., p. 59.

98 Amelia Jane Dickman (2008) *Key determinants of conflict between people and wildlife, particularly large carnivores around Ruaha National Park, Tanzania*, PhD thesis University College London (UCL) and Institute of Zoology, Zoological Society of London, p. 81.
99 Ibid., p. 81.
100 Ibid.
101 Hans Kruuk (2019) *The Call of Carnivores. Travels of a Field Biologist*, Exeter, UK: Pelagic Publishing, p. 60.
102 Ibid., pp. 59–60.
103 A. R. E. Sinclair (1995) Serengeti Past and Present, in A. R. E. Sinclair and Peter Arcese (eds) *Serengeti II Dynamics, Management and Conservation of an Ecosystem*, Chicago: University of Chicago Press, 3–33, p. 7.
104 Hans Kruuk (1972) *The Spotted Hyena A Study of Predation and Social Behaviour*, Chicago: University of Chicago Press, p. 13.
105 Ibid., pp. 18–9.
106 Anthony Sinclair (2012) *Serengeti Story*, Oxford: Oxford University Press, Kindle Edition, loc. 2630; personal communication from Dr Amy Dickman and Dr Laurence Frank.
107 Personal communication with game guard who wanted to remain anonymous, Lake Ndutu, Tanzania, May 2016.
108 K. M. Homewood and W. A Rodgers (1991) *Maasailand Ecology. Pastoralist development and wildlife conservation in Ngorongoro, Tanzania*, Cambridge: Cambridge University Press, p. 74.
109 Heribert Hofer and Marion East (1995) Population Dynamics, Population Size, and the Commuting Systems of Serengeti Spotted Hyenas, in A.R. E. Sinclair and Peter Arcese, *Serengeti II*, Chicago: University of Chicago Press 332–363, p. 340
110 Bernard M. Kissui et al. (2019) Patterns of livestock depredation and cost-effectiveness of fortified livestock enclosures in northern Tanzania, *Ecology and Evolution*, May 2019, 1–14. p. 2.
111 Ibid.
112 Ibid., p. 7.
113 Sinclair, 2012, loc. 649.
114 Hofer and East, 1995, p. 335.
115 Franco Peniel Mbise (2018) Livestock depredation by wild carnivores in the Eastern Serengeti Ecosystem, Tanzania, *International Journal of Biodiversity and Conservation*, 10, 3, 122–30, p. 122.
116 Ibid., p. 123.
117 Ibid., p. 126.
118 Tomas Holmern, Julius Nyahongo and Eivin Røskaft (2017) Livestock loss caused by predators outside the Serengeti National Park, Tanzania, *Biological Conservation*, 135, 4, 534–542, p. 534.
119 Ibid., p. 539.
120 Höner et al., 2005, p. 544.
121 Personal communication with Arjun Dheer.
122 Leandro Abade, David W. Macdonald and Amy J. Dickman (2014) Using Landscape and Bioclimatic Features to Predict the Distribution of Lions, Leopards and Spotted Hyaenas in Tanzania's Ruaha Landscape, *PLOS One*, 9, 5, 1–14, p. 2.
123 Leandro Alécio dos Santos Abade (2013) *Human-Carnivore Conflict in Tanzania: modelling the spatial distribution of lions (Panthera leo), leopards (Panthera pardus) and spotted hyaenas (Crocuta crocuta), and their attacks upon livestock, in Tanzania's Ruaha Landscape*, Masters of Science by Research Lady Margaret Hall/WildCRU, Trinity Term 2013, p. 28.
124 Ibid., p. 21.
125 Ibid., pp. 32–33.
126 Ibid.

127 Personal communication with Laurence Frank, who has worked with Laikipia pastoral communities.
128 Dickman, 2008, p. 186.
129 Ibid, p. 191.
130 Ibid., pp. 202–3.
131 Ibid., p. 216.
132 Rebuilding in Uganda, *Oryx*, 15, 3, April 1980, p. 220.
133 Keith Somerville (2019) *Ivory Power and Poaching in Africa*, London: Hurst and Co, pp. 126–30.
134 Edward Okot Omoya (2013) Estimating population sizes of lions Panthera leo and spotted hyaenas Crocuta crocuta in Uganda's savannah parks, using lure count methods, *Oryx*, 48, 3, 394–401, p. 394.
135 Ibid., p. 396.
136 Ibid., p. 398.
137 Ibid., p. 400.
138 Daniel W. Gade (2006) Hyenas and Humans in the Horn of Africa, *Geographical Review*, 96, 4, 609–632, p. 609.
139 Ibid.
140 Personal communication with Hans Bauer.
141 Gidey Yirga et al. (2016) Spotted hyena (Crocuta crocuta) concentrate around urban waste dumps across Tigray, northern Ethiopia, *Wildlife Research*, 42, 7, 563–569, pp. 566–7.
142 Ibid., p. 610.
143 Ibid.
144 Gidey Yirga and Hans Bauer (2010) Diet of the Spotted Hyena (Crocuta crocuta) in southern Tigray, northern Ethiopia, *World Journal of Science, Technology and Sustainable Development*, 7, 4, 391–6, p. 391.
145 Ibid., p. 393.
146 Ibid. p. 391.
147 Ibid., p. 392.
148 Gidey Yirga et al. (2011) Peri-urban spotted hyena (Crocuta crocuta) in Northern Ethiopia: diet, economic impact, and abundance, *European Journal of Wildlife Research*, 57, 759–765, p. 760.
149 Ibid., p. 762.
150 Gidey Yirga (2013) Spotted hyena (Crocuta crocuta) coexisting at high density with people in Wukro district, northern Ethiopia, *Mammalian Biology*, 78, 193–197, p. 194.
151 Ibid.
152 Ibid., p. 195.
153 Ibid., p. 197.
154 Ibid.
155 Ibid.
156 Gidey Yirga et al. (2012) Adaptability of large carnivores to changing anthropogenic food sources: diet change of spotted hyena (Crocuta crocuta) during Christian fasting period in northern Ethiopia, *Journal of Animal Ecology*, 81, 1052–1055, pp. 1052–3.
157 Yirga et al., 2012, pp. 1053–4.
158 Ibid.
159 Yirga et al., 2014, pp. 325–6.
160 Gade, 2006, pp. 215–20.
161 Ibid.
162 Ibid. pp. 627–8.
163 M. J. Fell et al. (2019) Facial injuries following hyena attack in rural eastern Ethiopia, *International Association of Oral and Maxillofacial Surgeons*, 43, 1459–1464, p. 1460.
164 Personal communication with Ludwig Siege.
165 Ibid.
166 Ibid.

167 Mills and Hofer, 1998, p. 75.
168 Ibid.
169 Kruuk, 1972, p. 114.
170 Ibid.
171 Marcus Baynes-Rock (2015) *Among the Bone-Eaters. Encounters with Hyenas in Harar*, Pennsylvania: University of Pennsylvania Press.
172 Camilla Gibb (1996) In the City of Saints: Religion, Politics and Gender in Harar, Ethiopia, PhD Thesis, University of Oxford, (BLLD 46–12470), p. 191.
173 Baynes-Rock, 2015, p. 14.
174 Marcus Baynes-Rock (2012) Hyenas like Us: Social Relations with an Urban Carnivore in Harar, Ethiopia, PhD thesis, Department of Anthropology Macquarie University, Sydney, p. 59.
175 Ibid., pp. 15–6.
176 Ibid., p. 22.
177 Ibid., pp. 23–4.
178 Marcus Baynes-Rock (2013) Local Tolerance of Hyena Attacks in East Hararge Region, Ethiopia, *Anthrozoös*, 26, 3, 421–433, p. 422.
179 Baynes-Rock, 2015, p. 103.
180 Ibid., p. 104.
181 Baynes-Rock, 2013, pp. 424–5.
182 Ibid., p. 425.
183 Ibid., p. 426.
184 Opportunity in Somalia, *Oryx*, 9, 3, December 1967, p. 180.
185 J. Cloudsley-Thompson (1973) Developments in the Sudan Parks, *Oryx*, 12, 1, 49–52, pp. 49–51.
186 Personal communication with Adrian Schenk.
187 Andy Catley, Tim Leyland and Suzan Bishop (2005) *Policies, Practice and Participation in Complex Emergencies: The Case of Livestock Interventions in South Sudan A case study for the Agriculture and Development Economics Division of the Food and Agriculture Organization*, Rome: FAO, accessed 29 July 2020, p. 3.
188 Hans Bauer et al. (2017) *Follow-up Visits to Alatash-Dinder Lion Conservation Unit, Ethiopia & Sudan*, Oxford: Wildlife Conservation Research Unit (WildCRU), p. 3.
189 James Clarke (2012) *Save me from the Lion's Mouth*, Johannesburg: Penguin Random House South Africa, Kindle Edition, loc. 2376.
190 See, for example, Rachel Nuwer (2017) The Rare African Park Where Elephants Are Thriving, National Geographic, 29 January, https://www.nationalgeographic.com/news/2017/01/wildlife-watch-chad-zakouma-elephants-poaching/ accessed 30 July 2020; African Parks (no date) Zakouma National Park, https://www.africanparks.org/the-parks/zakouma accessed 30 July 2020.
191 Ibid. – both sources mention the success in reducing poaching.
192 Nathalie Vanherle (2008) Report of the DAS/ROCAL pilot project in Zakouma National Park, Chad, in Barbara Croes et al. (ed), *Management and conservation of large carnivores in West and Central Africa*, Conservation of large carnivores in West and Central Africa. Proceedings of an International Seminar, CML/CEDC, 15 and 16 November 2006, Maroua, Cameroon, 53–72, p. 53.
193 Ibid.
194 Ibid., pp. 53–4.
195 Ibid., pp. 59 and 62.
196 Peter A. Lindsey et al. (2017) Relative efforts of countries to conserve world's megafauna, *Global Ecology and Conservation*, 10, pp. 243–252, figure S1.
197 African parks (no date) *Chinko*, https://www.africanparks.org/the-parks/chinko accessed 30 July 2020.
198 Thierry Aebischer (2020) Apex predators decline after an influx of pastoralists in former Central African Republic hunting zones, *Biological Conservation*, 241, 1–9, pp. 1–2.
199 Ibid., p. 2.

200 Jean Paul Kwabong (2008) Hunting of large carnivore in Cameroon over the past 20 years, in Barbara Croes et al. (ed), *Management and conservation of large carnivores in West and Central Africa*, Conservation of large carnivores in West and Central Africa. Proceedings of an International Seminar, CML/CEDC, 15 and 16 November 2006, Maroua, Cameroon, 103–8, p. 103.
201 H. Bauer and S. Kari (2001) Assessment of the people-predator conflict through thematic PRA in the surroundings of Waza National Park, Cameroon, *PLA Notes (IIED)*, 4, 1 August, p. 9.
202 Ibid., p. 11.
203 Ibid.; Pricella N. Tumenta et al. (2013) Livestock depredation and mitigation methods practised by resident and nomadic pastoralists around Waza National Park, Cameroon, *Oryx*, 47, 2, 237–242, p. 240.
204 Tumenta et al., 2013, p. 239.
205 Ibid., p. 241.
206 B. M. Croes et al. (2011) The impact of trophy hunting on lions (Panthera leo) and other large carnivores in the Bénoué Complex, northern Cameroon, *Biological Conservation*, 144, 3064–3072, p. 3069.
207 Ibid.
208 Ibid., pp. 3070.
209 Barbara Croes et al. (2008) Livestock-carnivore conflicts: results of an inventory around Bénoué National Park, Cameroon, in Croes et al., 29–40, p. 30 p. 33.
210 Croes et al., 2011, p. 3070.
211 Hans Bauer (2007) Status of large carnivores in Bouba Ndjida National Park, Cameroon, *African Journal of Ecology*, 45, 448–450, p. 449.
212 Iris Kirsten et al. (2018) Lion (Panthera leo) and spotted hyaena (Crocuta crocuta) abundance in Bouba Ndjida National Park, Cameroon, trends between 2005 and 2014, *African Journal of Ecology*, 56, 414–417, p. 413.
213 Ibid., pp. 415–6.
214 C. A. Spinage (1980) Parks and Reserves in Congo Brazzaville, *Oryx*, 15, 3, April, 292–5, p. 292.
215 Philipp Henschel, Guy-Aime Malanda and Luke Hunter (2014) The status of savanna carnivores in the Odzala-Kokoua National Park, northern Republic of Congo, *Journal of Mammalogy*, 95, 4, 22 August, pp. 882–892.
216 African Parks (2018) Annual Report, https://www.africanparks.org/sites/default/files/uploads/resources/2019-06/AFRICAN%20PARKS%20-%202018%20Annual%20Report%20-%20Full%20-%2029052019%20-%20Digital.pdf accessed 30 July 2020.
217 C. E. G. Tutin, L. J. T. White and A. Mackanga-Missandzou (1997) The Use by Rain Forest Mammals of Natural Forest Fragments in an Equatorial African Savanna, *Conservation Biology*, 11, 5, 1190–1203, p. 1191.
218 Personal communication with Phillip Henschel and Lee White.
219 I. Wiesel (2015) *Parahyaena brunnea. The IUCN Red List of Threatened Species*, https://www.iucnredlist.org/species/10276/82344448 accessed 5 August 2020.
220 R. L. Eaton (1976) The Brown Hyena: A Review Of Biology, Status And Conservation, *Mammalia*, 40, 3, 377–399, p. 377.
221 Ibid., p. 388.
222 Louisa Richmond-Coggan (2014) *Comparative abundance and ranging behaviour of brown hyaena (Parahyaena brunnea) inside and outside protected areas in South Africa*, a thesis submitted in partial fulfilment of the requirements of Nottingham Trent University for the degree of Doctor of Philosophy, February 2014, p. 19.
223 Ibid.
224 Wiesel, 2015.
225 Richmond-Coggan, 2014, p. 24.
226 Somerville, 2019, pp. 172–8.
227 Pedro Beja et al. (2019) The Mammals of Angola, in Brian J. Huntley et al., *The Biodiversity of Angola. Science & Conservation: A Modern Synthesis*, Springer Open, https://

link.springer.com/content/pdf/10.1007%2F978-3-030-03083-4.pdf accessed 13 July 2020, 357–444, p. 371–2.
228 Ibid.
229 Ibid.
230 Brian J. Huntley (2017) *Wildlife at War: The Rise and Fall of an African Eden*, Pretoria: Protea Book House, p. 69.
231 Ibid., p. 376.
232 Francisco M. P. Gonçalves (2019) A rapid assessment of hunting and bushmeat trade along the roadside between five Angolan major towns, *Nature Conservation*, 37, 151–160, p. 151.
233 Keith Somerville (2019) Angola: Demining key to conservation plans, *Global Geneva*, 21 November, accessed 5 August 2020.
234 Ibid.
235 Paul Funston (2017) *The distribution and status of Lions and other large carnivores in Luengue-Luiana and Mavinga National Parks*, Angola, KAZA TFCA Secretariat https://ichef.bbci.co.uk/news/660/cpsprodpb/0A83/production/_112019620_img-20200428-wa0027.jpg accessed 29 April 2020.
236 Ibid.
237 Ibid.
238 Personal communication with Angolan predator researcher Sara Fernandes Elizalde.
239 Ibid.
240 Jake Overton et al. (2017) A Large Mammal Survey of Bicuar and Mupa National Parks, Angola With Special Emphasis on the Presence and Status of Cheetah and African Wild Dogs, RWCP/Panthera/Ministerio de Ambiente, https://cheetahandwilddog.files.wordpress.com/2017/04/bicuarmupafinalreport_march2017.pdf accessed 28 April 2019, p. 7.
241 Ibid., p. 30.
242 Robert and June Kay (1962) Fauna preservation in N'Gamiland, *Oryx*, October, 5, p. 283.
243 L. N. Rich et al. (2017) Carnivore distributions in Botswana are shaped by resource availability and intraguild species, *Journal of Zoology*, 303, 90–98, p. 94.
244 Lauren C. Satterfield et al. (2017) Estimating Occurrence and Detectability of a Carnivore Community in Eastern Botswana using Baited Camera Traps, *African Journal of Wildlife Research*, 47, 1, 32–46, p. 36.
245 Brian F. Kuhn (2012) Bone Accumulations of Spotted Hyaenas (Crocuta crocuta, Erxleben, 1777) as Indicators of Diet and Human Conflict; Mashatu, Botswana, *International Journal of Ecology*, 2012, 1–6, p. 1.
246 Ibid., p. 2.
247 Ibid, p. 4.
248 Monika Schiess-Meier et al. (2006) Livestock Predation—Insights From Problem Animal Control Registers in Botswana, *Journal of Wildlife Management*, 71, 4, 1267–1274, p. 1267.
249 Schiess-Meier et al., 2006, p. 1270.
250 Vivien Tempest Kent (2011) The Status and Conservation Potential of Carnivores in Semi-Arid Rangelands, Botswana The Ghanzi Farmlands: A Case Study, Durham theses, Durham University, p. 4.
251 Ibid., p. 65.
252 Ibid.
253 Ibid.
254 Ibid., p. 53.
255 Ibid., p. 95.
256 Lorraine K. Boast, and Ann-Marie Houser (2012) Density of Large Predators on Commercial Farmland in Ghanzi, Botswana, *South African Journal of Wildlife Research*, 42, 2, 138–143, p. 141.
257 Ibid., p. 142.

258 Tempest Kent, 2011, p. 129.
259 Ibid., pp. 130–1.
260 Vivien T. Kent and Russell A. Hill (2013) The importance of farmland for the conservation of the brown hyaena Parahyaena brunnea, *Oryx*, 47, 3, 431–440, p. 431.
261 Ibid., pp. 435–6.
262 G. Cozzi et al. (2015) Effects of Trophy Hunting Leftovers on the Ranging Behaviour of Large Carnivores: A Case Study on Spotted Hyenas. *PLoS ONE*, 10–3, 1–15, p. 1.
263 Ibid., p. 4.
264 Ibid., p. 12.
265 G. Cozzi et al. (2013) Density and habitat use of lions and spotted hyenas in northern Botswana and the influence of survey and ecological variables on call-in survey estimation, *Biodiversity Conservation*, 22, 2937–2956, p. 2949.
266 M. Gussett et al. (2009) Human–wildlife conflict in northern Botswana: livestock predation by Endangered African wild dog Lycaon pictus and other carnivores, *Oryx*, 43, 1, 67–72, p. 67.
267 J. Weldon McNutt (2017) Living on the edge: characteristics of human–wildlife conflict in a traditional livestock community in Botswana, *Wildlife Research*, 44, 546–557, p. 547.
268 Ibid., pp. 549 and 550–1.
269 Personal communication with Peter Apps of Botswana Predator Conservation.
270 McNutt, 2017, p. 551.
271 Ibid.
272 Personal communication with Chief Timex Moalosi.
273 G. Maude, R. P. Reading and S. Harris (2019) Fluctuating food resources and home ranges in brown hyaenas living in a semi-arid environment, *Journal of Zoology*, 307, 53–60, p. 54.
274 Ibid., pp. 56–7.
275 Maude and Mills, 2005, p. 202.
276 Ibid., p. 205.
277 Ibid., p. 210.
278 Ibid., p. 211.
279 Ibid., p. 212.
280 Ibid., pp. 212–3.
281 Mark and Delia Owens (1985) *Cry of the Kalahari,* Glasgow: Fontana/Collins, p. 72.
282 M. G. L. Mills (1989) The Comparative Behavioral Ecology of Hyenas: The Importance of Diet and Food Dispersion, in John L. Gittelman (ed), *Carnivore Behavior, Ecology, and Evolution*, Ithaca, New York: Comstock Publishing, 125–142, p. 125; Michael G. L. Mills (2015) Living Near the Edge: A Review of the Ecological Relationships Between Large Carnivores in the Arid Kalahari, *African Journal of Wildlife Research*, 45, 2, 127–137, p. 130.
283 Willem D. Briers-Louw and Alison J. Leslie (2020) Dietary partitioning of three large carnivores in Majete Wildlife Reserve, Malawi, *African Journal of Ecology*, 2020, 00, June, 1–12, p. 2.
284 A. Balestra (1962) The man-eating hyeans of Mlanje, *African Wildlife*, 16, pp. 25–27.
285 Hans Kruuk (2002) *Hunter and Hunted. Relationships between carnivores and people*, Cambridge: Cambridge University Press, pp. 64–5.
286 BBC (2002) Man-eating hyenas spread fear in Malawi, 9 January, http://news.bbc.co.uk/1/hi/world/africa/1750825.stm accessed 19 August 2019.
287 Personal communication with Brennan PetersonWood.
288 Ibid.
289 Personal communication with Olivia Sievert, formerly of the Lilongwe Wildlife Trust.
290 Carnivore Research Malawi (2020) 2 January, https://www.facebook.com/CarnivoreResearchMalawi/ accessed 10 August 2020.
291 Conservation Research Malawi https://www.facebook.com/CarnivoreResearchMalawi/posts/3375312642497904 accessed 10 August 2020.

292 Madinga, 2019.
293 Brennan PetersonWood (2020) Importance of Spotted Hyena in Malawi and how to safely coexist with them, *Malawi 24*, 28 July, https://malawi24.com/2020/07/28/importance-of-spotted-hyena-in-malawi-and-how-to-safely-coexist-with-them/?fbclid=IwAR2a5aJoSH3vNQmdlxDhCe64ZLs09z8igfznOuP0td7kA56ZipqJo7GTV44 accessed 10 August 2020.
294 Hyenas kill woman in Ntcheu, *Malawi24*, 22 July 2020, https://malawi24.com/2020/07/22/hyenas-kill-woman-in-ntcheu/?fbclid=IwAR2 PrIsDdME0QJlXnfrJSGglnnP-0jEMZJa1iz9B9plQYJwtXKMSlfABGHY accessed 10 August 2020.
295 Niassa Lion Project, https://www.facebook.com/niassalionproject accessed 10 August 2020.
296 https://www.greatlimpopo.org/maps/project-maps/ accessed 10 August 2020.
297 A plane full of hyenas, *Africa Geographic*, 3 August 2020, https://africageographic.com/stories/a-plane-full-of-hyenas/ accessed 4 August 2020.
298 Wiesel, 2015.
299 Sarah Edwards (2015) Human-wildlife conflict issues on commercial farms bordering the Sperrgebiet and Namib-Naukluft National Parks borders, southern Namibia, PhD thesis, Royal Holloway, University of London, 8 September, pp. 56–7.
300 Results of killing "vermin", *Oryx*, 9, 3, December 1967, pp. 175–6.
301 Ibid.
302 Ibid.
303 Edwards, 2015, pp. 26–7.
304 Ibod., pp. 33–4.
305 L. Chris Weaver and Theunis Petersen (2008) Namibia Communal Area Conservancies, FAO, Best Practices in Sustainable Hunting, http://www.fao.org/tempref/docrep/fao/010/aj114e/aj114e10.pdf accessed 10 August 2020, pp. 48–52, p. 48.
306 Edwards, 2015, p. 35.
307 Ibid., pp. 35–6.
308 Mandy Schumann, Laurence H. Watson, and Bonnie D. Schumann (2008) Attitudes of Namibian commercial farmers toward large carnivores: The influence of conservancy membership, *South African Journal of Wildlife Research*, 38, 2, 123–132, p. 124.
309 N. A. Rust and L. Marker (2014) Cost of carnivore coexistence on communal and resettled land in Namibia, *Environmental Conservation*, 41, 1, 45–53, p. 45.
310 Ibid., p. 50.
311 Schumann et al., 2008, p. 124.
312 Ibid., p. 125.
313 Ibid.
314 Ibid., p. 130.
315 Edwards, 2015, pp. 175–9.
316 Ibid., p. 220.
317 Ibid., p. 250.
318 Edwards, 2015 p. 106.
319 Martina Trinkel et al. (2004) Spotted hyenas (Crocuta crocuta) follow migratory prey. Seasonal expansion of a clan territory in Etosha, Namibia, *Journal of Zoology*, 264, 125, 125–133, p. 125.
320 Martina Trinkel (2009) A keystone predator at risk? Density and distribution of the spotted hyena (Crocuta crocuta) in the Etosha National Park, Namibia, *Canadian Journal of Zoology*, 87, 941–947, p. 946.
321 Florian Weise, Stéphanie Périquet and Ken Stratford (2020) *Hyaena research and conservation efforts at the Ongava Research Centre*, http://conservationnamibia.com/articles/cnam2020-spotted-hyaenas.php?fbclid=IwAR1DenGRiEfpDng5gtG7tekH1q5HqGIJUZqANxbPNV4_onuAqtqi1gY1ibM accessed 14 August 2020.

322 Linda Barker (2019) Namibia culls hyenas to save its wild / feral horses, *Africa Geographic*, 26 February, https://africageographic.com/stories/namibia-culls-hyenas-to-save-its-wild-feral-horses/ accessed 11 August 2020.
323 Edwards, 2015, pp 35–6.
324 Brown hyena research project (2019) 29 December, https://www.facebook.com/BrownHyenaResearchProject/ accessed 1 January 2020.
325 Edwards, 2015, p. 50.
326 Andrew B. Stein, Todd K. Fuller and Laurie L. Marker (2013) Brown hyaena feeding ecology on Namibian farmlands, *South African Journal of Wildlife Research*, 43, 1, April 2013, 27–32, p. 27.
327 Ibid., p. 31.
328 Ibid.
329 Florian J. Weise et al. (2015) Evaluation of a Conflict-Related Brown Hyaena Translocation in Central Namibia, *African Journal of Wildlife Research*, 45, 2, 178–186, p. 182.
330 Ingrid Wiesel, Sabrina Karthun-Strijbos and Inga Jänecke (2019) The use of GPS telemetry data to study parturition, den location and occupancy in the brown hyaena, *African Journal of Wildlife Research*, 49, 1, 1–11, p. 2.
331 Ibid., pp. 2–3.
332 Personal communication from Dr Sarah Edwards at Okonjima.
333 Sarah Edwards (2019) Brown Hyena research Update, *Africat*, https://africat.org/africat-brown-hyaena-research-update-2019/?fbclid=IwAR0u-EfhBjy344iCFvyrO91b4TWzZU8VKoU0X4naNjHeWj3U5iNCQ86YZDE accessed 2 September 2019.
334 Ibid.
335 Sarah Edwards et al. (2020) Socioecology of a high-density brown hyaena population within an enclosed reserve, *Mammal Research*, February, https://link-springer-com.chain.kent.ac.uk/content/pdf/10.1007%2Fs13364-020-00477-z.pdf accessed 11 August 2020, no page numbers.
336 Ibid.
337 Sarah Edwards et al. (2020) First confirmed record of infanticide for wild brown hyaena, *African Journal of Ecology*, February, p. 2.
338 Lise Hanssen (no date) Spotted Hyena Dynamics in the Zambezi Region of Namibia, *Kwando Carnivore Project*, http://www.travelnewsnamibia.com/news/stories/conservation/conservation-spotted-hyena-in-the-zambezi/ accessed 20 August 2019.
339 Ibid.
340 Ibid.
341 Ibid.
342 Lise Hanssen (2013) Caprivi Carnivore Project, 2 July, https://tosco.org/2013/07/02/caprivi-carnivore-project/ accessed 20 August 2019.
343 Bohm and Höner, 2015.
344 Louisa Richmond-Coggan, 2014, p. 19.
345 Louisa Richmond-Coggan (2014) Comparative abundance and ranging behaviour of brown hyaena (Parahyaena brunnea) inside and outside protected areas in South Africa, a thesis submitted in partial fulfilment of the requirements of Nottingham Trent University for the degree of Doctor of Philosophy, February, p. 19.
346 Vivienne Linda Williams and John Whiting (2016) A picture of health? Animal use and the Faraday traditional medicine market, South Africa, *Journal of Etnophamracology*, 179, February, 265–73, p. 265.
347 A. B. Cunningham and A. S. Zond (1991). *Use of Animal Parts for the Commercial Trade in Traditional Medicines*. Institute of Natural Resources, Pietermaritzburg.
348 T. S. Simelane and G. I. H Kerley (1998) Conservation implications of the use of vertebrates by Xhosa traditional healers in South Africa. *South African Journal of Wildlife Research*, 28, 4, 121–126.

349 Sam M. Ferreira, and Paul J. Funston (2016) Population Estimates of Spotted Hyaenas in the Kruger National Park, South Africa, *African Journal of Wildlife Research*, 46, 2, 61–70, pp. 67–8.
350 G. L. Smuts (1982) *Lion*, Johannesburg: Macmillan, p. 167.
351 Ibid., pp. 214 and 220.
352 Lion Cropping in the Kruger, *Oryx*, 13, 3, February 1976, p. 228.
353 Ferreira and Funston, 2016, p. 67.
354 Mills and Hofer, 1998, p. 74.
355 Melissa Wray (no date) An Estimated 1600 Lions in Kruger, http://www.krugerpark.co.za/krugerpark-times-3-13-krugers-lions-23272.html#:~:text=It%20is%20estimated%20that%20there,nearly%20all%20in%20tiptop%20condition accessed 12 August 2020.
356 Ferreira and Funston, 2016, p. 69.
357 L. E. Belton, E. Z. Cameron, and F. Dalerum (2018) Spotted Hyaena Visitation at Anthropogenic Sites in the Kruger National Park, South Africa, *African Zoology*, 53, 3, 113–118, p. 114.
358 Ibid.
359 Ibid., p. 115.
360 Lydia E. Belton, Elissa Z. Cameron and Fredrik Dalerum (2018) Social networks of spotted hyaenas in areas of contrasting human activity and Infrastructure, *Animal Behaviour*, 135, 13–23, p. 16.
361 Bruce Bryden (2008) *A Game Ranger Remembers*, Johannesburg: Jonathan Ball Publishers, Kindle Edition, loc. 2742.
362 M. G. L Mills (1990) *Kalahari Hyenas Comparative Ecology of Two Species*, Caldwell, New Jersey: The Blackburn Press, p. 12.
363 Ibid., pp. 7 and 11.
364 Ibid., p. 32.
365 Ibid., pp. 86–92.
366 Michael G. L. Mills (1981) The Socio-Ecology and Social Behaviour of the Brown Hyaena Hyaena brunnea, Thunberg, 1820, in the Southern Kalahari, DSc thesis, University of Pretoria, srudy carried out in the Kalahari Gemsbok National Park South Africa and the adjacent Gemsbok National Park Botswana, p. 87.
367 Jan A. Graf (2009) Heterogeneity in the density of spotted hyaenas in Hluhluwe-iMfolozi Park, South Africa, *Acta Theriologica*, 54, 4, 333–343, p. 333.
368 Ibid., p. 334.
369 https://www.pilanesbergnationalpark.org/ accessed 12 August 2020.
370 M. Thorn et al. (2009). Estimating brown hyaena occupancy using baited camera traps, *South African Journal of Wildlife Research*, 39, 1, 10.p. 6.
371 Vivien T. Kent and Russell A. Hill (2013) The importance of farmland for the conservation of the brown hyaena Parahyaena brunnea, *Oryx*, 47, 3, 431–440, pp. 435–6.
372 Ibid.
373 Michelle Thorn et al. (2011) Brown hyaenas on roads: Estimating carnivore occupancy and abundance using spatially auto-correlated sign survey replicates, *Biological Conservation*, 144, 1799–1807, p. 801.
374 Richmond-Coggan, 2014, p. 19.
375 Ibid., p. 25.
376 Ibid., pp. 26–7.
377 Ibdi., p. 93.
378 Ibid., p. 99.
379 Philip B. Faure et al. (2019) Brown hyaena (*Parahyaena brunnea*) diet composition from Zingela Game Reserve, Limpopo Province, South Africa, *African Zoology*, 54, 2, 119–24, p. 121.
380 Ibid., p. 122.
381 Michelle Thorn et al. (2013) Characteristics and determinants of human-carnivore conflict in South African farmland, Biodiversity and Conservation, 22, 1715–1730, p. 1715.
382 Ibid., p. 1721.

383 Ibid., p. 1722.
384 Ibid., p. 1724.
385 Rebecca J. Welch and Daniel M. Parker (2016) Brown hyaena population explosion: rapid population growth in a small, fenced system, *Wildlife Research*, 43, 2, 178–187, p. 179.
386 Ibid.
387 Ibid. pp. 182–3.
388 Rebecca J. Welch (2016) Brown hyena habitat selection varies among sites in a semi-arid region of southern Africa, *Journal of Mammalogy*, 97, 2, 473–482, p. 474.
389 Welch, 2016, p. 474.
390 Ibid., p. 477.
391 John O'Brien (2012) The Ecology and Management of the Large Carnivore Guild on Shamwari Game Reserve, Eastern Cape, a thesis submitted in fulfilment of the requirements for the degree of Doctor of Philosophy of Rhodes University (copy supplied to me by author).
392 Personal communication from John O'Brien.
393 Ibid.
394 Ibid.
395 J. Comley et al. (2020) Do spotted hyaenas outcompete the big cats in a small, enclosed system in South Africa? *Journal of Zoology*, January 2020, doi:10.1111/jzo.12772, 1–9, p. 4.
396 Ibid, p. 4.
397 Chansa Chomba (2012) Patterns of human – wildlife conflicts in Zambia, causes, consequences and management responses, *Journal of Ecology and the Natural Environment*, 4, 12, 303–313, p. 303.
398 Jassiel M'soka (2016) Spotted hyaena survival and density in a lion depleted ecosystem: The effects of prey availability, humans and competition between large carnivores in African savannahs, *Biological Conservation*, 201, 348–355, p. 348.
399 Ibid.
400 African Parks (2020) https://www.facebook.com/AfricanParks/ accessed 8 April 2020; https://www.africanparks.org/the-parks/liuwa-plain accessed 13 August 2020.
401 M'soka, 2016, p. 353.
402 Peter Matthiessen and Bob Douthwaite (1985) The impacts of tsetse fly control campaigns on African wildlife, *Oryx*, 19, 4, 1985, 202–9, pp. 202–3.
403 Personal communication with Blondie Leatham, manager of the Bubye Valley Conservancy, Zimbabwe, and former Zimbabwean game ranger.
404 Somerville, 2020, p. 201.
405 Samuel T. Williams et al. (2016) The impact of land reform on the status of large carnivores in Zimbabwe. *PeerJ*, 4, e1537; DOI:10.7717/peerj.1537, 1–21, p. 6.
406 Ibid., p. 202.
407 Mills and Hoger, 1998, p. 72.
408 Ibid., p. 78.
409 S. Périquet (2015) Spotted hyaenas switch their foraging strategy as a response to changes in intraguild interactions with lions, *Journal of Zoology*, 297, 245–254, p. 247.
410 J. Loveridge et al. (2020) Evaluating the spatial intensity and demographic impacts of wire-snare bush-meat poaching on large carnivores, *Biological Conservation*, 244, 1–8, p. 4.
411 Mlamuleli Mhlanga et al. (2019) Influence of Settlement Type and Land Use on Public Attitudes towards Spotted Hyaenas (Crocuta crocuta) in Zimbabwe, *African Journal of Wildlife Research*, 49, 1, 142–154, pp. 142–3.
412 Ibid., p. 149.
413 Ibid.
414 Ibid. p. 988.
415 Moreangels M. Mbizah, Jorgelina Marino, and Rosemary J. Groom (2012) Diet of Four Sympatric Carnivores in Savé Valley Conservancy, Zimbabwe: Implications for

Conservation of the African Wild Dog (Lycaon pictus), *South African Journal of Wildlife Research*, 42, 2, 94–103, p. 94.
416 Samual T. Williams et al. (2016) The impact of land reform on the status of large carnivores in Zimbabwe. *PeerJ*, 4, 1537; 1–21, p. 6.
417 Ibid.
418 Ibid.
419 Darcy Ogada, 2014, p. 3.
420 Somerville, 2020, pp. 122–4.
421 Etotépé A. S. Sogbhossou et al. (2011) Human–carnivore conflict around Pendjari Biosphere Reserve, northern Benin, *Oryx*, 45, 4, 569–578, p. 569.
422 Ibid.
423 https://www.shakariconnection.com/hunting-benin.html; https://www.fauna-safari-club.com/destination/chasse-au-benin/; http://www.pyreneanoutfitters.com/portfolio-item/safari-benin-buffalo-roan-antelope-hunting/ all accessed 14 August 2020.
424 Somerville, 2020, pp. 122–4.
425 Sogbhossou et al., 2011, p. 570.
426 Hamissou H. Malam Garba & Ilaria Di Silvestre (2008) Conflicts between large carnivores and domestic livestock in the peripheral zone of the W transboundary Park in Niger, in Croes et al., 133–51, p.133.
427 Ibid.
428 John Hare (2013) *Last Man in The End of Empire in Northern Nigeria*, Benenden, Kent: Neville and Harding, p. 169.
429 Happold's account relayed by Andrew Dunn of WCS.
430 Personal communication from Andrew Dunn.
431 Mikita Brottman (2012) *Hyena*, London: Reaktion Books, pp. 85–7.
432 Pieter Hugo (no date) *'Gadawan Kura' - The Hyena Men*, http://archive.stevenson.info/exhibitions/hugo/nigeria_index2.htm accessed 13 August 2020.
433 China Global Television Network – CGTN (no date), Hyena Men, https://www.youtube.com/watch?v=–f79MZpW_A accessed 13 August 2020.
434 Ilaria Di Silvestre et al. (2000) Feeding habits of the spotted hyaena in the Niokolo Koba National Park, *African Journal of Ecology*, 38, 102–107, p. 102.
435 Jacques Verschuren (1985) Mauritania: Its wildlife and a coastal park, *Oryx*, 19, 4, October, 221–4, p. 223.
436 Mills and Hofer, 1998, p. 66.
437 Ibid.
438 Ibid., pp. 68–70.
439 Dale J. Osborn and Ibrahim Helmy (1980) The Contemporary Land Mammals of Egypt (including Sinai). Field Museum of Natural History, digitized version, https://archive.org/details/contemporaryland05osbo/page/422 accessed 24 July 2019, p. 422.
440 https://www.iucnredlist.org/search?landRegions=TN&searchType=species accessed 17 August 2020.
441 M. Masseti (2010) Holocene mammals of Libya: A biogeographical, historical and archaeozoological approach, *Journal of Arid Environments*, 74, 794–805, p. 799.
442 Osborn and Helmy, 1980, p. 422; Mills and Hofer, 1998, p. 69.
443 Mills and Hofer, 1998, p. 69.
444 Osborn and Helmy, 1980, pp. 426–8.
445 Mills and Hofer, 1998, p. 69.
446 Ibid., p. 429.
447 John Grainger and Francis Gilbert (2008) The St Katherine protectorate in South Sinai, Egypt, in IUCN, *Values of Protected Landscapes and Seascapes, Protected Landscapes and Cultural and Spiritual Values*, Vol 2, IUCN World Commission on Protected Areas, https://books.google.co.uk/books?hl=en&lr=&id=NXWxbvFEi8sC&oi=fnd&pg=PA21&dq=Striped+hyena&ots=wqymhsSeyL&sig=0sXuDLoz-Q1is-WW95I5aWQqVM0&redir_esc=y#v=onepage&q=Striped%20hyena&f=false accessed 24 January 2020.

448 Lisa V. Gecchele (2017) A pilot study to survey the carnivore community in the hyper-arid environment of South Sinai mountains, *Journal of Arid Environments*, 141, 16–24, p. 19.
449 Hans Kruuk (1976) Feeding and Social Behaviour of the Striped Hyaena (*Hyaena vulgaris Desmarest), East African Wildlife Journal*, 14, 2, 91–111, pp. 91–2.
450 Mikita Brottman (2012) *Hyena*, London: Reaktion Books, p. 135.
451 J. D. Skinner and G. Ilani (1979) The striped hyaena *Hyaena hyaena* of the Judean and Negev Dserts and a comparison with the brown hyaena *H. brunnea*, *Israel Journal of Zoology*, 28, 4, 229–232, p. 229.
452 D. W. Macdonald (1978) Observations on the behaviour and ecology of the striped hyaena, *Hyaena hyaena,* in Israel, *Israel Journal of Zoology*, 27, 4, 189–198, p. 189.
453 Julian C. Kerbis-Peterhans and Liora Kolska Horwitz (1992) A bone assemblage from a striped hyaena (*Hayena hyaena*) den in the Negev Desert Israel, *Israel Journal of Zoology*, 37, 4, pp. 225–245.
454 J. D. Skinner, S. Davis and G. Ilani (1980) Bone collecting by striped hyaenas, *Hyaena hyaena*, in Israel, *Paleontologica Africana*, 23, 99–104. p. 100.
455 Liora Kolska Horwitz and Patricia Smith (1988) The Effects of Striped Hyaena Activity on Human Remains, *Journal of Archaeological Science*, 1988, 15, 47, 471–481, p. 472.
456 Ibid., p. 473.
457 Ibid., pp. 473–4.
458 Mills and Hofer, 1998, p. 25.
459 Hila Shamoon and Idan Shapira (2019) Limiting factors of Striped Hyaena, Hyaena-hyaena, distribution and densities across climatic and geographical gradients (Mammalia: Carnivora), *Zoology in the Middle East*, 65, 189–200, pp. 193–6.
460 Ibid., p. 197.
461 Melissa Hogenboom (2016) The hyena that made its home in a wolf pack, http://www.bbc.co.uk/earth/story/20160324-the-hyena-that-lives-with-wolves accessed 25 June 2019.
462 Vladimir Dinets and Beniamin Eligulashvili (2016) Striped Hyaenas (Hyaena hyaena) in Grey Wolf (Canis lupus) packs: cooperation, commensalism or singular aberration? *Zoology in the Middle East*, 62, 1, 85–87, p. 85.
463 M. Albaba (2015) Primary survey of the striped hyaena, Hyaena hyanea (Linnaeus, 1758), (Carnivora: Hyaenidae) status in the West Bank Governorates, Palestine, *Global Scholastic Research Journal*, 1, 10, 39–44, p. 40.
464 Penny Johnson (2019) *Companions in Conflict. Animals in Occupied Palestine*, Brooklyn, New York: Melville House, p. xiii.
465 Albaba, 2015, pp. 41–2.
466 Ibid.
467 Mounir R. Abi-Said (2006) Reviled as a grave-robber: the ecology and conservation of striped hyaenas in the human-dominated landscapes of Lebanon, PhD thesis, Durrell Institute of Conservation and Ecology, University of Kent, EThOS ID: 443775, https://ethos.bl.uk/ProcessOrderDetailsDirect.do?documentId=1&thesisTitle=Reviled+as+a+grave-robber+%3A+the+ecology+and+conservation+of+striped+hyaenas+in+the+human-dominated+landscapes+of+Lebanon&eprintId=443775 accessed 25 June 2019, p. 14.
468 Ibid.
469 Ibid., p. 24.
470 Ibid., p. 47.
471 Ibid., p. 74.
472 Ibid., pp. 25–6.
473 Ibid., p. 83.
474 Mounir Abi-Said and Zuhair S. Amr (2012) Camera trapping in assessing diversity of mammals in Jabal Moussa Biosphere Reserve, Lebanon, *Vertebrate Zoology*, 62, 1, 145–152, p. 148.
475 Helen Sullivan (2019) In Lebanon, the hyena's main fear is fear itself, *New York Times*, 30 October, https://www.nytimes.com/2019/10/30/science/striped-hyena-lebanon.

html?fbclid=IwAR2oq5WiJUEM1cY_9eJyzE53apeHFnPxyJAZYk0iLVvjt4p6GyA_3DJ3qFY accessed 16 August 2020.
476 Abi-Said, 2006, p. 40.
477 M. Masseti (2009) Carnivores of Syria, in E. Neubert et al. (ed), *Animal Biodiversity in the Middle East. Proceedings of the First Middle Eastern Biodiversity Congress, Aqaba, Jordan, 20–23 October 2008*, ZooKeys 31, 229–252, p. 230.
478 Guy Mountfort (1964) Disappearing Wildlife and Growing Deserts in Jordan, *Oryx*, 7, 5, 229–232, p. 230.
479 Maisie Fitter (1967) New Hope for Wildlife in Jordan, *Oryx*, April, 9, 4, p. 35.
480 Mills and Hofer, 1998, p. 69.
481 Mayas A. Qarqaz, Mohammad A. Abu Baker and Zuhair S. Amr (2004) Status and ecology of the Striped Hyaena, *Hyaena hyaena*, in Jordan, *Zoology in the Middle East*, 33, 1, 87–92, p. 88.
482 Ibid., p. 89.
483 Ibid., p. 91.
484 Sarah Edwards et al. (2017) Mammalian activity at artificial water sources in Dana Biosphere Reserve, southern Jordan, *Journal of Arid Environments*, 141, 52–55, p. 53.
485 Omar Attum et al. (2017) Population size and artificial waterhole use by striped hyenas in the Dana Biosphere Reserve, Jordan, *Mammalia*, 81, 4, 415–419, p. 417.
486 Mills and Hofer, 1998, p. 69.
487 Peter SWchwartzstein (2017) Iraq's Unoique Wildlife Pushed to Brink by War, Hunting, *National Geographic*, 3 February, https://www.nationalgeographic.com/news/2017/02/iraq-wildlife-islamic-state-species-war/ accessed 20 August 2019.
488 Omar Attum et al. (2017) Population size and artificial waterhole use by striped hyenas in the Dana Biosphere Reserve, Jordan, *Mammalia*, 81, 4, 415–419, p. 417.
489 David L. Harrison (1968) The Large Mammals in Arabia, *Oryx*, 9, 5, September, 357–63, p. 357.
490 Chris Thouless (1992) Conservation in Saudi Arabia, *Oryx*, October, 222–228, p. 222.
491 P. L. Cunningham, T. Wronski, T and K. Al Ageel (2009) Predators persecuted in the Asir Region, South-western Saudi Arabia, *Wildlife Middle East News*, 4, 1, http://researchonline.ljmu.ac.uk/6893/ accessed 24 January 2020.
492 M. Zafar-ul Islami et al. (2014) Poisoning of endangered Arabian leopard in Saudi Arabia and its conservation efforts, *CAT news*, 60, Spring, http://nwrc.gov.sa/NWRC_ARB/mnshwrat_files/Islam%20et%20al.%20(2014)%20Poisoning%20of%20Arabian%20Leopard...%20Catnews%20(Spring%2060)%2016-17.pdf accessed 24 January 2020, p. 16.
493 M. Hall (2008) Arabia's last forest under threat: Plant biodiversity and conservation in the valley forest of Jabal Bura (Yemen), *Edinburgh Journal of Botany*, 65, 1, 113–135, p. 113.
494 Marcelo Mazzolli (2009) Arabian Leopard Panthera pardus nimr status and habitat assessment in northwest Dhofar, Oman (Mammalia: Felidae), *Zoology in the Middle East*, 47, 3–12, p. 5.
495 Lawrence Ball (2013) Observations in the Empty Quarter & A Rapid Biodiversity Assessment of Wadi Sayq, Dhofar, British Exploring Society, https://www.academia.edu/8085570/Observations_in_the_Empty_Quarter_and_A_Rapid_Biodiversity_Assessment_of_Wadi_Sayq_Dhofar?auto=download accessed 3 February 2020, p. 58.
496 *Wild Arabia,* series 1, episode 2, BBC iPlayer and Eden Channel.
497 Norman Ali Khalaf et al. (2020) The Arabian Striped Hyaena (Hyaena hyaena sultana, Pocock, 1934) (Carnivora: Hyaenidae) in the United Arab Emirates, *Gazelle: The Palestinian Biological Bulletin*, 38, 84, April 2020, 1–20, p. 5.
498 Hervé Monchot and Marjan Mashkour (2010). Hyenas around the cities. The case of Kaftarkhoun (Kashan-Iran), *Journal of Taphonomy*, 8, 1, 17–32, p. 18.
499 Mills and Hofer, 1998, p. 69.
500 Mehdi Chalani, Gholam Hosein Yusefi and Seyed Massoud Madjdzadeh (2012) *Ecology of striped hyena (Hyaena hyaena) L. 1758 (Mammalia: Carnivora) in Jajroud Complex Region (JCR)*, https://www.researchgate.net/profile/Seyed_Madjdzadeh/publication/

283090867_Ecology_of_striped_hyena_Hyaena_hyaena_L_1758_Mammalia_Carnivora_in_Jajroud_Complex_Region_JCR/links/562a898508ae04c2aeb1aafd/Ecology-of-striped-hyena-Hyaena-hyaena-L-1758-Mammalia-Carnivora-in-Jajroud-Complex-Region-JCR.pdf accessed 4 February 2020.

501 Flora Graham (2019) Daily briefing: Secretive Iran court sentences cheetah conservationists to prison, *Nature*, 22 November 2019, https://www.nature.com/articles/d41586-019-03619-9 accessed 17 August 2020.

502 Chalani et al., 2020.

503 Khaki Sahneh Saeid et al. (2016) Modelling habitat requirement of striped hyena (*hyaena hyaena*), in Lashgardar, Protected Area, Hamedan province, *Environmental Researches*, 7, 13, 11–20, https://www.sid.ir/en/journal/ViewPaper.aspx?id=528601 accessed 20 January 2020, p. 11.

504 Mahdieh Tourani, Ehsan M. Moqanaki and Bahram H. Kiabi (2012) Vulnerability of Striped Hyaenas, *Hyaena hyaena*, in a human-dominated landscape of Central Iran, *Zoology in the Middle East*, 56, 1, 133–136, p. 134.

505 Ibid., pp. 135–6.

506 Monchot and Mashkour, 2010 p. 19.

507 Karami Mahmood et al. (2009) The public perception of the striped hyena (Hyaena hyanea) in Khojir National Park, *Journal of Environmental Science and technology*, Fall 2009, 11, 3, 281–294, p. 281.

508 Ö. Emre Can and İnci Togan (2004) Status and Management of Brown Bears in Turkey, *Ursus*, 15, 1, pp. 48–53, p. 48.

509 A. M. De Marinis and M. Masseti (2009) Mammalian fauna of the Termessos National Park, Turkey, in E. Neubert et al. (eds), *Animal Biodiversity in the Middle East. Proceedings of the First Middle Eastern Biodiversity Congress, Aqaba, Jordan, 20–23 October 2008*, ZooKeys 31, 221–228, p. 225.

510 Wild World (no date) Natural history from around the world, http://iberianature.com/wildworld/guides/wildlife-and-nature-of-turkey/mammals-of-turkey/ accessed 25 June 2019.

511 Max Kasparek et al. (2004) On the status and distribution of the Striped Hyaena, Hyaena hyaena, in Turkey, *Zoology in the Middle East*, 33, 93–108, p. 96.

512 Ibid.

513 Ibid., pp. 101–2.

514 Ibid., pp. 103–4.

515 Abdullah Emin Akay, Selcuk Inac and Ismet Ceyhun Yildirim (2001) Monitoring the local distribution of striped hyenas (Hyaena hyaena L.) in the Eastern Mediterranean Region of Turkey (Hatay) by using GIS and remote sensing technologies, Environmental Monitoring Assessment, 181, 445–455, pp. 445–6.

516 Alec Nove (1972) *An Economic History of the USSR*, Harmondsworth, UK: Penguin, pp. 161–86 and 330–2.

517 Heptner, V. G. and Sludskii, A. A. (1992). Mammals of the Soviet Union: Carnivora (hyaenas and cats), Volume 2. Smithsonian Institution Libraries and National Science Foundation. https://archive.org/details/mammalsofsov221992gept/page/10 accessed 24 July 2019.

518 Ibid., pp. 18–23.

519 Mills and Hofer, 1998, pp. 68–70; Heptner and Sludskii, 1992, p. 19.

520 Heptner and Sludskii, 1992, p. 22.

521 Igor Khorozyan, Alexander Malkhasyan and Marine Murtskhvaladze (2011) The striped hyaena Hyaena hyaena (Hyaenidae, Carnivora) rediscovered in Armenia, *Folia Zoologica*, 60, 3, 253–261, p. 253.

522 Mills and Hofer, 1998, p. 88.

523 Jurgen W Frembgen (1998) The Magicality of the Hyena Beliefs and Practices in West and South Asia, *Asian Folklore Studies*, 57, 331–344, p. 336–7.

524 Wildlife of Pakistan (no date) Mammals, Carnivors. Stripped [sic] Hyena. *(Hyaena Hyaena)*, http://www.wildlifeofpakistan.com/stripped_hyena.html accessed 29 July 2020.

525 Ibid.
526 Personal communication from Arjun Dheer.
527 Frembgen, 1998, pp. 336–7.
528 @wildpakistan tweet 1/12/2019.
529 Mills and Hofer, 1998, p. 69.
530 E.P. Gee (1961) Report from India, *Oryx*, 1961, p. 99.
531 Oishimaya Sen Nag (2020) Time To Recognize That Hyenas Are Nice, Not Nasty, Worldatlas.com, https://www.worldatlas.com/news/time-to-recognize-that-hyenas-are-nice-not-nasty.html?fbclid=IwAR0-Ok9Yk3kBZ5ly66E4vCLD4yVrAC8F0VqJEVuC98Ns0-6cDH5xxi1hv5c accessed 5 May 2020.
532 Ibid.
533 A. Stephen, R Suresh and C. Livingstone (2015) Indian Biodiversity: Past, Present and Future, *International Journal of Environment and Natural Sciences*, 7, 13–28, https://www.researchgate.net/publication/276410026_Indian_Biodiversity_Past_Present_and_Future accessed 8 January 2020, p. 20.
534 Rachel Becker (2016) Cattle drug threatens thousands of vultures, *Nature*, 29 April 2016.
535 Julian Spillett (1967) Pesticide Poisoning of Tigers, *Oryx*, 9, 3, December, p. 183.
536 Side effects of Project Tiger, *Oryx*, 15, 1, June 1979, p. 3.
537 Shaurabh Anand and Sindhu Radhakrishna (2017) Investigating trends in human-wildlife conflict: is conflict escalation real or imagined? *Journal of Asia-Pacific Biodiversity*, 10, 2, 154–161.
538 Vivek Menon (2003) *A Field Guide to Indian Mammals*, London: Dorling Kidnersley, p. 77.
539 Divyabhanusinh (2008) *The Story of Asia's Lions*, Mumbai: Marg Publications, 2nd ed., p. 18.
540 Sudipta Mitra (2005) *Gir Forest and the Saga of the Asiatic Lion*, New Delhi: Indus, p. 145.
541 M. Shamshad Alam, Jamal A. Khani and Bharat J. Pathak (2015) Striped hyena (Hyaena hyaena) status and factors affecting its distribution in the Gir National Park and Sanctuary, India, *Folia Zoologica*, 64, 1, May 2015, 32–39, p. 32.
542 Mitra, 2005, p. 170.
543 Nikunj Gajera, S. M. Dave and Dharaiya Nishith (2009) Feeding patterns and den ecology of striped hyena (Haeyena haeyena) in North Gujarat, India, *The Tiger Paper*, January–March 2009, 36, 1, 12–7, p. 15.
544 V. Athreya (2013) Big Cats in Our Backyards: Persistence of Large Carnivores in a Human Dominated Landscape in India, PLoS ONE 8, 3, e57872, doi:10.1371/journal.pone.0057872, 1–8, p. 2.
545 Priya Singh (2008) Population density and feeding ecology of the striped hyena (Hyaena hyaena) in relation to land use patterns in an arid region of Rajasthan, Thesis Submitted to The Manipal University In partial fulfilment for the degree of Master of Science in Wildlife Biology and Conservation, p. 1.
546 Priya Singh, Arjun M. Gopalaswamy and K. Ullas Karanth (2010), Factors influencing densities of striped hyenas (*Hyaena hyaena*) in arid regions of India, *Journal of Mammalogy*, 91, 5, 1152–1159, p. 1153.
547 Ibid.
548 Ibid., p. 1155.
549 Ibid., p. 1157.
550 L.S. Rajpurohit (2011) Status of five species of predators in Thar Desert, Jodhpur District, Rajasthan (India), *Zoo's Print*, XXVI, 8, 18–20, p. 19.
551 Divya Karnad (2017) The persistence of the Striped Hyena Hyaena hyaena Linnaeus, 1758 (Mammalia: Carnivora: Hyaenidae) as a predator of Olive Ridley Sea Turtle Lepidochelys olivacea Eschscholtz, 1829 (Reptilia: Testudines: Cheloniidae) eggs, *Journal of Threatened Taxa*, 9, 12, 1–3, p. 1.

552 S. R. Jnawali (2011) *The Status of Nepal's Mammals: The National Red List Series*, IUCN Species Survival Commission, https://www.researchgate.net/publication/273896557_The_Status_of_Nepal's_Mammals_The_National_Red_List_Series accessed 22 January 2020, pp. 1–2.

553 Ibid., p. 11.

554 Ibid., p. 66.

555 Ibid.

8

MYTHS AND REPRESENTATIONS FROM EARLY HUMANS TO *THE LION KING*

No other animal is quite so bound up in people's minds with death as the hyena, and across practically all of its range in Africa, West Asia and South Asia it features in mythology and folktales, with strong associations with evil, cowardice and witchcraft. European accounts of hyenas, secondhand and garbled, follow many of the same patterns and incorporate the myths about hermaphroditism derived from the unusual external form of the female genital organs of spotted hyenas. In folktales and myths, there is a thread of genuine fear of an animal that killed stock, even killed and ate people, and which digs up graves and eats corpses. This provides a starting point for the imagery of the hyena, but with inaccuracies, superstition, and outright falsehoods woven into the narratives. Over millennia this has created an image of hyenas that is common globally and taken up with gusto in films and the media, and even in wildlife documentaries.

In his account of studying carnivores, Kruuk wrote of attitudes towards hyenas in comparison with other predators:

> there is one totally different aspect to the relation between hyenas and people, something very irrational and weird, hugely supernaturally important in people's lives out here in the bush. Witchcraft. Hyenas are deeply involved in this, much more so than any animal, and over the years I come across strange beliefs many a time.[1]

Having studied them and written the seminal work on spotted hyenas, he concluded that "hyenas are no more supernatural than any other creature…They are difficult, destructive sometimes, but they are not evil".[2] He recounted how hyenas attacked his vehicle biting the wing mirrors, bumper and tyres:

> it reminds me how human they seem to be, these hyenas, how much like a gang of destructive teenagers on the rampage in town. No other animal has ever struck me like that. I think this nigh human streak is one of the reasons why hyenas feature so prominently and uniquely in African witchcraft.[3]

This parallel to human bad behaviour, combined with the amazing range of vocalizations of spotted hyenas, the scavenging by spotted and striped hyenas from middens and graveyards, collections of bones at striped and brown hyena dens and their silent nocturnal foraging all come together to produce an image of the hyena as cunning, devious, contemptible and linked with the occult. Spotted hyenas howl, whoop, growl and have an almost insane laughing vocalisation that can strike fear into people, reinforced by the frenzied behaviour and sounds of spotted hyenas on a carcass, their ability to crack bones and, in the case of brown and striped hyenas, the carrying of bones and body parts back to their dens. That they could appear soundlessly at night and disappear as dawn came, not to be seen again until the next night, all added to the myths and misunderstandings. As Kruuk added in his attempt to explain why people associated them with magic and witches:

> …Their loud, staccato 'giggles' are like those of a mad person, and they are mixed with deep growls and howls, all together the cacophony of a very aggressive orgy…for a frightened human in the dark, many such sounds together are hauntingly human, supernatural, a witches Sabbath.[4]

Such images conjured up by the sound of hyenas at night and beliefs about their threat to humans – physical or supernatural – have given rise to the evolution of myths and revulsion for hyenas of all species among humans. These myths and inaccuracies become exaggerated as they are repeated to people who have never seen or heard a hyena.

Fear and loathing in the African bush

Many cultures and communities with long experience of coexisting with hyenas have translated myths into stories, superstitions and belief in supernatural roles, and have used their skins and body parts in traditional medical. They developed characterisations of hyenas varying from stupid and greedy to magical, cunning and calculating. The supernatural and occult remain important in many African and Asian societies, with enduring beliefs in spiritual power and the efficacy of traditional healers using medicines derived from plants and animals and magic to cure people of illness or deal with curses. This is not endemic to Africa or Asia. Europe, for example, has a long history of belief in the supernatural among many different cultures and communities. These may have evolved in the face of education, technology and other manifestations of the modern world, but they have not disappeared. Thomas has pointed out that in Europe and the Americas, organised religions still embody beliefs that are supernatural, such as spiritual healing, efficacy of pilgrimages and confession as forgiveness of sin that leads to immortality after death.[5]

In Africa, witchcraft and sorcery are inextricably linked with social relationships and tensions, with accusations of the use of witchcraft for evil ends often relating to conflicts or pressure within or upon communities.[6] Beliefs in the supernatural and uses of witchcraft are ubiquitous. As Nyamnjoh points out:

> They are as practised in urban spaces as in rural ones, as much by the most cosmopolitan as by the most localist. They may assume new forms and new uses, but witchcraft and occult practices form part of everyday life in urban Africa.[7]

In contemporary South Africa, "Hardly a week passes in South Africa without press reports of witches being killed".[8] In Tanzania, approximately 400 people in the western part of the country were killed after being accused of witchcraft between 1997 and 2000, and between June and July 2001, over 800 people were killed in a witch hunt in the north-eastern Democratic Republic of Congo.[9]

Other forms of violence are associated with witchcraft or sorcery across Africa. Foremost among these are what are called "*muti* murders", in which people (often children, and in some countries albino children)[10] are killed for their body parts which are used in *muti* or magic, often to lay curses or to produce charms for protection against natural of supernatural forces.[11] The use of animal body parts is an important part of *muti*, with hyena body parts often an ingredient and believed to have powers to cure illnesses or ward off evil.[12] Hyena scent marking – leaving a strong-smelling paste on grasses and low branches – has given rise to the concept of "hyena butter" being a magical substance. This paste has a strong odour and in many cultures in Africa is called witches' butter. There is the belief it has magical properties and is used by witches as fuel for torches carried at night, when they search for victims.[13] The view of hyenas as occult creatures is behind the practice of taking the body parts of hyenas for use in medicine and rituals. There are also widespread beliefs that witches can turn themselves into animals,[14] hyenas, leopards and lions being common forms of this shape-shifting. Examples of this occur during times of great social or community tension and during outbreaks of predation on humans by hyenas, lions and other predators – the man-eaters being believed to be witches taking on animal forms and killing for evil reasons, including getting body parts for *muti*.

There are many examples of hyenas being associated with witches, the occult and magic. I will present just a few to demonstrate their role in belief systems. Among the Kaguru people of the Kilosa and Mpwapwa districts between Dodoma and Dar es Salaam in Tanzania, beliefs in witches and sorcerers who can bring or avert misfortune are persistent and involve hyenas. Witches are said to travel at night in the company of owls and use ground hornbills and hyenas as their familiars. At night they can cover great distances "by means of hyena familiars which they hug by the belly as they race through the air".[15] They also rely on hyenas to dig up bodies from graves. If a body is fresh it is used by the witch or eaten by them. If it is decomposed the hyena is allowed to eat it. For the Valangi of central Tanzania, hyenas are associated with witches and with thieves, and they believe that "hyenas are at the furthest conceivable point from everything which is wholesome and decent".[16] Dunham noted that hyenas were the most common predator in the region and raided stock enclosures and even huts at night to take goats, but didn't feel this was sufficient "to account for the fascinated hatred" with which the people regarded them, and that traditional supernatural beliefs were involved.[17] They believed that if owls or hyenas came near a house at night, someone in that house would die. A further belief among the Valangi is that when an apprentice witch graduates, he or she is given a hyena, which they learn to ride at night to find bodies or live victims. The hyenas supposedly sleep during the day in the witches' houses.[18] Gray recorded similar beliefs among the Wambugwe community, living at the southern end of Lake Manyara, in a game-rich area with a population of predators including spotted and striped hyenas. They believed hyenas were owned by witches and referred to them as "night cattle", which bore a brand to identify who owned them.[19] The witches ride on the hyenas carrying torches of flaming hyena marking paste.[20]

An aspect of southern African supernatural/occult beliefs is the ape-like creature the tokoloshe or tokolotši, found in Zulu, Xhosa, Sotho, Tsonga and other belief systems.[21] The belief is that it is a witch that turns itself into a tokoloshe or that the tokoloshe is a familiar of the witch, which can ride on the back of a hyena. The Gusii people of Kenya also associate hyenas with witches operating at night and tell their children they must not leave their huts at night as the witches and hyenas will catch and devour them.[22] In Malawi, Chewa hyena stories are based on the belief that hyenas are familiars of sorcerers who know when someone has died, helping the sorcerers find the corpses to take body parts for occult use. These sorcerers are held responsible for their hyena-familiars killing livestock.[23] Linked with this is the belief that wild animals embody the dark forests beyond human habitations.[24] Game animals are sources of meat for humans, while animals like hyenas and other carnivores are, in the Chewa language, *chirombo* or wild beasts. Wild beasts are viewed as intrinsically bound up with the spirit world of the forests.[25] Kruuk says that in East Africa and across most of the spotted hyena's range in Africa, the hyena is viewed as a contemptible and sinister creature and is the animal most widely credited with being involved in witchcraft. One African account, not further identified by Kruuk, refers to hyenas as the "living mausoleum of the dead"[26]. Among the beliefs he encountered in many years studying spotted and striped hyenas was one that if you fed your livestock ground up hyena skins they would be protected from attacks by hyenas.

In addition to the beliefs in the occult aspects of the hyena, there are many folk stories from Africa concerning its character, especially its greed and alleged stupidity. A typical example comes from a folktale from Malawi. The Yao and Chewa peoples there have a story similar to greed-linked stories about hyenas from across Africa.[27] It starts with the premise that hyenas rely on lions for their food – the traditional view of the spotted hyena as a scavenger. The tale starts with the hyena asking the lion for food. The lion gives him food. The next day, the hyena asks for more. The lion refuses, saying he gave him food yesterday. The lion decides to play a trick on the greedy hyena. The lion gets an antelope to climb a tree in the middle of a pool of water, so that its reflection is in the water. The hyena arrives and the lion asks him to fetch the dead antelope that is in the water. The hyena tries to grab the reflection of the antelope, getting stuck in the mud and swallowing so much water he dies.[28] Another tale collected by Macdonald and Doke tells of a hyena repeatedly tricking a fox out of food. Once more, the hyena is lured into deep water and dies because of his greed.[29]

A folktale from Tanzania talks of a lion, hyena and rabbit setting up a farm. On the way to see their crops at harvest time, the rabbit tricks both animals and the hyena is eaten by the lion and then the lion by the hare. The thrust of the story is that the rabbit is intelligent and the two predators stupid and gullible.[30] A not dissimilar story is to be found among the Maasai and involves a hare, elephant, snake, tortoise and hyena owning a herd of cattle and donkeys. Four of them fear the elephant will use his size to cheat them of their animals and resolve to kill him. The snake kills him but the other three plot to kill the snake, which the tortoise does. The hare and the hyena move the cattle and donkeys to stop the tortoise getting them, but the hare tricks the hyena out of the cattle leaving it with the donkeys, which the hare somehow lures into thick mud, where they are stuck. Eventually, the hare persuades the hyena to jump over a fire. The hyena falls into the fire and dies.[31]

In Namibia, home to spotted and brown hyenas, the Himba pastoralists have stories that follow this pattern of greed and idiocy on the part of hyenas. Their characterisations show the tortoise as wise, the jackal as cunning and clever, but the hyena is "hapless and slow-witted, as an animal easily exploited and bested by its fellows". This is curious, as Crandall notes, as most Himba will have lost stock to the hyena, which is a powerful predator as well as a scavenger. But their characterisation is of the hyena as "dull and foolish persists".[32] In every Himba story, the jackal outwits the hyena. This is also found in San, Khoikhoi and Venda stories from South Africa. One tells that the jackal persuaded the hyena to jump up and eat a cloud. The hyena jumped ever higher, with the promise that the jackal would catch him. The jackal lets the hyena fall and it breaks its back legs. This, according to the story, is why hyenas' hind legs are shorter than their front ones.[33]

There is one story, derived from San folklore, which unusually has the hyena outwitting a lion by offering it soup made from an ostrich, pouring soup down the lion's throat, then ramming the soup pot on to the lion's head and killing it.[34] But most San and Khoikhoi stories of the hyena are in line with the general view of the stupid hyena. Many were taken up by the Afrikaner writer G.R. von Weilligh in his children's stories in Afrikaans. He calls the hyena wolf throughout – a common Afrikaner usage, with the brown hyena known as a strandwolf (beach wolf). Once more the jackal continually outwits the greedy and stupid hyena. One story has the jackal cheating the hyena of a hartebeest he has killed and is taking home to his mother. Over several days, the jackal gets the hartebeest, an eland and a small antelope by tricking the hyena and his mother.[35] Other stories involve the hyena being tricked into challenging lions to a fight, or being persuaded to fight a human, who shoots the hyena and hits him with a sabre.[36]

Stupidity and "moral bankruptcy" are characteristics ascribed to the hyena by the Beng people of Ivory Coast.[37] Their folklore abounds with tales of the hare outwitting the hyena and of the hyena's huge greed, which always clouds its judgment and leads it into stupid actions, often encouraged by the hare or other animals. Gottlieb concludes that one of the key aspects of these stories is the disastrous result for the hyena (and by extension greedy people) of the temptation to satisfy unreasonable desires at the expense of others.[38] The Beng also are revolted and angered by the hyena's attempt to dig up and consume human corpses, seen as a crime against the spirits of the ancestors.[39]

Global myths

In Greek and Roman myths, the genital peculiarities of the spotted hyena gave rise to myths about the ability of hyenas to change sex or their hermaphroditism.[40] In Aesop's fables there is reference to hyenas which involve the change of sex and implied perversion, anal intercourse and homosexuality.[41] The Christian theologian, Clement of Alexandria, perpetuated the sexual myth and used hyena as a metaphor for excessive sexuality.[42] The sexual ambiguity figures prominently in the work known as the *Physiologus*, believed to have been written in Alexandria between the 2nd and 5th centuries CE and which is very similar to Clement's work. The hyena is referred to as a brute and Christians are told to obey the biblical rule that "'Thou shalt not eat the brute, nor anything similar to it.' This animal is an *arenotelicon*, that is, an alternating male-female. At one time it becomes male, at another female, and it is unclean because it has two natures".[43] It compares the hyena to "double-minded men" who take on a "womanly nature", strongly suggesting homosexuality.[44]

In the early fourteenth century, an Italian theologian, Cecco d'Ascoli, wrote in his poem *L'Acerba* that the hyena is a sodomitic beast and it could change its sex: "Muta 'l sexo, animal sodomito".[45] These myths survived in the European medieval period and became part of the revulsion with which hyenas were presented in bestiaries and viewed by those who had the slightest knowledge of them. They were transformed by the bestiaries into "a signifier of the abominable par excellence".[46] In some bestiaries:

> …the hyena is depicted as a deceitful, dirty creature known to dwell in graveyards and dig up the recently deceased to eat their corpses…The hyena's story in the bestiary is an allegory for the temptations of the devil and the importance of living a Christian life.

Illustrations often show hyenas devouring human bodies.[47] In a typical hate and prejudice-filled comparison, some bestiaries compared hyenas to Jews, saying "The Children of Israel, who served the living God at first, are compared to this brute".[48] Less derogatory references to the hyena Christian mythology come in the accounts of the Christian monk Macarius, one of the "desert fathers" who became hermits living in remote parts of the Egyptian desert in the 4th century CE. He is reputed to have had a hyena (presumably a striped one) enter his cave and lick his feet. The hyena then pulled at his clothes and pulled him towards the door. The monk followed the hyena and found her cub, which was blind. He prayed over the cub and its sight was restored.[49] Hardly a miracle, of course, as striped hyena cubs are born blind, like many mammals. The hyena then returned one day to his cave and brought him a sheepskin for him to sleep on, according to the legend.

In the Middle East and West Asia, the striped hyena is viewed as a contemptible creature embodying vices and occult powers. In some Arab cultures, the hyena is depicted as having the power to put spells on people, so they can subdue them, drag them back to their caves to devour them.[50] The Israeli hyena researcher, Jonathan Tichon, wrote that "I had a Bedouin guy swear to God that he was enchanted by a hyena and that his friend saved him by slapping him on the head at the last moment".[51] In his doctoral dissertation on the striped hyena in Lebanon, Abi-Said wrote that:

> Stories and myths are present wherever striped hyaenas exist. Many stories and myths about striped hyaena can be found in old Arabic literature. which might in turn have helped to perpetuate these stories, and to have given rise to the negative images that the striped hyaena generally attracts.[52]

He goes on to say that many Arabic poets and writers have perpetuated myths, such as hyenas attacking and killing or sucking the blood of sleeping people, while others dwell on the hyena's habit of foraging in graveyards and digging up corpses, an accurate depiction of one of many sources of food for striped hyenas, but one that is the basis of exaggerations, misrepresentations and myths about the hyena being a ghoulish grave-robber above all else.[53] They also refer to the supposed ability of the hyena to mimic human sounds and lure men and dogs to their deaths. Arabic proverbs refer to the stupidity and supposed treachery of the hyena.[54] Another myth is based on the belief that hyenas can bewitch people by exuding a strong scent, which makes the person follow the hyena to its den to be eaten.[55]

In West and South Asia, a variety of myths and connections with magic are attached to the striped hyena, particularly the use of teeth and other body parts to ward off evil.[56] In Afghanistan, hair from the mane or tail is used as a love charm or as a cure for the sick, while in India hyena fat is said to cure rheumatism.[57] Hyenas occur in the Hindu Vedic texts and in works of religion and mythology. In the Rig Veda (dating from 1500–1200BCE) and other Vedic texts, the myths state that a male god, Indra, in the form of a female hyena, hands over a group of Jain monks to her cubs to eat. In some versions, several of the monks survive; in others, there is no mention of survivors.[58] In the ancient epic poem, the *Mahabharata,* there are periodic references to hyenas, often in the context of scavengers consuming the dead on battlefields. One character, Bhima, threatens a devil that he will leave him for the vultures and hyenas.[59]

Hyena images in literature, film and the media

Hyenas also occupied the attention of writers whose major focus was not on natural history. Sir Walter Raleigh, in his *History of the World* (1614, Book I), dealt with hyenas in the context of Noah's Ark. In one passage, he wrestles with how all the existing animal species were accommodated within the limited space provided by the ark. Raleigh suggests that hyenas were not on the ark because they were the unnatural offspring of foxes and wolves.[60] Shakespeare has Rosalind in *As You Like It* telling Orlando that after they are married she will laugh like a hyena.[61] In the mid-17th century, the English poet John Milton in his epic poem *Samson Agonistes* has Samson reviling Delilah after she betrays him: "Out, out Hyæna; these are thy wonted arts, And arts of every woman".[62] At another point in the poem, as Bewell noted, "When Dalila pleads for forgiveness, he reviles her as a Hyaena feeding on the wreckage, feigning remorse for her rash…unfortunate misdeed".[63] By the late 18th century, strong, intelligent women, like the writer and feminist Mary Wollstonecraft, were seen by men as threatening and troublesome, and compared with hyenas. Horace Walpole called her a hyena in petticoats. The equality of women being viewed as dangerous and labelled as unnatural and as destroying the basic nature of women – a probable hyena allusion being its supposed ability to change sex.[64] The conservative British political writer, Edmund Burke, described his republican opponents as hyenas which wished to prey on the carcass of the conservative political order.[65]

The use of hyena images as political abuse or criticism has been carried forward into the modern era, with the South African trade unionist and politician Zwelinzima Vavi referring to the corrupt elements of the ruling African National Congress under President Jacob Zuma turning the South African government and the party into "a predatory state. The hyenas are gathering".[66] After Zuma left office, his successor warned, somewhat lamely, about corruption in government and the ANC, with people profiting from the responses to the COVID-19 pandemic like a pack of hyenas circling wounded prey.[67]

The Lion King

While hyena references, almost always derogatory ones, crop up throughout literature, taking in Dryden, Hemingway (in his books like the *Green Hills of Africa* and *Snows of Kilimanjaro*) and a host of other writers, the most striking and perhaps influential contemporary depiction is in the two Disney films, the 1994 cartoon *The Lion King* and the 2019 computer-animated remake. The characterisation of hyenas in the two films has been hugely influential in reinforcing and advancing the concept of hyenas as evil, skulking, cowardly, destructive and cunning but stupid animals. There are intelligent people I know, but who themselves know little about natural history beyond films and TV depictions, who reacted with surprise when I told them I was writing about hyenas, which I believed were fascinating, intelligent and worthy of respect. A typical reaction was, "Haven't you seen *The Lion King*, that shows what hyenas are like, they are ghastly". It was and remains highly influential in a bad way, because it may be the first and perhaps only depiction many people will see of hyenas. They may not take it as absolute truth but it creates a lasting image and one that does not help hyena conservation.

In the films, hyenas destroy the wildlife in an unsustainable way, reducing the environment to a grey, bone-filled wasteland – in stark contrast to the biodiversity and fecundity of the so-called pridelands, ruled by lions. The hyenas are typified by the three characters – the vile, scheming, vicious Shenzi; the mad, violent Banzai; and the cringing, giggling, stupid Ed, who gnaws on his own leg, thinking it's a bone. When the lion cubs enter the bone-strewn, hyena territory, slinking grey shapes can be seen surrounding them and then the three lead hyenas appear to attack them, only to be seen off by the dominant male lion, Mufasa. The plot between the weak but ambitious lion Scar and the hyenas to kill and succeed Mufasa lead to the pridelands becoming another empty, bone-strewn wasteland. Even Scar, himself presented as totally evil and cowardly, accuses the hyenas of destroying their own land and stripping it of every living thing. When Scar is overthrown by Simba and the lion pride, Scar is killed by the hyenas – which is perhaps the one accurate part of the film, as hyenas do kill old, sick lions forced from their prides.

Laurence Frank and Stephen Glickman tried to advise Disney when the company's cartoonists were deciding how to depict the hyenas – but their advice was ignored and the very negative representations were chosen to have impact and establish a clear divide between the good, noble lions and the destructive, greedy and stupid hyenas. Glickman wrote that several artists from the Disney studios were invited by Frank to spend two days at the Berkeley university campus observing and sketching hyenas maintained in a colony at the Field Station for Behavioral Research. He went on to say:

> Frank, and other scientists working at the colony, asked how hyenas were to be portrayed in the film and expressed a strong request that it be positive. The artists explained that the script was written and that a trio of hyenas (Banzai, Shenzi, and Ed) were to be the allies of an evil older lion, who would eventually lose to the hero, a noble young lion. However, they seemed very appreciative of the animals and said they would do their best to make them appear comical instead of evil.[68]

The result was different, and evil was the main motif for the hyenas. The same depiction occurs in spin-off computer games and other cartoons that Disney has produced as shorter TV films, like *The Lion Guard*, though in the latter there are some hyenas who have broken away from the evil majority and the term themselves the hyena resistance, working with the lions.

The *Lion King* remake was Disney's highest earning film and at $1.435 billion, it's now the ninth highest-grossing movie of all time, according to Box Office Mojo, and that doesn't count the tens or hundreds of millions who have watched it on DVD, Blu-Ray or digital TV channels. The original cartoon grossed $968 million in 1994, which converts to about $1.676 billion in 2019 money.[69] This level of earnings reflects the huge number of viewers and indicates the extent of propagation of a very negative, damaging image of the hyena.

One of the few images of striped hyenas is in another film, *Mowgli: Legend of the Jungle*, directed by the actor Andy Serkis. This turns the jackal character Tabaqui, from Rudyard Kilping's story *The Jungle Book*, into a smelly, cowardly, fly-blown striped hyena, who spies for the tiger Shere Khan. Tabaqui appears in the original book but was omitted from the famous 1967 Disney version. Another film featuring a hyena is the version of Yann Martel's novel, *The Life of Pi*. The human character Pi is adrift in a lifeboat with a spotted hyena, a zebra, and orang-utan and a tiger. The hyena is depicted as cowardly – it eats the zebra as it is dying but still alive, and then eats the orang-utan.[70]

Images and descriptions of hyenas on TV are often only marginally better than the fictional representations. With only a very few exceptions, natural history documentaries rarely concentrate on hyenas as animals worthy of examination let alone respect in their own right. They are shown as gory killers that rip their victims apart in a frenzy or try to use superior numbers or strength to chase leopards, cheetahs and lions from their kills. In the long running BBC series *Big Cat Diary* and *Big Cast Week*, hyenas are present as B-movie villains that threaten the young of the big cats or steal their kills. Colin Jackson, a producer of many editions of the programmes, told me that they put together the footage to make what were effectively wildlife soap operas. He said they did not falsify or fake anything but they did have storyboards for each episode, and these almost always ended with a note of jeopardy for a cat or a group of cats, and usually involving a threatened or missing cub.[71] Spotted hyenas were often the threatening animal depicted as about to kill a cub.

In the BBC documentary series *Dynasties*, first broadcast in December 2018, in the episodes about wild dogs and lions, hyenas were again shown as threats to the featured, more charismatic animals, with no other context supplied. Another example was on broadcast on 3 November 2019, when Sky TV ran an ad for the National Geographic Wild HD Channel for new series of *Savage Kingdom* with first episode *Dawn of Darkness*: with the description, "Rivals have always clashed on the African Savannah. But now a new age of terror emerges as a ruthless hyena commander seizes power from the lions". Another documentary, *Scavengers of the Savanna*, describes hyenas as carrion-eaters rather than hunters and although their scavenging role is described as "commendable", they are called "gluttons" with "a taste for blood"; lions are described as "peerless" predators.[72]

One of the results of the film and documentary vilification of hyenas is that even affluent, educated Europeans and North Americans going on wildlife safaris, while loving lions, elephants, giraffes, leopards, cheetahs and other charismatic mammals, have the negative images of hyenas drawn from the dominant historical, literary, film and TV representations. An informal survey carried out for me by Wilderness Safaris in southern Africa drew sadly predictable responses about hyenas from safaris goers. Guides said that many guests came up with the same three points about hyenas – they are only scavengers, they steal rather than hunt and they are selfish and only look after themselves. Neil Midlane and Henry Parsons of Wilderness told me they do all they can to combat these images and tell of the hyena's hunting abilities and their strong family/clan oriented social behaviour.[73]

Emsie Verwey, Skeleton Coast Brown Hyena Project and Research Manager, Wilderness Safaris, told me she was saddened by the poor images people had of them and that they took the popular, bad image given to spotted hyenas as true and applied it to all hyenas. She said she regretted that:

> ...there are way too little popular articles out there, too little cute and cuddly hyaena books, videos and soft toys for children, too few serious documentaries about their great attributes and very little portraying them for all that they are. They are just not 'charismatic' or 'sexy' enough. They don't come across as vulnerable.[74]

It is perhaps for these reasons that few zoos have them. There are 48 zoos and six safari parks in Britain. In Britain, Colchester Zoo, Longleat and Yorkshire Wildlife Park have hyenas in their collections. Malcolm Fitzgerald, the chief Curator at the Zoological Society of London, told me that the zoo had hyenas in the past (the *Proceedings of the ZSL* quoted in previous chapters, lists animals acquired or presented to the zoo, including striped and spotted hyenas), but there is no record of them being exhibited in the last 20 years. He said this was not related to charisma or visitor-appeal:

> In terms of species choices...we use the soon to be launched, ZSL Species Planning tool which links with the IUCN Red List...and European Association of Zoos and Aquaria (EAZA) regional collection plans. This allows us to prioritise on species choices in our zoos that have a conservation breeding (insurance) role and ZSL have an involvement in research or conservation...It may be in the future that we will have a hyena species at one of our zoos. Populations in the wild are declining, with striped hyena in particular likely to move into the threatened IUCN category.[75]

Prague Zoo has brown hyenas, Berlin Zoo in 2012 had all four species, but now only lists spotted hyenas, and Barben, near Aix in France, has striped hyenas. Dierenpark Zoo in the Netherlands has striped and spotted hyenas. One list of animals in European zoos[76] suggests there are striped, spotted and brown hyenas in a number of zoos in Belgium, France, Estonia and a few other East European countries, but searching on the zoos' sites does not yield results.

In the United States a few more zoos have hyenas. Frank says that some of the offspring of his Berkeley spotted hyena study group were sent to zoos, and he thinks a number of European zoos have them.[77] On the US Zoochat site, contributors say spotted are more common in zoos and Denver, Milwaukee and St. Louisand Franklin Park are among zoos that have them. It is not recorded if there are brown hyenas in any zoos. The Living Desert, Denver, Fort Worth and Utica zoos have striped hyenas.[78]

There are very few positive representations in documentaries that could change these views. One exception is National Geographic's *Ghosts of the Great Salt Lake*, about brown hyenas in Botswana. The film features carnivore researcher, Dr Glyn Maude at Mkgadikgadi Pan NP. There is no hyperbole or misrepresentation and it shows how important hyenas are in the food chain.[79] Another National Geographic film follows the hyena specialist Ingrid Wiesel studying brown hyenas of the Nambian coast and Namib desert. It is factual with no propagation of myths.[80] A similarly positive representation of brown hyenas occurs in the BBC's *Seven Worlds One Planet* documentary, with a section on a brown hyena living in an abandoned town in the Namib Desert – again with a straight, factual presentation that shows how the female hyena has survived and raised nine generations of cubs.[81]

What is important about these films, which give an accurate, factual account of hyenas, their feeding, foraging, breeding and other behaviours, is that striking images and a lack of sensationalism give an insight into the real lives or hyenas without the mythological, panic-mongering baggage of millennia of demonising hyenas. They will have informed people and perhaps even have endeared a normally maligned species to millions of viewers. Such portrayals are in stark contrast to the usual depiction in wildlife documentaries, media reporting and movies. It is worth remembering that in their 1998 *Status Survey and Action* for hyenas for the IUCN, Mills and Hofer stressed that representations are important because the perceptions of people towards hyenas are central to their conservation and management; and they urge those involved in research and conservation of them to undertake a "concerted education campaign" to educate people who live alongside hyenas and Western audiences that may influence global approaches to conservation that at present ignore hyenas – "by gaining sympathy and respect the status of Hyaenas will improve. This will be reflected in the willingness of people to make contributions to hyaena conservation".[82] That was written 12 years ago. Although there have been a few positive documentaries and only a little positive media coverage of research (which has been substantial and vital for hyena conservation), there is a long way to go before hyenas are accorded the positive coverage, concern and ability to elicit donations for conservation currently enjoyed by more obviously charismatic creatures like elephants, rhinos lions, tigers, pandas, whales, gorillas and orang-utans.

What next for the hated hyenas?

The spotted, striped and brown hyenas, very different from each other in many ways, face a broadly similar range of threats, some derived from enduring attitudes:

- Human hostility – mixing ignorance, fear derived from age-old beliefs about the cowardly, evil, perverse and occult nature of hyenas.
- The development of those myths through storytelling, literature, films and documentaries into a generalised demonisation of hyenas. They are ugly, cowardly but cunning, they steal and scavenge and are grave robbers on a major scale.
- Hyenas do scavenge, they do eat human remains and even forage from graves, but this is no different to what other more charismatic carnivores do without the resulting opprobrium – to repeat Kruuk's observation: "hyenas are no more supernatural than any other creature…They are difficult, destructive sometimes, but they are not evil".[83]
- They are blamed disproportionately for livestock predation, especially brown and striped hyenas, and as threats to humans – especially children – and so are persecuted, with widespread killing of hyenas.
- Prey loss in many parts of the range – parts of Tigray in Ethiopia being the strongest examples – forces hyenas into conflict with humans through livestock killing, and over-dependence on anthropogenic sources of food. This threatens their long-term survival as their future is totally in the hands of people's land-use and waste disposal methods.
- The illegal bushmeat trade reduces wild prey species and leads to widespread snaring of hyenas.
- Hyenas can adapt to different environments, foraging techniques, food sources and to living close to people, but continued habitat loss across the range of the four species is a threat and enhances the dangers of increased conflict as hyenas and people are thrown more closely together.
- Ultimately, they are not charismatic, and so not worthy of great consideration when it comes to conservation priorities.

What is needed is a recalibration of our view of conservation, moving away from overemphasis on charismatic, headlining species and towards emphasis on habitat and biodiversity conservation, in which a is hyena no less important and worthy of survival than a feral horse, lion, tiger, panda or elephant.

Many researchers are devoting their lives to researching the ecology and behaviour of hyenas. Ways need to be found to get this valuable, in-depth research into the public domain, so the demonisation ends and hyenas become valued for what they are – complex, clever, adaptable and social animals with a vital role to play in maintaining biodiversity and healthy ecosystems.

Notes

1 Hans Kruuk (2019) *The Call of Carnivores. Travels of a Field Biologist*, Exeter, UK: Pelagic Publishing, p. 57.
2 Ibid., p. 58.
3 Ibid.
4 Ibid., p. 59.

5 Keith Thomas (1971) *Religion and the Decline of Magic*, London: Penguin, p. 27.
6 Ibid., p. 21.
7 Frances Nyamnjoh (2001) Delusions of Development and the Enrichment of Witchcraft Discourses in Cameroon in Moore and Sanders (eds), 47–73 p. 52; see also, Khaukanani Mavhungu (2012) *Witchcraft in Post-Colonial Africa. Beliefs, Techniques and Containment Strategies*, Bamenda, Cameroon: Langaa Research and Publishing, p. 1.
8 John Alan Cohan (2011) The Problem of Witchcraft Violence in Africa, *Suffolk University Law Review*, XLIV, 4, 803–872, p. 804.
9 Ibid., p. 839.
10 Andrew Blonfield and Henry Mhango (2019) Malawi albinos kidnapped and sacrificed by witchdoctor gangs on the hunt for election charms, *Daily Telegraph*, 6 April, https://www.telegraph.co.uk/global-health/terror-and-security/malawi-albinos-kidnapped-sacrificed-witchdoctor-gangs-hunt-election/ accessed 26 February 2020; BBC (2019) Tanzania arrests 65 'witchdoctors' over killings, 4 March, https://www.bbc.co.uk/news/world-africa-47447103 accessed 26 February 2020.
11 Ibid., pp. 806–7
12 Personal communication with Dr Vivienne Williams, School of Animal, Plant and Environmental Sciences, University of Witwatersrand, concerning *muti* markets in South Africa and use of hyena skins and body parts.
13 Robert Gray (1963) Some Structural Aspects of Mbugwe Witchcraft, in John Middleton and E.H. Winter, 143–74, p. 166.
14 Middleton and Winter, 1963, p. 1.
15 Margaret Dunham (2006) *The Hyena: Witch's Auxiliary or Nature's Fool?* Hal Archives Ouverte, halshs-00009734, https://halshs.archives-ouvertes.fr/halshs-00009734 accessed 4 June 2019, p. 5.
16 Ibid.
17 Ibid.
18 Ibid., pp. 9–10.
19 Gray, 1963, pp. 164–5.
20 Ibid., p. 166.
21 Isak A. Niehaus (2012) *Witchcraft and a Life in the New South Africa*, Cambridge: Cambridge University Press, p. 12.
22 Robert Le Vine (1963) Witchcraft and sorcery in a Gusii community, in Middleton and Winter, 221–255, p. 250.
23 G. Marwick (1963) The sociology of sorcery in a Central African tribe, *African Studies*, 22, 1, 1–21, p. 8.
24 Brian Morris (1995) Woodland and Village: Reflections on the 'Animal Estate' in Rural Malawi, *Journal of the Royal Anthropological Institute*, 1, 2, 301–315, p. 302.
25 Ibid.
26 Kruuk, 2002, pp. 185–7.
27 Duff Macdonald and C. M. Doke (1938) Yao and Nyanja Tales, *Bantu Studies*, 12, 1, 251–285, p. 264–5.
28 Ibid.
29 Ibid., p. 267–71.
30 Tanzanian Folktale (no date) The Lion, the hyena and the rabbit, https://www.worldoftales.com/African_folktales/African_Folktale_41.html accessed 4 June 2019.
31 Naomi Kipury (1983) *Oral Literature of the Maasai*, Nairobi: East African Educational Publishers, pp. 68–9.
32 P. Crandall (2002) Himba Animal Classification and the Strange Case of the Hyena, *Africa*, 72, 2, 293–311, p. 293.
33 James A. Honey (1910) *South African Folk Tales*, New York: Baker and Taylor, http://www.gutenberg.org/files/38339/38339-h/38339-h.htm p. 106.
34 W. H. I. Bleek and L. C. Lloyd, *Bushman Folklore*, Kindle Edition, loc. 349; see also, Folklore and Symbolism, http://www.heartofsnow.net/hyena/hyenalore.php accessed

24 February 2020; and http://www.sacred-texts.com/afr/sbf/sbf16.htm accessed 24 February 2020.

35 G. R. von Wielligh (2011) *Animal Tales 1*, Pretoria: Protea Book House, pp. 55–8.

36 Ibid., p. 77.

37 Alma Gottlieb (1989) Hyenas and Heteroglossia: Myth and Ritual Among the Beng of Côte d'Ivoire, *American Ethnologist*, 16, 3, 487–501, pp. 488–9.

38 Ibid.

39 Ibid., p. 492.

40 Andreas Krass (2018) The Hyena's Cave Jeremiah 12.9 in Premodern Bestiaries, *Interfaces*, 5, 111–128, p. 111.

41 Ibid.

42 Ibid., pp. 114–5.

43 Glickman, 1995, 515–6.

44 Ibid.

45 Krass, 2018, pp. 116–7.

46 Krass, 2018, p. 111.

47 Jessica Sheppard-Reynolds (2018) The Corpse-Devouring Hyena of the Medieval Bestiary: In a world of good versus evil, the hyena plays the role of the bad guy, 10 May, http://blogs.getty.edu/iris/the-corpse-devouring-hyena-of-the-medieval-bestiary/ accessed 4 June 2018.

48 T. H. White (ed) (1984) *The Book of Beasts, being a translation from a Latin bestiary of the twelfth century*, New York: Dover Publications, p. 31.

49 Macarius and the Hyena, Written for English Language and History, based on 'The Lives of the Desert fathers', https://englishlanguageandhistory.com/?id=macarius-hyena accessed 26 February 2020.

50 Jason Bittel (2018) Striped Hyenas Don't Have Magical Powers, But Their Disappearing Act Is Real, 25 January, https://www.ecowatch.com/hyenas-in-india-2528561632.html accessed 21 August 2020.

51 Ibid.

52 Mounir R. Abi-Said (2006) Reviled as a grave-robber: the ecology and conservation of striped hyaenas in the human-dominated landscapes of Lebanon, PhD thesis, Durrell Institute of Conservation and Ecology, University of Kent, EThOS ID: 443775, https://ethos.bl.uk/ProcessOrderDetailsDirect.do?documentId=1&thesisTitle=Reviled+as+a+grave-robber+%3A+the+ecology+and+conservation+of+striped+hyaenas+in+the+human-dominated+landscapes+of+Lebanon&eprintId=443775 accessed 25 June 2019, pp. 33–6.

53 Ibid.

54 Ibid, pp. 36–7.

55 Dan Boneh (1987) Mystical Powers of Hyenas: Interpreting a Bedouin Belief, *Folklore*, 98, 1, 57–64, p. 58.

56 See Jurgen W Frembgen (1998) The Magicality of the Hyena Beliefs and Practices in West and South Asia, *Asian Folklore Studies*, 57, 331–344, p. 339.

57 Ibid.

58 Joel P. Brereton (1993) The Ravenous Hyenas and the Wounded Sun: Myth and Ritual in Ancient India by Stephanie W. Jamison Review, *Journal of the American Oriental Society*, 113, 4, 601–602, p. 601.

59 R. K. Narayan (1978) *The Mahabharata. A Shortened Modern prose Version of the Indian Epic*, London: Penguin, pp. 30–31.

60 Stephen E. Glickman (1995) The Spotted Hyena from Aristotle to the Lion King: Reputation is Everything, *Social Research*, 62, 3, 502–37, p. 4522; Alan Bewell (2014) Hyena Trouble, *Studies in Romanticism*, Fall 2014, 369–398, p. 369.

61 Glickman, 1995, p. 521.

62 Milton, *Samson Agonistes*, https://www.dartmouth.edu/~milton/reading_room/samson/drama/text.shtml accessed 26 February 2020.

63 Bewell, 2014, p. 373.

64 Ibid., p. 375.

65 Edmund Burke (1796) Burke and his pension, In the House of Lords, by the Duke of Bedford and the Earl of Lauderdale in the present session of Parliament, in Edmund Burke (2015) *Edmund Burke. The Complete Works*, Pergamon Media (Highlights of World Literature), Kindle Edition, loc. 30124–30129.
66 Zwelinzima Vavi (2016) We must stop South Africa from descending into a mafia-controlled predatory state, *Daily Maverick*, 25 March, https://www.dailymaverick.co.za/article/2016-03-25-op-ed-we-must-stop-south-africa-from-descending-into-a-mafia-controlled-predatory-state/?__cf_chl_captcha_tk__=63f71daa697af27096ede9732e1050c4c72c129a-1598020554-0-AcgSYUfxIHXjLZ8E7BpqJ7BUdr3cbf81w3H8PW8Xhba9Z4BjMWl2D4H8RRnP_K6Mae3LNmVMfB0blF2Gtt5-e311iskz8Rw9IqHVmbjuKngBqe4GxOSivndXZhfr5yE_ZF4s6Gi60CIVjjJQpMH6-jM7zkP2hrB9vC9UVhOgEp3ple7j45lNEmdhw79JHtlCh1pXHoZ6wCyX01BkU4hlNF63aC_lRcdJosleJCPv7uZiCg4VhRWS-xMcSexk3VdRsEynYmYpYbHp_ibN6i86NfMElHZARTzEed9FW7rwhqTFL_U0_h8QScCrme2YwEwOgSnU1qPmzK8673ca4vweQGSp2uYnHLNGb5uRMHQrq2VK3ob36pxFlYH1-n4vjCybsF5Tpel363VF8eODoZVyhj0CIRTfz7XcEPXIQSLu10gvwLeeIfGkjS2DIUOwwggje5nhlJD3x_e6w35dANMHOwsZG9I9HjFWiRUltH29upShMGWwlgGZq4GjaH97atj-txaBKA5SM_4SIJuOoYhxQAMkv7mHZ4lfOduNpWWp2sioywiIKlh5CzSHraERtmxlTrRrgRKkWfd98AnczqGd9S-QZ9IKyAXBsvM9Mbw5lF4B-Zb10R8YEBzny0Etm0Yhsg8uxN0rtuDUJCQm8G9r5LKH3v-71KGlBuV6geUHS0_dleD1XsdVQPZjV3L8m2wY0g accessed 21 August 2020.
67 News24, 3 August 2020, https://www.news24.com/news24/Columnists/CyrilRamaphosa/cyril-ramaphosa-profiting-from-a-pandemic-is-like-a-pack-of-hyenas-circling-a-prey-20200803?isapp=true&fbclid=IwAR1rfbRGFf8FbJc2u8RzIY83q0hLnAl38IJcAq6NlTZxwl8L78AZdfAw4-E accessed 3 August 2020.
68 Glickman, 1995, pp. 504–5.
69 Daniel Van Boom (2019) Lion King is the 9th highest grossing film ever, *CNET*, 18 August 2019, https://www.cnet.com/news/lion-king-remake-is-the-9th-highest-grossing-film-ever/#:~:text=The%20Lion%20King%20grossed%20%24968,%24496%20million%20in%20the%20US accessed 21 August 2020.
70 Mikita Brottman (2012) *Hyena*, London: Reaktion Books, p, 103.
71 Personal communication with Colin Jackson.
72 Canal+ and Saint Thomas productions, YouTube - https://www.youtube.com/watch?v=Xrt8KlPtX0w accessed 12 July 2020.
73 Personal communication with Derek de la Harpe, Neil Midlane and Henry Parsons of Wilderness Safaris.
74 Personal communication with Emsie Verwey.
75 Personal communication with Malcolm Fitzgerald, Chief Curator, ZSL.
76 http://www.zootierliste.de/?klasse=1&ordnung=115&familie=11506&art=1120704 accessed 3 September 2020.
77 Personal communication with Laurence Frank.
78 https://www.zoochat.com/community/threads/hyenas-in-the-usa.468521/ accessed 31 August 2020.
79 National Geographic's *Ghosts of the Great Salt Lake*, https://www.youtube.com/watch?v=NSmBwIXldpU accessed 1 August 2020.
80 National Geographic, *Hyena Coast*, https://www.youtube.com/watch?v=If_IlSaAgcY accessed 21 August 2020.
81 BBC, *Seven Worlds One Planet*, https://www.youtube.com/watch?v=OwwsMTgAHzM accessed 21 August 2020.
82 Gus Mills and Heribert Hofer (1998) *Status Survey and Conservation Action Plan. Hyaenas*, Gland, Switzerland: IUCN, pp. 6 and 104.
83 Ibid., p. 58.

INDEX

For Product Safety Concerns and Information please contact our EU
representative GPSR@taylorandfrancis.com
Taylor & Francis Verlag GmbH, Kaufingerstraße 24, 80331 München, Germany

www.ingramcontent.com/pod-product-compliance
Ingram Content Group UK Ltd.
Pitfield, Milton Keynes, MK11 3LW, UK
UKHW020143200926
475702UK00005B/888
* 9 7 8 0 3 6 7 4 3 6 4 2 1 *